BOB MIDDLETON'S HANDBOOK OF ELECTRONIC TIME-SAVERS AND SHORTCUTS

BOB MIDDLETON'S HANDBOOK OF ELECTRONIC TIME-SAVERS AND SHORTCUTS

Robert G. Middleton

PRENTICE-HALL
Englewood Cliffs, New Jersey

Prentice-Hall International, Inc., *London*
Prentice-Hall of Australia, Pty. Ltd., *Sydney*
Prentice-Hall Canada, Inc., *Toronto*
Prentice-Hall of India Private Ltd., *New Delhi*
Prentice-Hall of Japan, Inc., *Tokyo*
Prentice-Hall of Southeast Asia Pte. Ltd., *Singapore*
Whitehall Books, Ltd., Wellington, *New Zealand*
Editora Prentice-Hall do Brasil Ltda., *Rio de Janeiro*
Prentice-Hall Hispanoamericana, S.A., *Mexico*

Library of Congress Catalog Card Number: 86-062121

Printed in the United States of America

A WORD FROM THE AUTHOR ON THE UNIQUE, PRACTICAL VALUE THIS BOOK OFFERS

You will save time, increase your effectiveness, and have fun applying the electronic time-savers and shortcuts illustrated in the following pages. All of the novel procedures and techniques described in this book are based on many years of practical experience.

You will see a number of never-before-published quick tests, shortcuts, test tips, new electronic testers, new ways to use conventional test instruments, and new servicing techniques. Many of these topics have never been published before because they were developed especially for this new book.

- How to build a simple voltage-controlled audio oscillator that permits you to use your tape recorder as a dc voltage monitor.
- When supplemented by an "endless-loop" reference level indicator, you can use your dc voltage monitor as a practical recording voltmeter.
- You will see practical examples of how dc voltages can be added or subtracted with a voltmeter.
- A technique of measuring dc voltages in very high-impedance circuitry by means of a two-DVM method that draws zero current from the circuit under test.
- When supplemented by a bias box, you can make a sensitive test for amplifier distortion with a dc voltmeter.

- You can use your dc voltage monitor as a dc current monitor when supplemented by a current shunt.
- An appendix section illustrates and details "Getting It All Together" with a modern workbench/shelf/storage facility that you can easily build.
- When supplemented by a few simple components, your dc voltmeter becomes a high-performance dynamic ohmmeter that automatically measures the internal resistance of "live" circuits.
- With the addition of a pair of diodes, your dynamic ohmmeter becomes a high-performance dynamic impedance meter that automatically measures the internal impedance of "live" circuits.
- It is shown how to build an improved AF/RF/IF/VF signal tracer/level checker that speeds up preliminary troubleshooting procedures.
- When slightly modified, a conventional emitter-follower arrangement has zero insertion loss, facilitating voltage measurements in high-impedance circuitry.
- You can monitor and record voltage levels in low-level circuitry with a double-Darlington emitter-follower tester.
- New and useful applications are shown for your workbench test panel.
- Your ohmmeter can be put to good use as an IC "gauntlet" tester in preliminary digital troubleshooting procedures.
- You will see some brand-new digital troubleshooting quick checks called the 7/11 and 8/12 ground rules, along with the helpful "NC" quick check.
- When you encounter larger ICs, you can speed up quick checks with the aid of the 10/12/14 ground rule.
- Also, you will observe how to use a couple of proto-clips with indicator LEDs for an unusually informative digital quick checker in comparison procedures.
- Another helpful topic explains how to use a logic pulser and a logic probe in an "exerciser" procedure with the digital quick checker to uncover "tough-dog" faults.
- You will see how your DVM temperature probe can do yeoman service in analysis of digital IC temperature "signatures."
- . . . plus other analog and digital time-savers and shortcuts.

Various instructive projects are described and illustrated in the following chapters. For example, you will see how to construct a simple arrangement for using your tape recorder as a digital data memory. You can also use an RG LED with a resistor for preliminary data-stream analysis to determine whether two nodes have the same data flow or whether they may have complementary data flows. New types of test probes are explained for speeding up trouble localization in both analog and digital circuitry. New applications for conventional test probes are shown.

Down-to-earth procedures are described and illustrated for troubleshooting audio, radio, television, CB, tape-recorder, intercom, CCTV, telephonic, and digitally-controlled equipment. You will observe how to use the triggered-sweep oscilloscope in speedy quick checks. You will also see how to identify unmarked components and devices with conventional test equipment, and with specialized testers that you can easily build.

Note that Chapter 1 is primarily concerned with low-frequency circuitry and amplifiers. Subsequent chapters explain high-frequency tests.

This handbook is written for the beginning to avanced troubleshooter, as well as the hobbyist, experimenter, technician, and student. For example, the beginner will see what causes circuit loading when a meter is applied in a comparatively high resistance circuit. Again, the beginner will see what frequencies should be used in distortion tests, and what signal amplitudes are suitable. Practical examples with test results are cited. The more advanced troubleshooter will find discussions of controlled timbre tests and digital-logic troubleshooting ground rules, with examples of oscilloscope application. Similarly, the more advanced troubleshooter will find descriptions of automatic internal-resistance ohmmeters and modified emitter followers with zero insertion loss that can be easily constructed.

Hobbyists and experimenters will find discussions of novel circuitry in this handbook, such as simple RC networks that develop an output voltage which exceeds the input voltage (pseudo-active RC configurations), and of dual-DVM test techniques that provide dc voltage measurement with zero current demand from the circuit under test. Technicians will find various case histories cited that can speed up typical test-data evaluation. Students can profit from the circuit-action insights that are provided by various practical experiments. Both beginners and advanced troubleshooters will find the labor-saving computer programs of continuing utility.

Just about everyone agrees on the basic principles that time is money and that knowledge is power. Your personal satisfaction and your success in the electronics profession is limited only by the horizons of your technical know-how. The novel tricks of the trade, new techniques, and shortcut methods described and illustrated in this unusual book provide key steppingstones for you to take from your present position to your goal.

Robert G. Middleton

CONTENTS

A WORD FROM THE AUTHOR iii

Analog Section

CHAPTER 1 DC TESTS AND MEASUREMENTS 3

Addition and Subtraction of DC Voltages *3*
DC Voltage Measurement in Very High Impedance Circuits *4*
Quick Check for Circuit Loading *7*
Sensitive Amplifier Distortion Test *13*
DC Voltage Monitor with Audible Indication *21*
Recording DC Voltage Monitor *22*
Emitter Follower for DC Voltage Monitor *25*
Reference Level Indicator for DC Voltage Monitor *28*
DC Current Monitor with Audible Indication
Turn-off and Turn-on Quick Checks *31*
Critical versus Noncritical DC Voltages *37*
DC Current "Sniffer" *40*
Double-Darlington Emitter Follower *40*
Tolerances in Comparison Quick Checks *42*
Tape-Recorder Recycling Quick Test *43*

CHAPTER 2 RESISTANCE CHECKS AND MEASUREMENTS 49

Ohmmeter Operation *49*
Cautionary Note *53*
E/I Resistance Measurements *55*
Measured Resistance of Stacked Diodes *55*
Low-Power Ohmmeters *61*
Bipolar Transistor Checkout with Ohmmeter *61*
In-Circuit Resistance Measurement with Conventional VOM *64*
Automatic Internal Resistance Ohmmeter *66*
Workbench Test-Equipment Requirements *68*

CHAPTER 3 AC TESTS AND MEASUREMENTS 73

Overview *73*
Basic AC Waveforms *74*
Components of Pulsating DC Waveforms *77*
Apparent RMS Values of Basic Pulsating DC Waveforms on a Half-Wave AC Voltmeter *77*
If Your AC Voltmeter Has a Full-Wave Instrument Rectifier . . . *81*
Peak-Response AC Voltmeters *81*
True RMS Voltmeter *81*
AC Voltmeter Response to Pure DC *82*
The Turnover Check *83*
Distortion Checks with an AC Voltmeter *83*
Measurement of Stage Gain *84*
AC Voltage Monitor/Recorder Arrangement *89*
Diode Probe with Zero Barrier Potential *91*
Positive-Peak/Negative-Peak/Peak-to-Peak High-Frequency AC Probe 91
Audio Current "Sniffer" *95*
Improvement of Audio-Oscillator Waveform *98*
AC Voltmeter with Double-Darlington Emitter Follower *100*
Modified Emitter Follower with Zero Insertion Loss *100*
Shortcuts in Transceiver Troubleshooting *102*

CHAPTER 4 IMPEDANCE MEASUREMENTS AND QUICK CHECKERS 105

General Considerations *105*
Audio Impedance Checker *108*
Impedance at Battery Clip Terminals *108*
Measurement of Internal Impedance *114*
Measurement of Amplifier Input Impedance *117*
Measurement of Amplifier Output Impedance *119*
Measurement of Speaker Impedance *120*
Measurement of Microphone Output Impedance *120*
Measurement of Impedance at Radio Battery Clip Terminals *121*
Impedance of Mini-Speaker *126*
Crossover Operation *126*
Measuring Inductance with a DVM *130*

CHAPTER 5 SIGNAL TRACERS AND ANALYZERS 134

Overview *134*
Audio Signal-Tracer/Residue Analyzer *135*
Audio Signal-Tracer/Level Indicator *143*
Signal Tracing at Higher Frequencies *145*
Unmodulated Signals Made Audible *149*

CHAPTER 6 SIGNAL INJECTORS AND ANALYZERS 163

General Discussion *163*
Two-Tone Signal Injector *164*
IFM Signal Injector *166*
Speedy Stage Indentifier *166*
Practical Examples *168*
Transistor Click Test *169*
Stage-Gain Measurement by Signal Injection *170*
Out-of-Place Signal Voltages *171*
Quick Check for AM Rejection by FM Detector *172*
TV Analyzer *173*

Digital Section

CHAPTER 1 ***LOOKING AHEAD*** *181*

Gate Recognition *181*
High, Low, and Bad Logic Levels *183*
Generalized Troubleshooting Procedures and Data Storage *185*
Fan-In and Fan-Out *195*
The 7/11 Ground Rule *198*
The 8/12 Ground Rule *201*
NC Pins *202*
Unexpected Grounds *203*
"Normal" Bad Level *203*

CHAPTER 2 ***PROGRESSIVE DIGITAL TROUBLESHOOTING PROCEDURES*** *209*

Troubleshooting Latches *209*
Troubleshooting the Basic *D* Latch *215*
Checkout of the Gated (Transparent) Latch *215*
AND-OR Gate Operation *219*
AND-OR-INVERT Gate Operation *219*
Junction Characteristics *221*
Ohmmeter Quick Check of *D* Latch *225*
Basic Latches in 14-Pin IC Packages *225*
Basic Latches in 16-Pin IC Packages *225*
Gates in 16-Pin IC Packages *225*
The 10/12/14 Ground Rule *225*
"False Alarm" Nodes *227*
Common Faults *227*
Intermittent Monitoring *227*

CHAPTER 3 ***COMPARISON AND QUASI-COMPARISON QUICK TESTS*** *234*

Troubleshooting Flip-Flops and Clocks *234*
Single-Shot Pulsers *242*
Continuity "Wipe" Test *242*

NAND Implementation *242*
Tracing Circuit-Board Wiring *243*
Troubleshooting CMOS Gates *245*
Comparative TTL and CMOS Current Demands *247*
Inverter Inplementation *251*
Practical Note *255*
Principles of Data Tracing *256*

CHAPTER 4 VOLTAGE-BASED, CURRENT-BASED, AND RESISTANCE-BASED TROUBLESHOOTING PROCEDURES 260

Troubleshooting Digital Counters *260*
Checkout of 4-Bit Asynchronous Binary Up Counter *262*
Running the "Garbage" Out *265*
Clearing the "Garbage" Out *265*
Checkout of Asynchronous 4-Bit Down Counter *266*
Experimental IC Asynchronous Up Counter *266*
Troubleshooting with the Logic Comparator *269*
Diode Switch Experiment *276*
Negative Fan-out Test *276*
Cumulative Barrier Potentials *278*

CHAPTER 5 DIGITAL MAPOUT TROUBLESHOOTING PROCEDURES 279

Troubleshooting Synchronous Counters *279*
Programmable Down Counters *286*
Frequency Counter *288*
Direction of Data Flow *293*
Binary Coded Decimal Counter *293*
Series Operation of Devices *296*
Handy IC Comparison Quick Checker *297*

CHAPTER 6 ENCODER AND DECODER IDENTIFICATION AND TROUBLESHOOTING 299

Digital Stethoscope *299*
2-to-4 Line Decoder *300*

BCD-to-Decimal Decoder *301*
Binary Number Decoder *302*
1-of-16 Decoder *306*
2′421-to-8421 Code Converter *306*
Keyboard-to-BCD Encoder *306*
Binary-to-Seven-Segment Encoder *309*
Priority Encoder *312*
The Darlington Probe *313*

***APPENDIX I* Time-Saving Computer Programs for the Experimenter, Hobbyist, Technician, Student, and Circuit Designer 315**

***APPENDIX II* Time-Saving Program Conversions 329**

***APPENDIX III* Time-Saving Program for Computation of RC High-Pass Filter Bootstrap Circuit Output Voltage and Phase versus F 341**

***APPENDIX IV* Getting it All Together and Workbench Time-Savers 343**

BOB MIDDLETON'S HANDBOOK OF ELECTRONIC TIME-SAVERS AND SHORTCUTS

ANALOG SECTION

1

DC Tests and Measurements

ADDITION AND SUBTRACTION OF DC VOLTAGES

Subtraction of one dc voltage from another as shown in Fig. 1–1 is a well-known procedure. The addition of two or more dc voltages is less familiar, although it is a basically simple procedure, as exemplified in Fig. 1–2.* Observe that a VOM such as the Micronta (Radio Shack) 50,000 Ohms/Volt Multitester provides dc current ranges of 0–25–50 microamperes. In other words, the 0–25 microampere range can be used as shown in Fig. 1–2 as a 0–25 volts range, and the 0–50 microampere range can be utilized as a 0–50 volts range. Notice that service-type DVMs do not provide microampere ranges; only milliampere ranges are available. (See also Chart 1–1.)

Addition of three dc voltages is depicted in Fig. 1–3. This arrangement also can be used to measure the sum of two dc voltages, if desired—one of the branch test leads is merely left unconnected in this case. Observe that this arrangement can be employed to add negative voltages, if the polarity indication of the VOM is reversed. A combination of positive and negative voltages also may be added—the algebraic sum of the applied voltages will be indicated.

*Pulsating dc voltage tests and measurements are described and illustrated in Chapter 3.

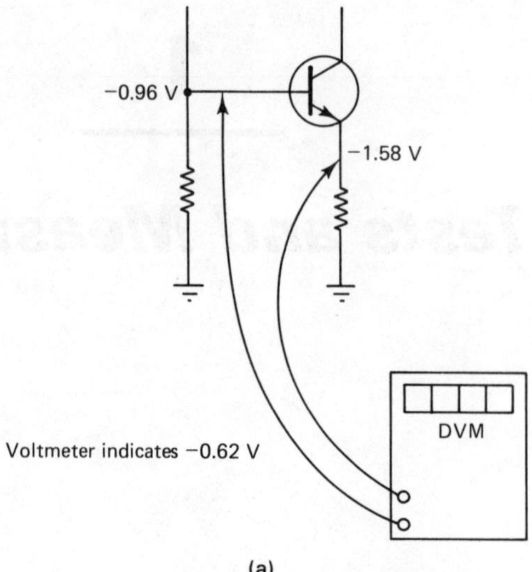

Note 1: The dc voltmeter effectively subtracts the base-to-ground voltage from the emitter-to-ground voltage, and indicates the difference between these two voltages. In other words, −0.96 V subtracted from −1.58 V equals −0.62 V. Thus, the emitter is biased −0.62 V negative with respect to the base, or the transistor is forward-biased 0.62 V.

Note 2: When the test leads to a DVM are not connected to a circuit (test leads are "floating" on open-circuit), the DVM usually indicates some dc voltage value such as 50 mV, or more. This "false" indication is due to stray-field pickup by the open test leads. The stray-field voltage tends to over-drive the DVM input circuit, resulting in partial rectification and a "false" voltage indication.

Figure 1–1 A familiar application of the dc voltmeter in subtraction of one voltage from another. (a) Test connections. (b) An auto-ranging digital multimeter. (Courtesy Radio Shack, a Tandy Corporation Company.)

DC VOLTAGE MEASUREMENT IN VERY HIGH IMPEDANCE CIRCUITS

Experienced technicians know that a very wide range of impedance (resistance) will be encountered in various electronic circuits. This is the internal resistance of the circuit under test—the resistance that the voltmeter "sees" from the point under test to ground. It is evident from Chart 1–1 that when

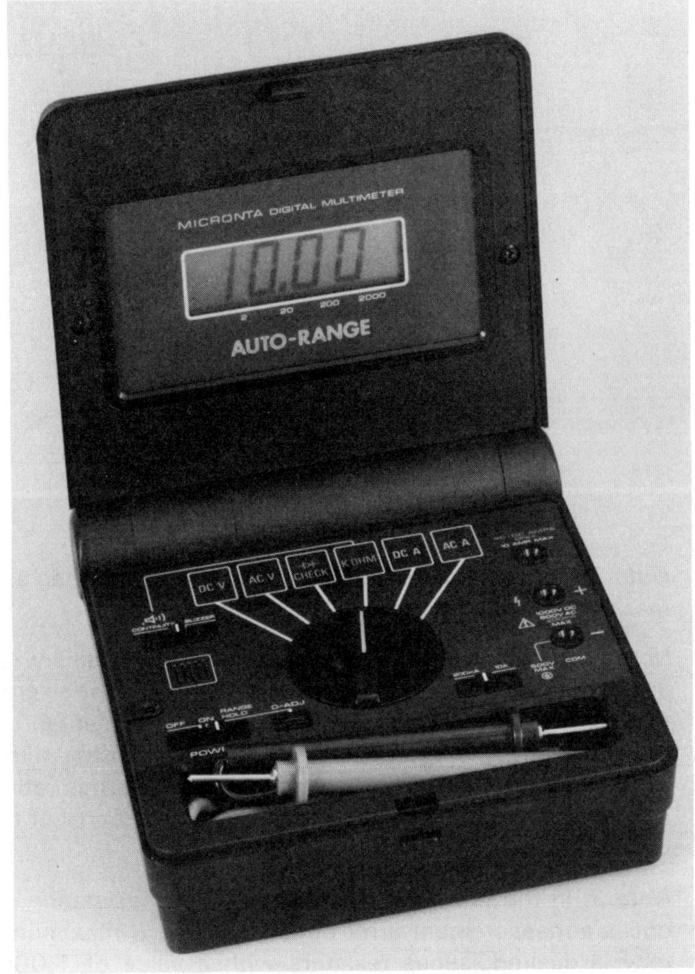

(b)

Figure 1–1 (*cont.*)

a voltmeter is applied across a resistor in a circuit the input resistance of the voltmeter is shunted across the resistor under test. Inasmuch as the input resistance of the voltmeter is thereby connected in parallel with the resistor under test, the result is that the effective resistance of the resistor under test is decreased. In turn, the circuit action is disturbed, and the effective voltage across the resistor under test is decreased.

As an illustration, if the voltmeter has an input resistance of 200,000 Ω, and the voltmeter is applied across a 200,000-Ω resistor in the

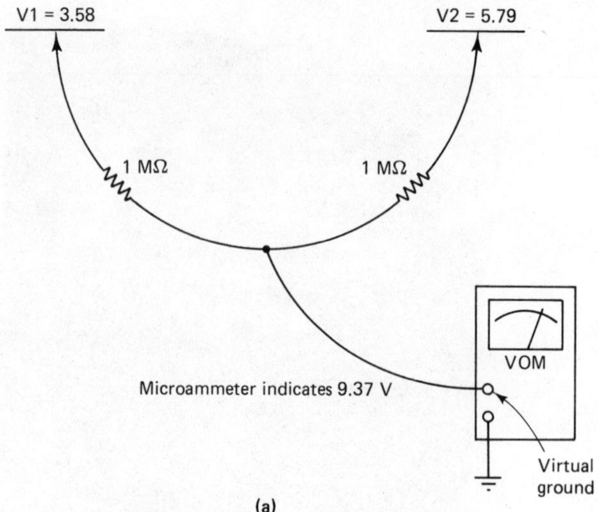

Note 1: This experiment shows how a VOM operates as an op-amp summer without an op-amp.

Note 2: When a 1-MΩ resistor is connected in series with the test lead to a VOM operated on its microampere function, the microampere scale indicates volts. If a pair of 1-MΩ resistors is connected to the test lead as shown above, the microampere scale indicates the sum of the two voltages. Observe that the microammeter has a comparatively low input resistance, such as 2,500 Ω. Thus, the input terminal of the microammeter is a virtual ground with respect to 1 MΩ.

Note 3: In the foregoing example, an input resistance of 2,500 Ω introduces a measurement error of 2.5 percent. If maximum indication accuracy is desired, series resistors with a value of 1,000,000–2,500 Ω (997,500 Ω) may be used instead of a value of 1,000,000 Ω.

Figure 1–2 (a) Example of addition of two dc voltages with a VOM. (b) Addition of two dc voltages with a VOM is simpler than employment of an op amp in analog-computer arrangements.

circuit under test, this paralleling of resistances reduces the effective resistance of the resistor under test to 100,000 Ω. It is evident that this circuit-loading action by the voltmeter will change the dc-voltage distribution throughout the circuit. Accordingly, an incorrect voltage value will be indicated by the voltmeter—the magnitude of the error depends upon the detailed configuration of the particular circuit.

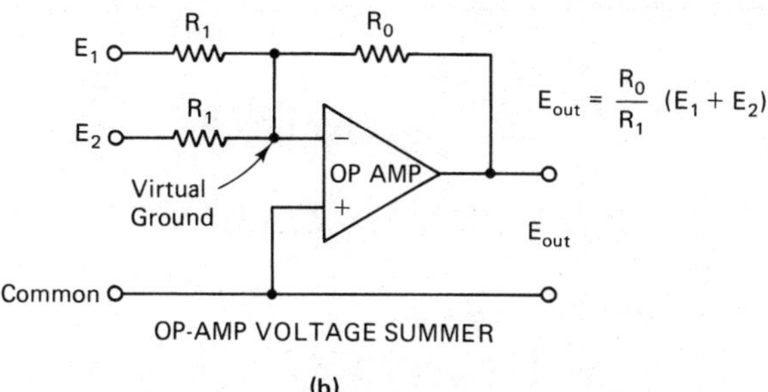

OP-AMP VOLTAGE SUMMER

(b)

Note 1: This is an example of a conventional op-amp voltage summer employed in various types of analog computers. If $R_O = R_1$, E_{out} will equal the sum of *E*1 and *E*2. E_{out} can be indicated by a VOM connected at the E_{out} terminals.

Comparison of this arrangement with that in Fig. 1–2 (a) shows that it is simpler to add voltages directly with a VOM, instead of utilizing an op amp. The only advantage of an op-amp configuration is that it provides somewhat greater precision than if the voltages are added directly with a VOM.

Note 2: If you are interested in op-amp analog-computer arrangements, you can use 1-MΩ resistors in the op-amp voltage summer shown above, with a Type 741 op-amp as detailed in the author's *New Ways To Use Test Meters.*

Figure 1–2 (*cont.*)

Typical DVMs have an input resistance of 10 MΩ on their dc-voltage ranges. This comparatively high value of input resistance minimizes indication error due to circuit loading in most situations. However, objectionable circuit loading will be occasionally encountered, as in various MOSFET circuitry. In such a case, a practical method of dc voltage measurement can be used in which a bias box and two DVMs are employed, as illustrated in Fig. 1–4.

QUICK CHECK FOR CIRCUIT LOADING

When measuring dc voltages in unfamiliar circuitry, doubt may occur concerning whether the dc voltmeter may be significantly loading the circuit under test. To quickly determine this possibility, note the DVM reading and then note the reading when a 10-MΩ resistor is connected in series with the

Chart 1–1

VOLT-OHM-MILLIAMMETER (VOM)

(Reproduced by special permission of Reston Publishing Co. and Miles Ritter-Sanders from "Electronic Meters.")

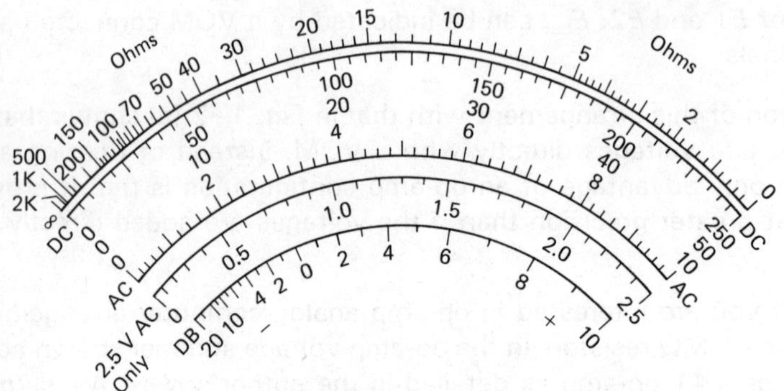

Scales Used on Simpson VOM

The VOM is also called a *multitester* or *multimeter*. Almost everyone who is into electronics uses this basic instrument from time to time. Like any test instrument, the VOM has its advantages and its limitations. Thus, some of its advantages are:

1. Comparatively low cost.
2. Portability (battery operated).
3. Low-range current function.
4. Decibel function.

Some of its disadvantages may be noted:

1. Comparatively low input resistance on its lower voltage ranges.
2. Lack of low-voltage ranges such as millivolt or microvolt ranges (except that an occasional VOM may provide a 250-mV range).
3. Lack of low-power ohms ranges.

VOMs are available in 1,000, 10,000, 20,000, 30,000, and 50,000 ohms-per-volt types. The input resistance of a VOM on its dc voltage ranges is determined as follows: Multiply the full-scale voltage of the selected range by the ohms-per-volt rating of your VOM. For example:

20,000 OHMS–PER–VOLT VOM

Full-scale Range (dc volts)	*Input Resistance (Ω)*
2.5	50,000
5.0	100,000
10.0	200,000
50.0	1,000,000

We need to be alert to the possibility of objectionable circuit loading when using a VOM (particularly a 1,000 ohms-per-volt meter). Observe the following examples of measurement errors; in each case, the voltmeter indicates less than the actual voltage (the actual voltage is 5 V):

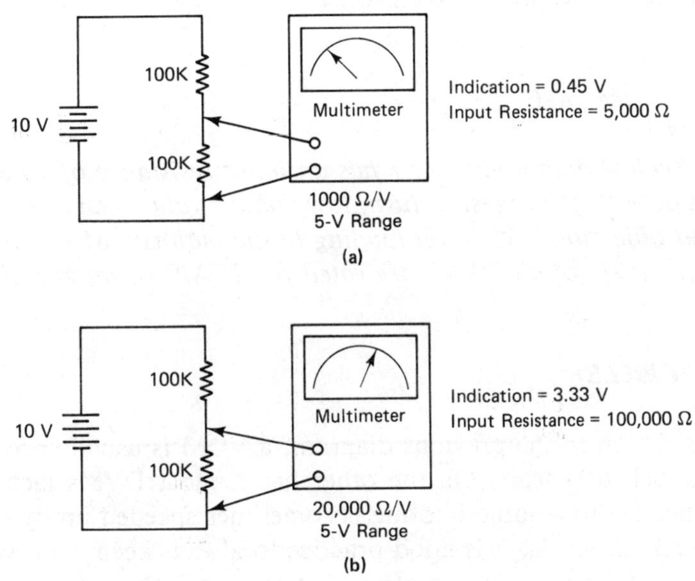

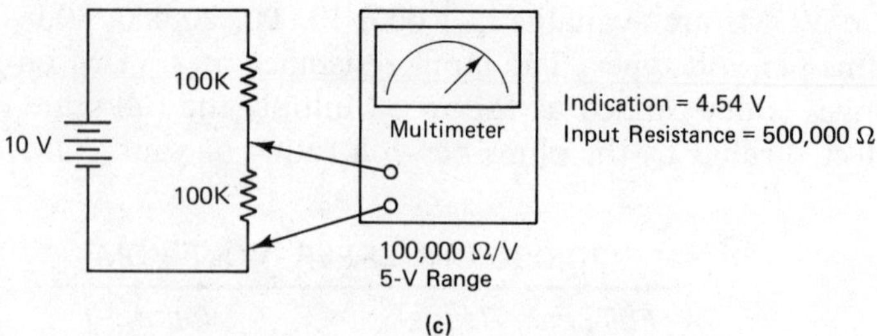

Typical voltage measurement errors due to loading of the circuit under test by the input resistance of the VOM. (a) 1,000 ohms-per-volt meter indicates 0.45 V instead of 5 V; (b) 20,000 ohms-per-volt meter indicates 3.33 V; (c) 100,000 ohms-per-volt meter indicates 4.54V.

Here's a useful trick of the trade: If you are in doubt whether objectionable circuit loading may be present, switch to the next higher range; then, if you read a higher voltage, significant circuit loading is present. On the other hand, if you read the same voltage on both ranges, circuit loading can be neglected.

Although circuit loading can be minimized by switching the VOM to a much higher voltage range, it then becomes impractical to read the scale with acceptable precision.

BOTTOM LINE

A typical digital voltmeter has an input resistance of 10 MΩ on all of its dc voltage ranges. In turn, low voltage values can be measured without objectionable circuit loading in the majority of familiar electronic circuits. (Not all DVMs are rated for 10 MΩ of input resistance.)

DECIBELS

As shown in the previous diagram, a VOM is usually provided with a decibel (dB) scale. On the other hand, most DVMs lack a dB scale. Since audio troubleshooting is sometimes speeded up by the ability to check dB levels, it is good practice to always keep a VOM at hand.

The decibel unit is often preferred over the voltage unit by audio experimenters because the decibel unit is proportional to ear response, whereas the voltage unit is not. As an illustration, if the sound level is increased by 6 dB, we judge that the sound is about twice as loud. Or, if the sound level is decreased by 6 dB, we judge that the sound level is about half as loud. On the other hand, if the voltage level were increased by 6 times, for example, we would not necessarily judge that the sound level has been doubled, or tripled, or increased by any specific amount. This is because a small voltage increase is much more effective with regard to ear response at low sound levels than it is at high sound levels. (Notice that an increase of 1 dB in the sound level is about the smallest change that the ear can detect.)

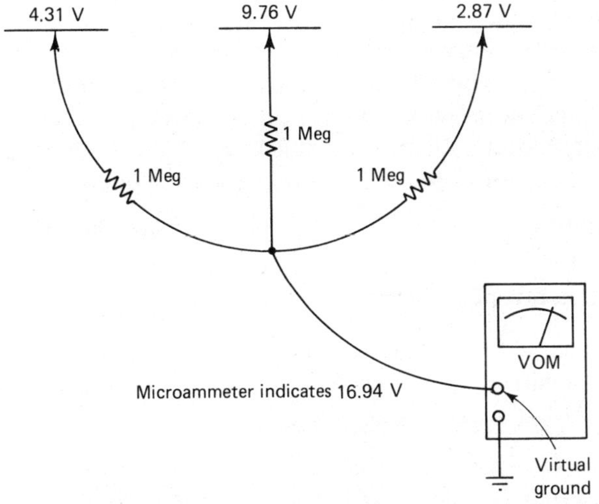

Note 1: In this example, three dc voltages are added and their sum is indicated on the microammeter scale. As previously observed, maximum accuracy of measurement can be obtained if the value of the input resistance to the VOM on its microampere function is subtracted from each of the 1-MΩ series resistors.

Note 2: Although the value of the input resistance is slightly reduced by the shunting effect of the three series resistors, this is such a small change in the effective input resistance that it may be disregarded from a practical viewpoint.

Figure 1–3 Example of addition of three dc voltages with a VOM.

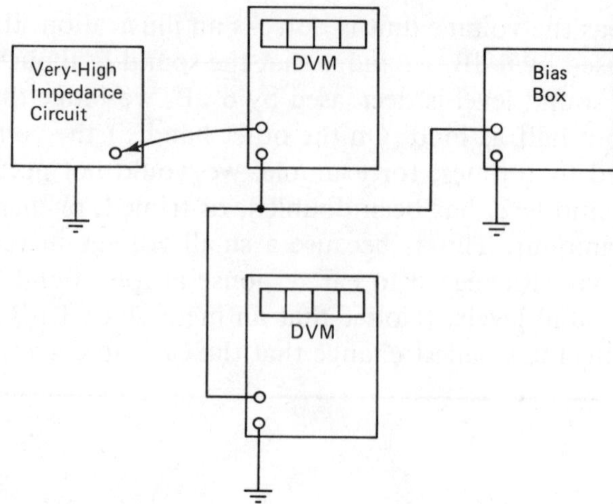

Note 1: This is a method for measurement of dc voltage in very-high-impedance circuitry. The upper DVM is set to its lowest voltage range and operates as a sensitive null indicator. When the bias box is adjusted for null indication, zero current is then flowing into or out of the very-high-impedance circuit. In turn, the lower DVM indicates the true voltage at the test point in the very-high-impedance circuit.

Note 2: A typical service-type DVM can indicate 0.1 mV (100μV). It also indicates the polarity of the readout (+ or −). Observe that 0 V is neither positive nor negative.

Figure 1–4 Measurement of dc voltage in very-high-impedance circuitry.

test lead. If the circuit under test is not being noticeably loaded, the second reading will be precisely one-half of the first reading. On the other hand, if the circuit under test is being significantly loaded, the second reading will be greater than one-half of the first reading.

The 10-MΩ resistor used in the foregoing procedure should be a high-precision type. Usually, you can select an accurate resistor from an assortment. Or, use potentiometer carefully set to a value of 10 MΩ. This adjustment can be easily determined by measuring the voltage of a battery, and then repeating the measurement with the potentiometer connected in series with the test lead. The potentiometer is adjusted for a reading equal to one-half of the first reading.

Notice that various types of bias boxes may be devised for use in electronic test procedures. If a variable-voltage power supply is available, it can serve the purpose. It is desirable to employ a power supply that has smooth adjustment of the output voltage level. As a practical consideration, it is good practice to use a regulated power supply, so that possible line-voltage fluctuation will have a minimum effect on the output voltage level. A variable-voltage power supply may be provided with a pointer and scale to indicate the output voltage value. However, this form of indication is often far from precise, and it is always advisable to connect a DVM across the power-supply output terminals.

Nearly all commercial power supplies provide transformer coupling to the power line. This is an important consideration in order to avoid shock hazard and possible equipment damage due to "hot chassis." Observe that if your power supply does not provide variable-voltage output, you can connect a suitable potentiometer across the power-supply output terminals. The potentiometer should have a value that will draw a reasonable amount of current from the power supply, so that the internal resistance of the bias arrangement will not be excessive. In turn, the potentiometer must have an appropriate power rating to avoid damage from overheating.

Hobbyists, experimenters, and technicians sometimes use a battery and potentiometer arrangement with a DVM as a bias box. This is a very satisfactory arrangement—it has no coupling to the power line, and it provides maximum voltage-level stability. The only disadvantage of a battery bias box is the necessity for occasional replacement of the battery (as in the case of an ohmmeter). Also, if you forget to switch the battery circuit off, and leave the battery connected to the potentiometer for a considerable length of time, the battery will be "dead" when you return.

SENSITIVE AMPLIFIER DISTORTION TEST

A sensitive test for even-harmonic amplifier distortion can be made with a dc voltmeter and bias box as shown in Fig. 1-5. This is a particularly helpful test when using comparatively low-sensitivity voltmeters. Stated otherwise, a voltmeter has maximum sensitivity when operated on its lowest range. For example, a typical DVM has a 200-mV range; however, this range cannot be used to check collector voltage unless a bias box is utilized, as depicted in Fig. 1-5.

This test method will indicate the presence of any form of even-harmonic distortion, and will also indicate the presence of various forms of

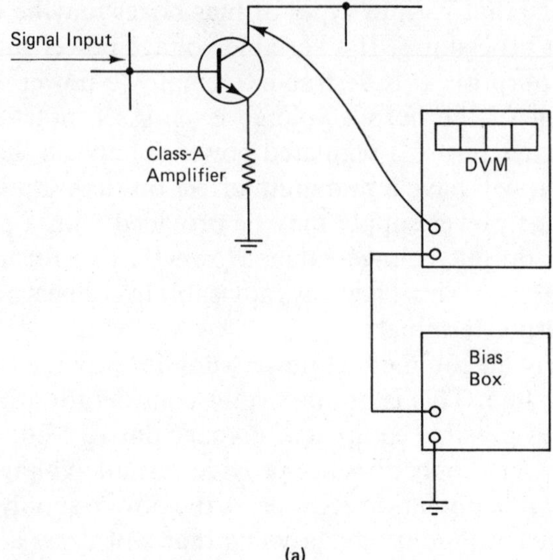

Note 1: This even-harmonic distortion test is based on shift of the dc-voltage operating point in case the transistor develops a nonlinear transfer characteristic as the input signal level is varied. In other words, the dc collector voltage will remain constant as the signal level is varied if the transistor is operating on a linear transfer characteristic. For maximum sensitivity of indication, the bias box voltage should be adjusted for a reference reading of zero on the DVM. (This permits the voltmeter to indicate the minimum feasible change in collector voltage as the input signal level is varied.)

Note 2: Signal input is usually obtained from an audio oscillator, but may be from any source, such as the 60-Hz power line. In other words, we are checking for distortion produced in the amplifier, and the test does not indicate distortion that may be present in the source waveform.

Figure 1–5 Sensitive test for even-harmonic distortion in a class-A amplifier stage. (a) Test setup. (b) Sine waves with various percentages of distortion. (c) Classes of amplifier operation. (Reproduced by special permission of Reston Publishing Co. and Derek Cameron from *Audio Technology Systems.*)

odd-harmonic distortion. However, an exception occurs in the event that the transistor develops a transfer characteristic that is equally and oppositely curved at both ends, with the dc voltage operating point placed at the center of the characteristic. To check for this possibility, the distortion test can be slightly elaborated as illustrated in Fig. 1–6. Note that single-ended class-A

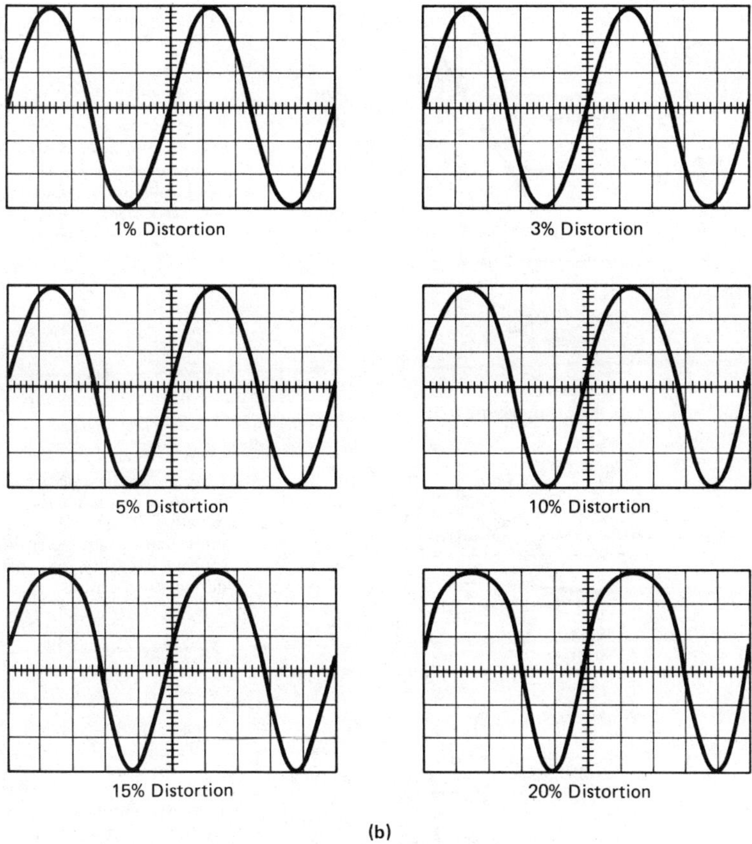

1% Distortion

3% Distortion

5% Distortion

10% Distortion

15% Distortion

20% Distortion

(b)

Figure 1–5 (*cont.*)

amplifier stages are truly linear only over a comparatively restricted input signal-level range.

> The foregoing tests for distortion (amplitude distortion) are based on the fact that a true class-A amplifier is completely linear and does not rectify any signal voltage whatsoever. On the other hand, a distorting amplifier (nonlinear amplifier) will rectify at least a small fraction of the signal voltage.*

*This chapter is primarily concerned with low-frequency circuitry and amplifiers. Subsequent chapters explain tests in high-frequency circuits and amplifiers.

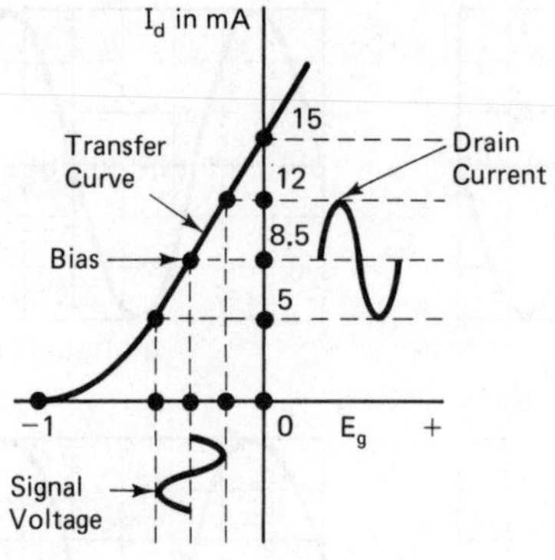

Class-A amplification provides an output waveform that is a precise replica of the input waveform. Amplifier efficiency is low, particularly during idling periods.

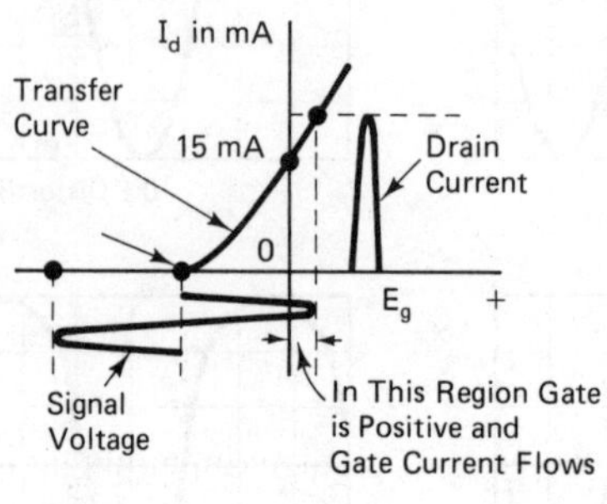

Class-B amplification provides an output waveform that is a half cycle of the input waveform; these half cycles are compressed in the low-current region. Amplifier efficiency is fairly good, particularly during idling periods.

(JFET Stage)

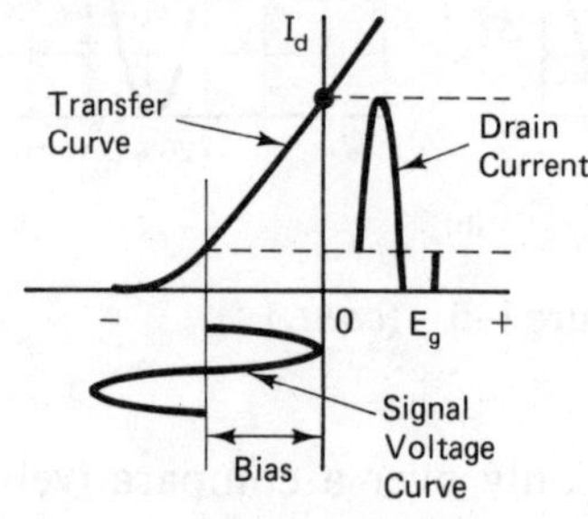

Class-AB amplification provides an output waveform that is more than a half cycle, but less than a full cycle of the input waveform. Amplifier efficiency is intermediate to that of Class-A and Class-B. amplification.

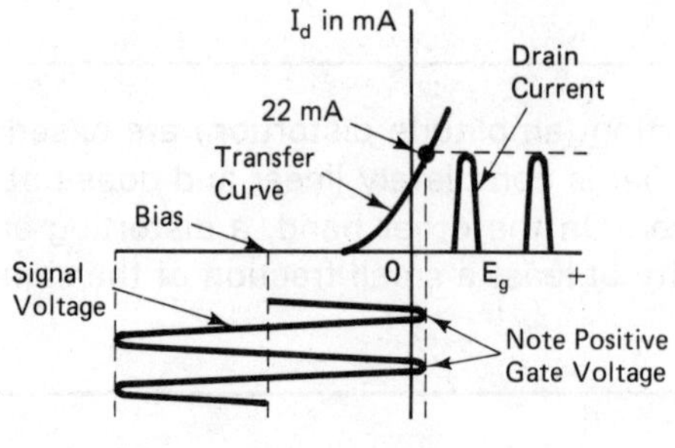

Class-C amplification provides an output waveform that is less than a half cycle of the input waveform. In audio technology, Class-C amplification represents amplifier malfunction. Amplifier efficiency is higher than for Class-B operation.

(JFET Stage)

(c)

Figure 1–5 *(cont.)*

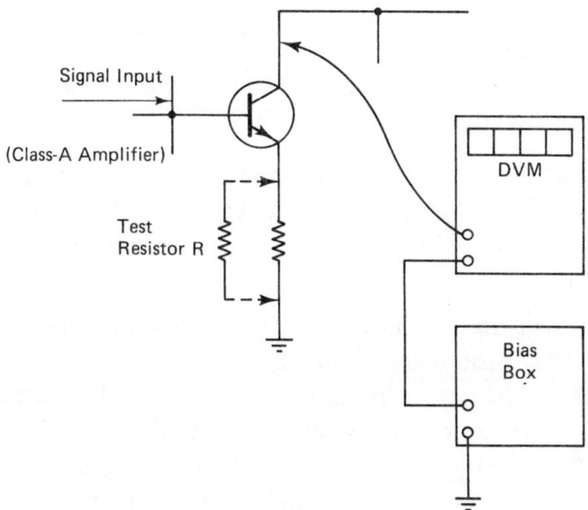

Note 1: This odd-harmonic distortion test employs a test resistor *R* for obtaining a small shift in the dc-voltage operating point of the transistor. Resistor *R* may have a value approximately five times the value of the emitter resistor. The test is first made as in Fig. 1–5. Then, if no change in DVM reading occurs as the input signal level is varied over its normal range, a follow-up test is made. The test resistor *R* is shunted across the emitter resistor, and the input signal level is again varied over its normal range. If any change in DVM reading now occurs, it is concluded that odd-harmonic distortion is present.

Note 2: The practical distinction between even-harmonic distortion and odd-harmonic distortion is that even-harmonic distortion almost always produces a waveform with unequal peak voltages, whereas odd-harmonic distortion produces a waveform with equal peak voltages.

Figure 1–6 A follow-up test for odd-harmonic distortion in a class-A amplifier stage.

Experienced technicians and troubleshooters know that commercial amplifiers are rated for frequency response, maximum power output, and total harmonic distortion. Distortion is basically measured at 1 kHz in audio amplifiers, although distortion ratings may also be specified at some frequency in the bass region and at some frequency in the treble region. Routine distortion tests of audio amplifiers are customarily made at 1 kHz. Notice that percentage distortion usually increases somewhat at very low and at very high operating frequencies. A typical hi-fi amplifier is often rated for less than 1 percent total harmonic distortion from 20 Hz to 20 kHz at its rated maximum power output level.

Beginners may ask how to measure the power output from an amplifier. This procedure is as follows:

1. Connect a power-type resistor across the amplifier output terminals in place of the speaker (if you wish to avoid distracting noise during tests). The resistor should have a value equal to the rated speaker impedance for the amplifier under test.
2. Connect an ac voltmeter across the amplifier output terminals. When a 1-kHz sine-wave input is applied to the input terminals of the amplifier, the ac voltmeter will indicate the corresponding root-mean-square (rms) output voltage that is developed.
3. Note that the output power in watts is equal to the output voltage squared, divided by the value of the load resistance. For example, if you measure 3 V rms across 4 Ω, the power output is equal to 9/4, or 2.25 W.

If you have had no previous experience with amplifier construction and test procedures, you will probably be interested in the project described in Chart 1-2. This simple audio amplifier can be constructed easily from parts that you are likely to have in your "junk box." Of course, if you prefer, you can use a proto-board socket for breadboarding. Or, if you wish to keep the completed amplifier on your bench as a simple utility preamp, you can construct it in a permanent version by utilizing a perf board, or a pre-etched circuit board (both available from Radio Shack).

Chart 1-2

EXPERIMENTAL PROJECT: CONSTRUCTION AND CHECKOUT OF A ONE-TRANSISTOR CLASS-A AMPLIFIER

The configuration of a one-transistor Class-A amplifier used in this project follows. The amplifier is constructed from the following parts:

Wood or plastic base 4″ × 1-1/4″ × 1/8″
(2) 8-lug terminal strips

(2) 3/8″ 6-32 flat-head machine screws and nuts
(2) 100-μF, 35-V electrolytic capacitors
20K 1/8-W resistor
100K 1/8-W resistor
150K 1/8-W resistor
65K 1/8-W resistor
NPN silicon transistor
9-V battery
Battery connector
Insulated hookup wire (6 in.)

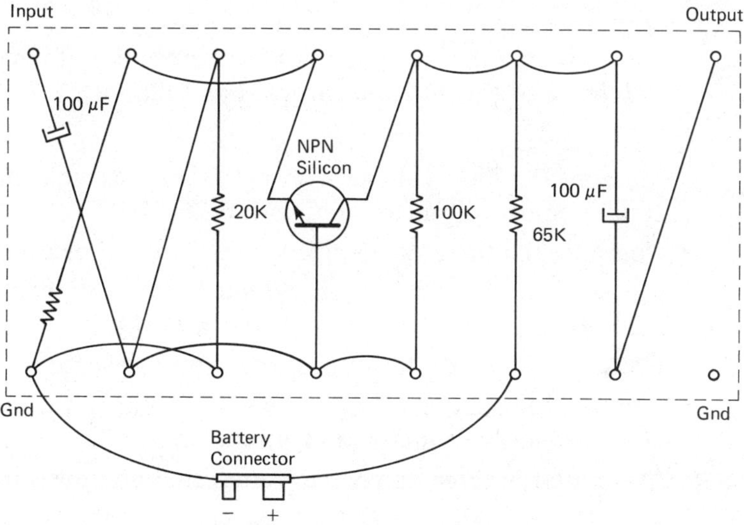

The two terminal strips are mounted symmetrically on the base with their angle brackets overlapping, and secured by the machine screws. In turn, the "breadboard" consists of two rows of terminal lugs with 1/2 in. spacing between them (see diagram). Parts are mounted between the terminal lugs as shown in the diagram, and are supported by their leads. Interconnections are made with the hookup wire.

When construction is completed, measure the dc voltages in the amplifier circuit. Your measured values will depend to some extent on the resistor tolerances, the transistor tolerances, the battery terminal voltage under load, and also upon the type of dc voltmeter that you use. Typical measured values are in the following. Widely different values would point to an error in circuit wiring (or to a seriously off-value component).

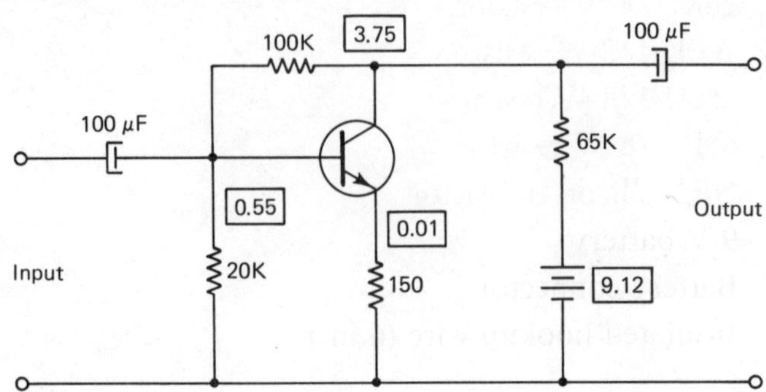

Now, observe the following practical examples of *circuit loading*. With reference to the diagram, proceed as follows:

1. Measure the base voltage with a 1,000 ohms/volt meter, with a 20,000 ohms/volt meter, and with a DVM.
2. Compare the three dc voltage values that you measured.
3. Measure the collector voltage with a 1,000 ohms/volt meter, with a 20,000 ohms/volt meter, and with a DVM.
4. Compare the three dc voltages that you measured.
5. Measure the emitter voltage with a 1,000 ohms/volt meter, with a 20,000 ohms/volt meter, and with a DVM.
6. Compare the three dc voltage values that you measured.

Next, observe the following practical examples of *in-circuit resistance measurement*. With reference to the diagram, proceed as follows:

1. Measure the resistance from base to ground with a high-power ohmmeter; connect the positive lead of the ohmmeter to the base terminal. (The 9-V battery *must* be disconnected.)
2. Repeat the resistance measurement with the negative lead of the ohmmeter connected to the base terminal.
3. Measure the resistance from base to ground with a low-power ohmmeter.
4. Compare the three resistance values that you measured.

5. Measure the resistance from collector to ground with a high-power ohmmeter; connect the positive terminal of the ohmmeter to the collector terminal.
6. Repeat the resistance measurement with the negative lead of the ohmmeter connected to the collector terminal.
7. Compare the three resistance values that you measured.
8. Measure the resistance from emitter to ground with a high-power ohmmeter; connect the positive terminal of the ohmmeter to the emitter terminal.
9. Repeat the resistance measurement with the negative lead of the ohmmeter connected to the emitter terminal.
10. Compare the three resistance values that you measured.

DC VOLTAGE MONITOR WITH AUDIBLE INDICATION

A useful dc voltage monitor is illustrated in Fig. 1-7. This tester frees the user from repeatedly glancing at a meter over an extended period of time while making circuit adjustments, or experimental replacement of components or devices.* This monitor is particularly helpful when an intermittent unit must be monitored over an indefinite interval, awaiting onset of the intermittent malfunction. Thus, the operator can connect the monitor at a key point in the intermittent unit, and then go about another project. Later, when the intermittent condition occurs, the operator is immediately alerted by the change in pitch of the sound output from the monitor.

This particular configuration is intended for monitoring of negative voltages. Thus, a −1 V potential produces a higher pitch in the sound output than does a 0 V potential. Again, a −2 V potential produces a higher pitch in the sound output than does a −1 V potential, and so on. Positive voltages may be monitored by a slightly different configuration, as explained subsequently. Observe that the input impedance to this basic monitor is comparatively low, and may tend to load high-impedance circuitry. The effective input impedance to the monitor can be greatly increased with the addition of an emitter follower, as will be explained.

*The ac signal level can be simultaneously monitored as explained in Chapter 3.

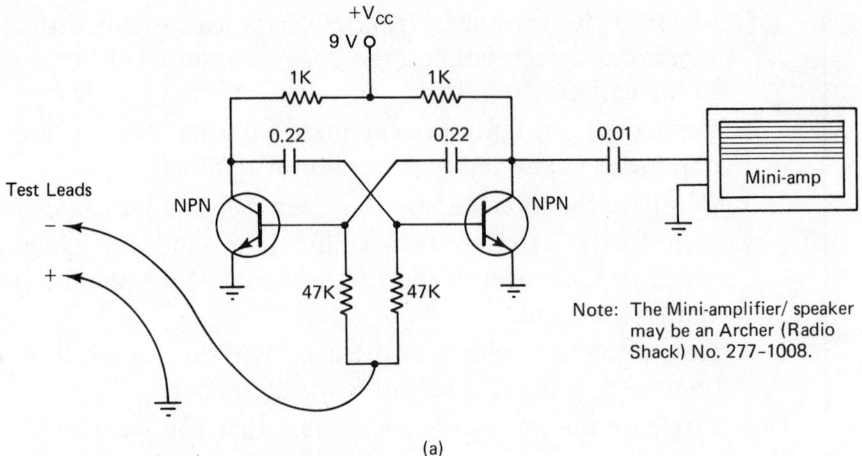

Note 1: The astable multivibrator operates as a voltage-controlled oscillator; its output frequency is reproduced as an audible tone by the mini-amp. The pitch of the tone varies over a wide range as the voltage applied to the test leads ranges from ground potential to −6 V, for example. As the negative input voltage increases, the pitch of the tone increases. In turn, the arrangement serves as a practical voltage monitor that liberates the user from watching a meter over an extended period of time.

Note 2: The transistors used in this experimental arrangement were Archer (Radio Shack) No. 276-1617. However, any general-purpose NPN transistors will operate satisfactorily. Observe that when the test leads are short-circuited together, a very low frequency is generated. On the other hand, if the test leads are connected across a 9-V battery, a very high frequency is generated.

Figure 1–7 (a) A dc voltage monitor with audible indication. (b) Experimenter socket is a major time-saver when breadboarding experimental circuitry: arrangement of breadboard and external view. (Courtesy Radio Shack, a Tandy Co. Corp.)

RECORDING DC VOLTAGE MONITOR

The monitor arrangement shown in Fig. 1–7 may be operated as a recording voltmeter when supplemented by a tape recorder. In other words, a tape recorder may be placed near the monitor while a voltage level is being checked over an extended interval. This is a particularly useful procedure when an intermittent unit is being checked, for example. Stated otherwise,

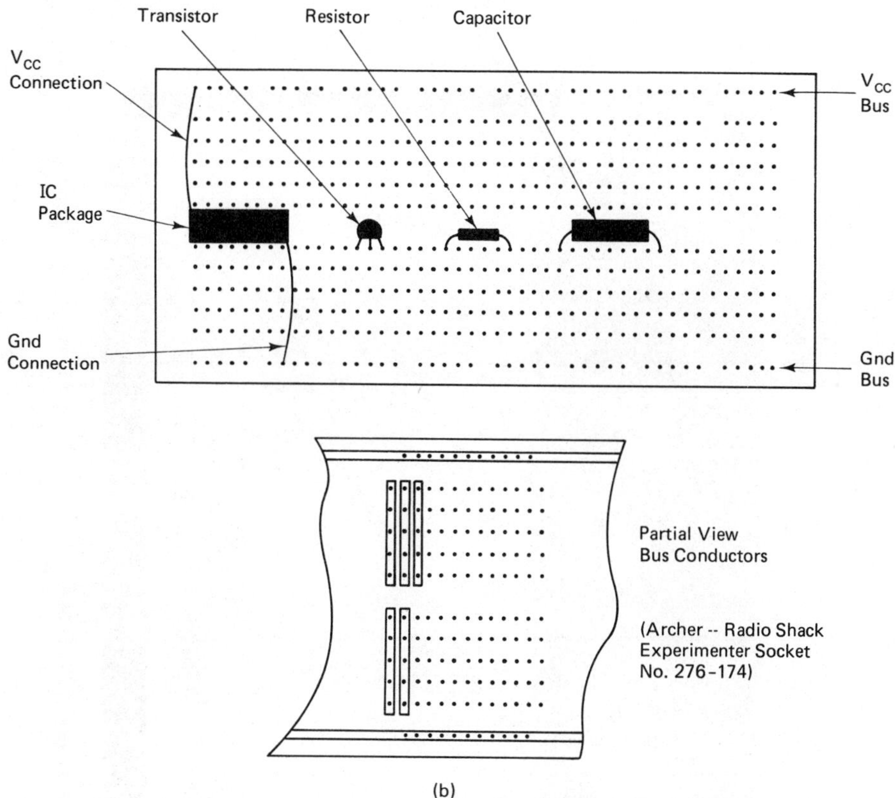

(b)

Note: This experimenter's socket board provides 560 holes into which device and component pigtails may be plugged for interconnection. Underneath the socket board, conductors are provided as shown in the partial view. These conductors provide interconnection of each column of five holes, and also provide interconnection of each row of holes along the top and bottom of the socket board.

Figure 1–7 (*cont.*)

the operator does not need to remain within hearing distance of the monitor inasmuch as the recorded tone track can be played back at any later time for analysis.

Silent recording may be employed, if desired, by connecting a patch cord between the output of the mini-amp and the input of the tape recorder. Observe also, that if a stereo tape recorder is utilized, two dc voltages can be simultaneously monitored and recorded on the *L* and *R* tracks. This is a particularly helpful procedure when a comparison test is to be made of a

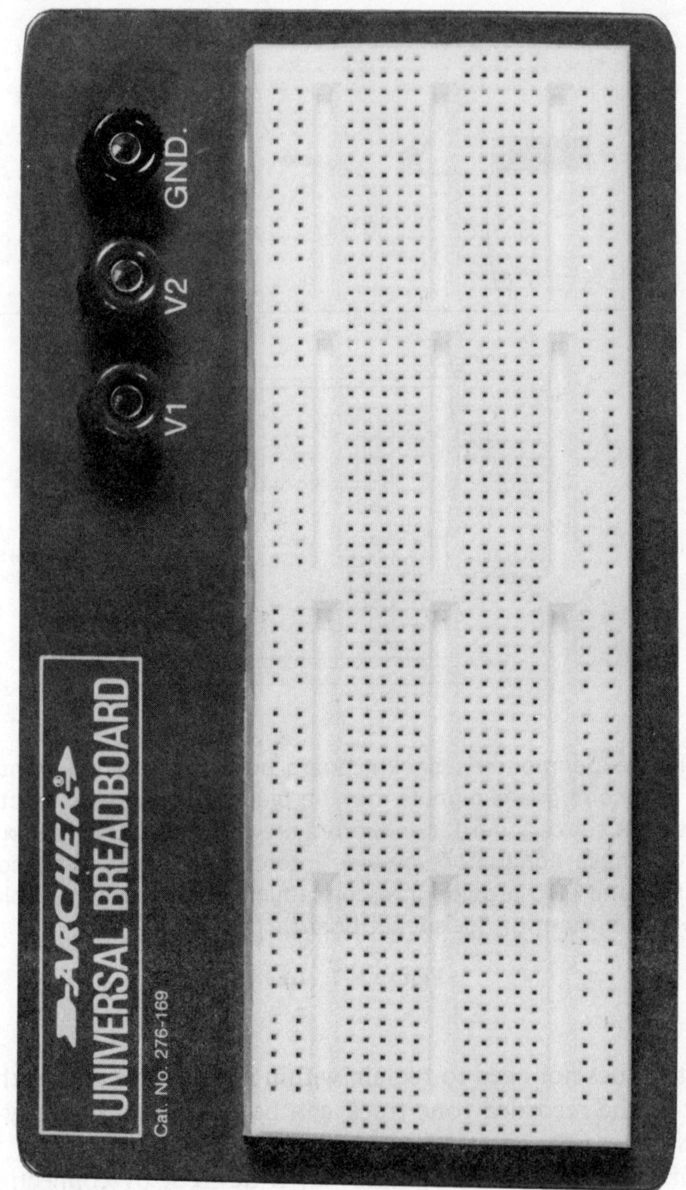

Figure 1–7 *(cont.)*

malfunctioning unit and a similar comparison unit in normal operating condition. Then, during playback at some later time, any discrepancy in pitch between the two monitor outputs becomes clearly evident.

EMITTER FOLLOWER FOR DC VOLTAGE MONITOR

As previously noted, the input resistance to the basic dc voltage monitor is comparatively low. Accordingly, when checking high-impedance circuitry, it may become desirable to increase the effective input resistance of the monitor with an emitter follower as shown in Fig. 1–8. This arrangement is suitable for monitoring negative dc voltage levels. On the other hand, to monitor

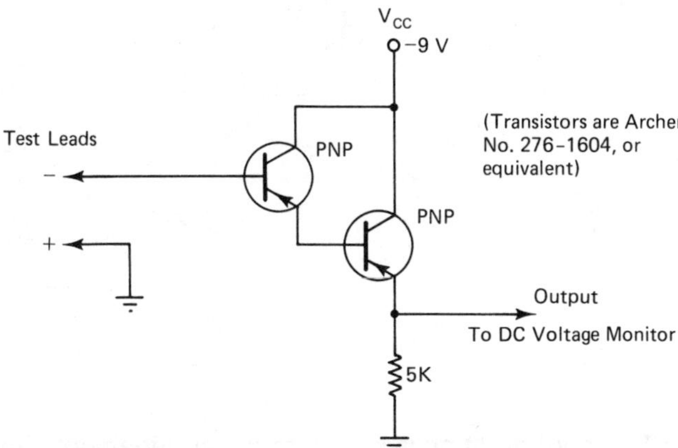

Note 1: An emitter follower operates as a current amplifier, and its output voltage is somewhat less than its input voltage. This is an example of the Darlington connection in an emitter follower, which provides very high current gain. In turn, the input impedance to the emitter follower is very high. Thus, when the dc voltage monitor is preceded by this emitter-follower section, tests can be made in very high impedance circuits without objectionable loading. Observe that this arrangement is suitable for monitoring negative dc voltages.

Note 2: If you are monitoring voltages in low-level circuitry, you may prefer to use the double-Darlington emitter follower detailed at the end of the chapter. If the output from the double-Darlington emitter follower has incorrect polarity, reverse the input test leads.

Figure 1–8 An emitter follower with a pair of Darlington-connected transistors provides very high input resistance.

positive dc voltage levels, the emitter-follower configuration depicted in Fig. 1-9(a) should be employed. Observe that the ground bus in the emitter follower will be at the same potential as the ground bus in the unit under test. However, the ground bus in the dc voltage monitor will be at the emitter potential of the lower transistor in the emitter follower.

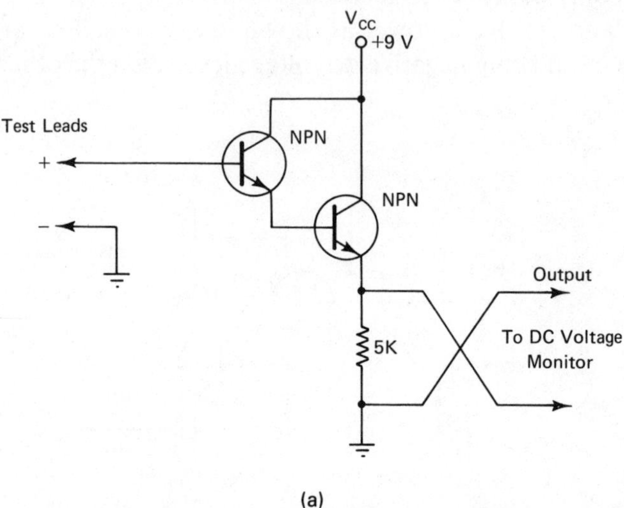

(a)

Note: This configuration is similar to the arrangement shown in Fig. 1–8., except that NPN transistors are utilized. In turn, this emitter follower provides for monitoring positive dc voltage levels. It provides very high input impedance as a result of the Darlington-connected transistors. Observe that the emitter of the lower transistor becomes more positive as the input is driven more positive. Accordingly, the ground lead of the emitter follower becomes more negative as the input is driven more positive. The practical significance of this relationship is that this emitter follower must utilize a separate collector supply battery, and that its ground connects to the 47 *k* resistors in Fig. 1–7. Consequently, the ground in the dc voltage monitor must "float" (a common ground must not be used between the emitter follower and the dc voltage monitor).

Figure 1-9 (a) Alternate emitter-follower configuration for monitoring positive dc voltage levels with the basic monitor arrangement shown in Fig. 1-7. (b) Emitter follower demonstration. (c) Another emitter follower experiment.

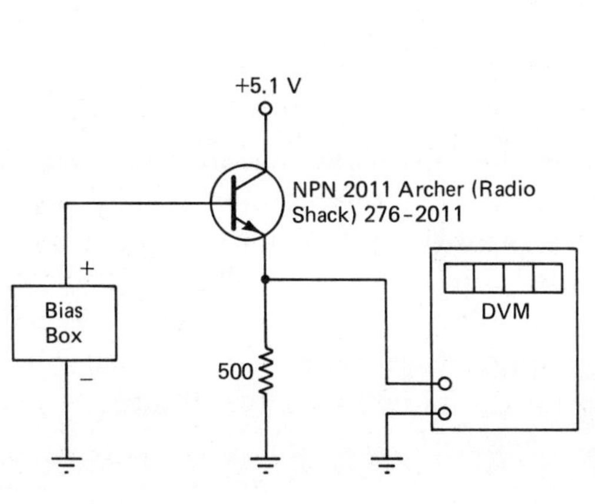

Bias Box Voltage	DVM Reading	V_{out}/V_{in} Percentage	$V_{in} - V_{out}$ Voltage
0.00			
0.60			
0.70			
0.80			
0.90			
1.00			
1.50			
2.00			
3.00			
4.00			

(b)

Note: This is a very helpful experiment for the beginning troubleshooter. It demonstrates the voltage input/output relations for a typical emitter-follower circuit. In turn, it gives an overview of the device bias values that are appropriate in various tests that employ an emitter follower.

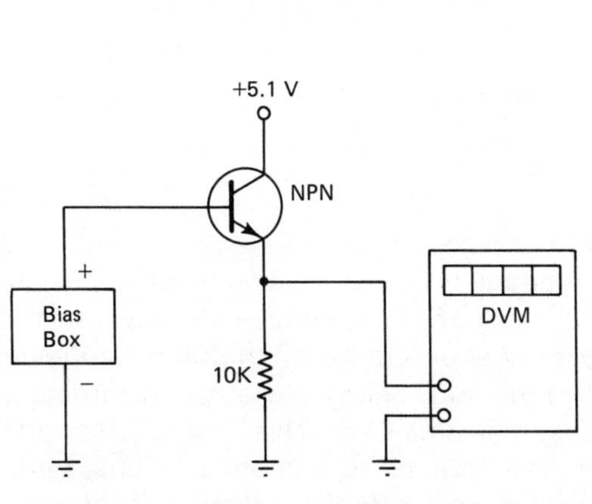

Bias Box Voltage	DVM Reading	V_{out}/V_{in} Percentage	$V_{in} - V_{out}$ Voltage
0.00			
0.60			
0.70			
0.80			
0.90			
1.00			
1.50			
2.00			
3.00			
4.00			

(c)

Note: This is another very helpful experiment for the beginning troubleshooter. It provides a practical comparison with the output obtained from a lower value of load, such as exemplified in Fig. 1–9(b). It gives an overview of the device bias values that are appropriate in various tests that employ an emitter follower.

Figure 1–9 (*cont.*)

REFERENCE LEVEL INDICATOR FOR DC VOLTAGE MONITOR

When analyzing a recorded tape from the dc voltage monitor, particularly a lengthy tape, it is often helpful to use a reference level indicator. A reference level indicator may only provide the initial tone pitch at 10-s intervals for comparison with the recorded tone. Again, a reference level indicator may provide a series of tone "bursts" corresponding to 1-V, 2-V, 3-V, 4-V, 5-V, or 6-V input levels for comparison with the recorded tone. Or, a reference level indicator may provide the initial tone pitch, followed by a series of voltage-level "bursts."

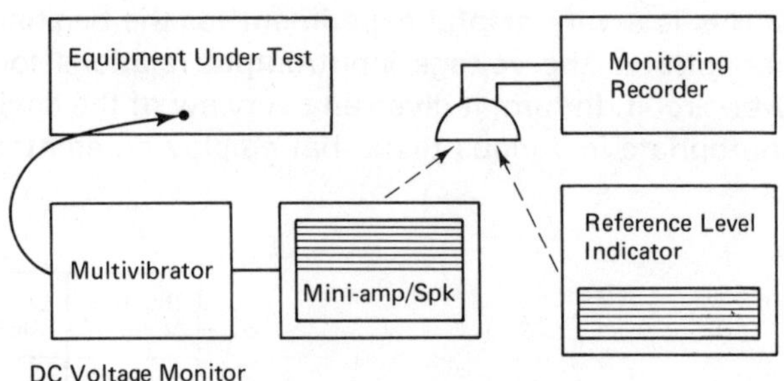

(The microphone in the monitoring recorder responds to sound from both the mini-amp/spk and the reference level indicator)

(a)

Note: In this example, the dc voltage from a chosen point in the equipment under test is fed to the multivibrator of the dc voltage monitor, and in turn a corresponding sound frequency is outputted by the mini-amp/speaker. The initial sound-frequency tone has been previously recorded as a "burst" on the reference-level indicator tape. This "burst" tone is outputted by the reference level indicator at 10-s intervals. Thus, both the output from the mini-amp/speaker and from the reference level indicator are simultaneously recorded on the monitoring recorder tape. This recurrent reference tone can be helpful in evaluating the monitoring recorder sound track at a later time, to determine whether any significant change in dc voltage at the test point may have occurred.

Figure 1–10 (a) Monitoring recorder may be energized by the outputs from both a dc voltage monitor and a reference level indicator. (b) Recorded tape frequency evaluation.

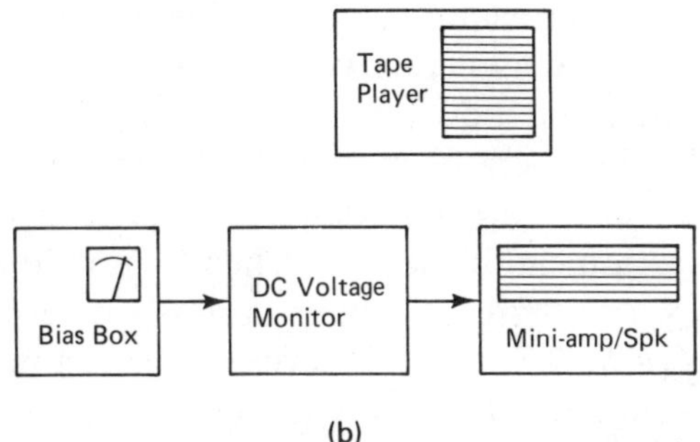

(b)

Note 1: This arrangement permits you to precisely identify a dc voltage value that corresponds to any arbitrary tone on a recorded tape (tape previously recorded from the output of the dc voltage monitor). The procedure is to advance the output from the bias box until the tone outputted by the tape player exactly matches the tone outputted by the mini-amp/speaker. Then, the voltage value indicated on the bias box is the same as the dc voltage value corresponding to the particular tone on the recorded tape.

Note 2: This method of voltage recording and measurement is only as accurate as the units that are utilized. The multivibrator output frequency can be made more reliable by using a regulated power supply. Hi-fi tape recorders are more stable than utility-type tape recorders. Tape stretch and drop-out are typical factors that can alter the pitch of the recorded tone, and thereby impair the precision of measurement.

Figure 1–10 *(cont.)*

The reference level indicator is a supplementary tape recorder that operates with a 10-s endless-loop tape such as the Realistic (Radio Shack) No. 43-404. On playback, the endless loop tape will repeat any "burst" at 10-s intervals. If only the initial tone pitch is to be provided for reference, the operator starts by recording a half-second (or 1-second) tone burst using the initial tone pitch as a source. Then, when the monitoring recorder is started, the reference level indicator is also started and both tones are simultaneously recorded, as illustrated in Fig. 1–10.

Any slow drift in dc voltage from the equipment under test becomes readily apparent on playback of the monitor tape, inasmuch as the recurrent "burst" has a noticeably different pitch and beats with the pitch from the mini-amp/speaker. Of course, an intermittent malfunction may develop a rapid drift or rapid fluctuation in dc voltage at the test point. In any case, the availability of the "burst" tone from the reference level indicator facilitates analysis of the monitoring-recorder tape at any later time.

The previous procedure is qualitative rather than quantitative. In other words, it will indicate whether voltage changes have taken place during the monitored period, but it does not indicate the values of these voltage changes. Consequently, the operator may desire to operate the arrangement as a recording voltmeter in a semi-quantitative procedure. As noted above, this mode of operation can be accomplished by providing the reference level indicator tape with a series of tone "bursts" corresponding, for example, to 1-V, 2-V, 3-V, 4-V, 5-V, and 6-V levels at the test point. These "bursts" can be sourced from a dc voltage divider used to initially energize the reference level indicator.

When the monitoring recorder is started, the reference level indicator is also started, and the sound outputs from both the mini-amp/speaker and the reference level indicator are picked up by the monitoring recorder. Thus, the tone corresponding to the dc voltage value at the test point and the reference-voltage "bursts" are simultaneously recorded by the arrangement depicted in Fig. 1–10. At some later time when the monitoring recorder tape is played back for analysis, the operator can form a reasonable judgment concerning the prevailing voltage-level value with respect to the reference-voltage "bursts."

Notice that a somewhat elaborated reference level indicator tape can be devised using a 45-s endless-loop tape such as the Realistic (Radio Shack) No. 43-404. For example, the 1-V level may be recorded with one "burst," followed by the 2-V level recorded with two shorter "bursts," followed by the 3-V level recorded with three still shorter "bursts," and so on. This method provides easier voltage-level identification on playback.

DC CURRENT MONITOR WITH AUDIBLE INDICATION

When working with low-impedance electronic equipment such as regulated power supplies, a dc current monitor such as shown in Fig. 1–11 is often helpful. This indicator frees the user from repeatedly glancing at a meter over an extended interval of time while making circuit adjustments, or experimental replacements of components or devices. A dc current monitor can be a major time-saver when an intermittent unit must be monitored over an unpredictable period of time, waiting for the intermittent malfunction to occur. The operator may connect the monitor into a suitable circuit, and then turn his attention to another task. At some later time when the intermittent malfunction occurs, the operator is immediately alerted by the change in pitch of sound output from the dc current monitor.

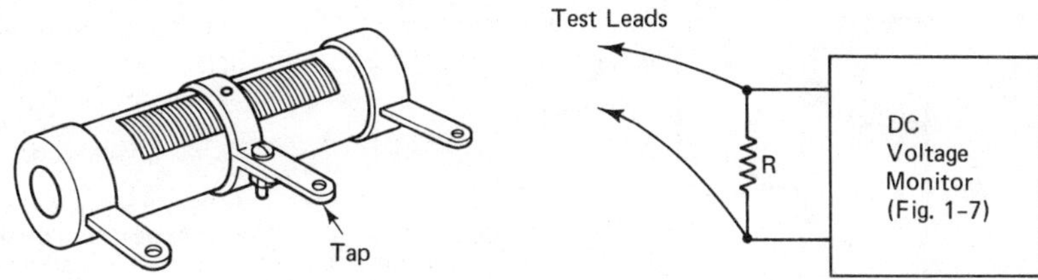

Note: Resistor *R* is a current shunt; it may be a wirewound power resistor, for example. Test-current flow through the shunt produces a voltage drop that energizes the dc voltage monitor. In turn, the pitch of the sound output from the monitor varies with the input current level.

Observe that the current shunt is always connected in series with the circuit under test. The voltage drop across the shunt is given by Ohm's law. Thus, if the shunt has a resistance of 2 Ω, a current flow of 3 A will produce a voltage drop of 6 V across the shunt.

A current shunt must necessarily have an ample power rating. For example, if the shunt is intended to conduct a maximum current value of 4 A, and the shunt has a resistance of 2 Ω, it must be able to dissipate 32 W. Conservative operation would typically dictate the use of a 50-W resistor in this example.

Figure 1–11 Basic dc current monitor arrangement.

As seen in Fig. 1–11, a basic dc current monitor is similar to a dc voltage monitor, except that a current shunt is connected across the input of the monitor. Operation is similar in current-monitoring and voltage-monitoring procedures. In other words, as the current flow through the shunt increases, the pitch of the sound output from the monitor increases accordingly. Notice that a dc current monitor is useful when working with low-impedance electronic equipment because the rate of change of current with respect to voltage is large in a low-impedance circuit.

TURN-OFF AND TURN-ON QUICK TESTS

Although this topic is not really new, it is a prime time-saver and merits mention here. A turn-off test is an in-circuit transistor quick-check. As illustrated in Fig. 1–12, a turn-off test is made as follows.

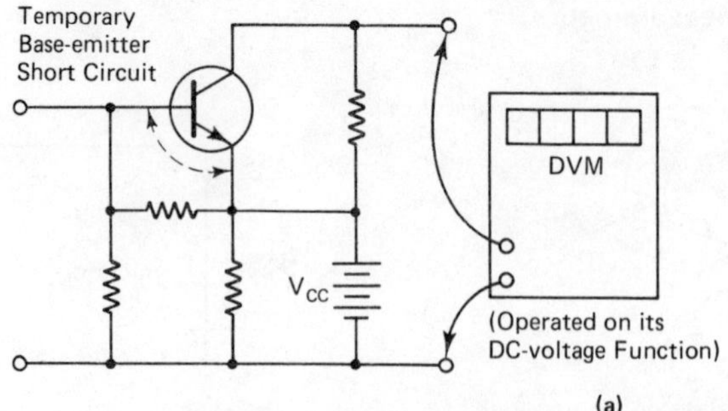

Note: Failure to pass the turn-off test usually indicates a defective transistor. However, a faulty component such as an open resistor or a poor connection can also result in failure to pass the turn-off test.

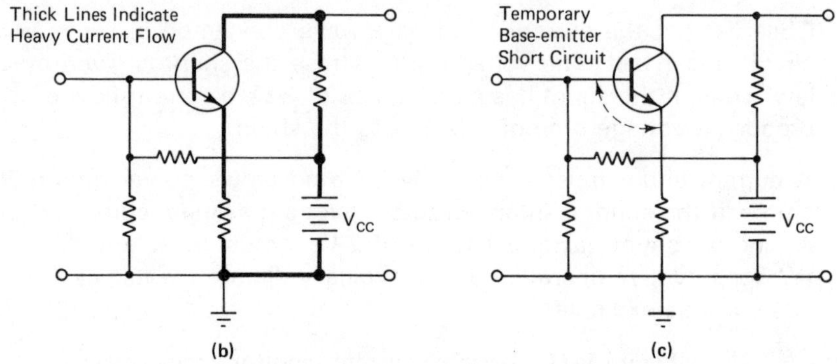

Note 1: A normally operating stage biased for class-A operation. The collector voltage is 3.5 V, and the collector current is approximately 0.1 mA. The base draws a comparatively small current.

Note 2: When the temporary short-circuit is applied from base to emitter, the base normally draws no current. In turn, the collector voltage rises to practically the V_{CC} value, and the collector current falls to virtually zero.

Note 3: Failure to pass the turn-off quick check does not necessarily indicate that the transistor has collector-junction leakage. For example, the transistor might be direct-coupled to a following stage—in such a case, a turn-off quick check is not valid (a turn-on quick check should be made, instead). Again, although the transistor may be capacitively coupled to a following stage, the coupling capacitor could be leaky. Accordingly, the troubleshooter needs to keep these possibilities in mind when evaluating test results.

Figure 1–12 Test connections for an in-circuit turn-off quick check. (a) DVM reading is noted before and after short-circuit is applied. (b) Current flow before short-circuit is applied. (c) Current flow after short-circuit is applied.

1. Apply a temporary short-circuit between the base and emitter terminals of the transistor, with V_{CC} present.
2. Note the collector voltage value indicated on the DVM.
3. Compare the indicated voltage with the voltage directly measured across the V_{CC} terminals.
4. If the indicated voltage is practically the same as the voltage measured across the V_{CC} terminals, the transistor passes the turn-off test.
5. On the other hand, an indicated voltage that is significantly less than the measured V_{CC} voltage means that the transistor fails the turn-off test.

The principle underlying the turn-off test is that when a temporary short-circuit is applied between the base and emitter terminals of a bipolar transistor in a class-A stage, the transistor is normally cut off due to effective reverse-biasing of the emitter-base junction by the internal barrier potential of the transistor. Since the transistor "cuts itself off" under this test condition, the collector voltage will normally "jump up" to the V_{CC} value. In turn, the technician concludes that the transistor is definitely workable. On the other hand, if there is only a small (or no) increase in collector voltage when the turn-off test is made, the technician concludes that the transistor (or an associated circuit component) is defective.

Next, observe the turn-on test shown in Fig. 1–13. A turn-on test provides a time-saving quick check of transistor condition in situations that are unsuitable for a turn-off test. Note that a turn-off test is not feasible in this particular circuit inasmuch as a bias resistor is connected between collector and base of the transistor. Since this bias resistor simulates collector-junction leakage in a quick test, a turn-off test would be inclusive. Accordingly, the technician proceeds with a turn-on test, as follows:

1. A temporary "bleeder" resistor is applied between the battery terminal and the base terminal of the transistor.
2. The DVM reading is noted.
3. Then the "bleeder" resistor is disconnected and the DVM reading is again noted.
4. If the collector voltage increases when the "bleeder" resistor is disconnected, the technician concludes that the transistor is workable.
5. However, if the collector voltage remains unchanged when the "bleeder" resistor is disconnected, the technician concludes that the transistor is unworkable. (It is possible for a component defect in the associated circuit to make the transistor "look bad.")

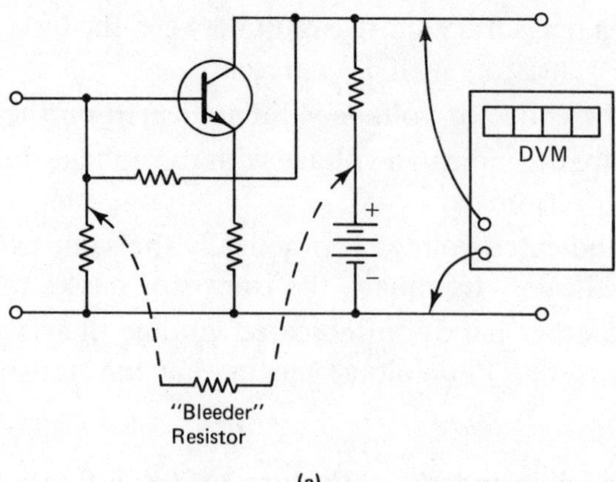

(a)

Note: Failure to pass the turn-on test usually indicates a defective transistor. However, a component defect such as an open resistor or a poor connection can also result in failure to pass the turn-on test.

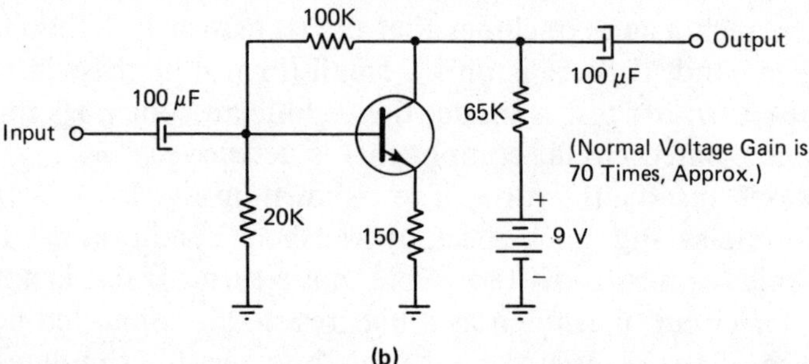

(b)

Case History: Although the utility audio amplifier shown here was operative, and although it responded to a turn-on test, its output was comparatively weak. A turn-off test was not feasible because of the 100-kΩ bias resistor between collector and base. When the transistor was checked out-of-circuit with an ohmmeter, its collector-base reverse resistance measured approximately 30 kΩ (transistor had collector-junction leakage).

Figure 1–13 (a) A 100-kΩ resistor is applied as shown by the dashed lines to make a turn-on test. (b) An instructive case history. (c) Practical example of dc voltage changes caused by common defects in an amplifier stage. (d) The transistor in a normally operating stage may appear to be reverse-biased in measurements of dc voltages.

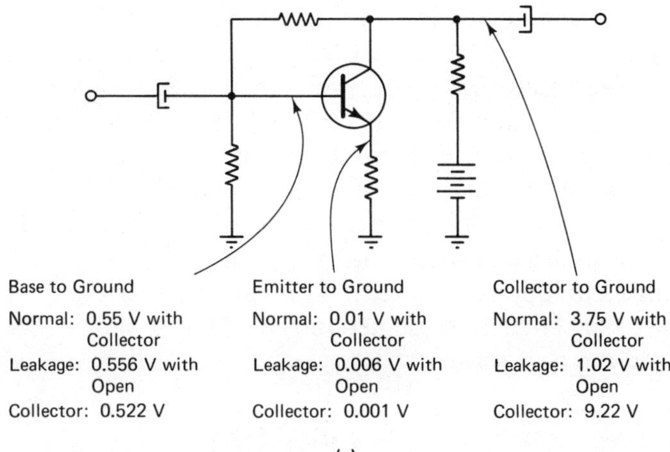

(c)

Bottom Line: Trouble analysis of the basic common-emitter amplifier on the basis of dc voltage measurements is made to best advantage on the basis of the collector-to-ground voltage.

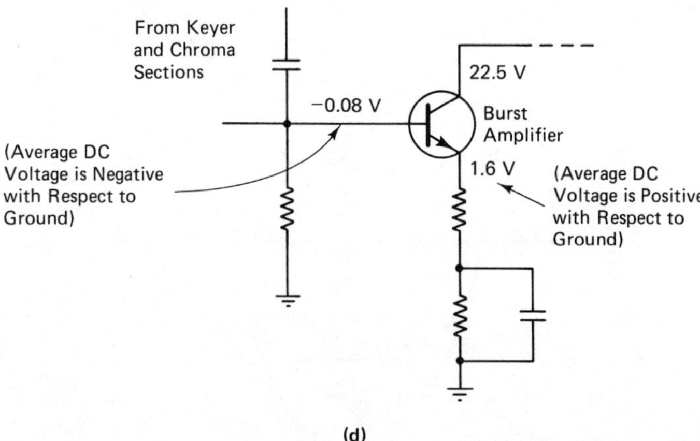

(d)

Note: The burst amplifier in a color-TV receiver is pulsed positively, but the signal-developed bias on the base of the transistor is negative due to the heavy base-current flow on the peaks of the keyer pulse. In this example, base-current pulses charge the base capacitor to an average value of −0.08 V.

Figure 1–13 *(cont.)*

In the example of Fig. 1–13, the bleeder resistor had a measured value of 108 kΩ. With the bleeder resistor connected into the circuit, the collector voltage measured 0.10 V. Next, when the bleeder resistor was disconnected, the collector voltage jumped up to 3.71 V. Accordingly, it was indicated that the transistor was definitely workable. (The transistor appeared to be defective because of a fault in the preceding stage.)

Next, the question is sometimes asked whether a field-effect transistor can be quick-checked in-circuit. The answer is "yes," although a slightly different technique is required. With reference to Fig. 1–14, an amplifier stage with an N-channel depletion-type FET is shown. A turn-off test is shown on the next page.

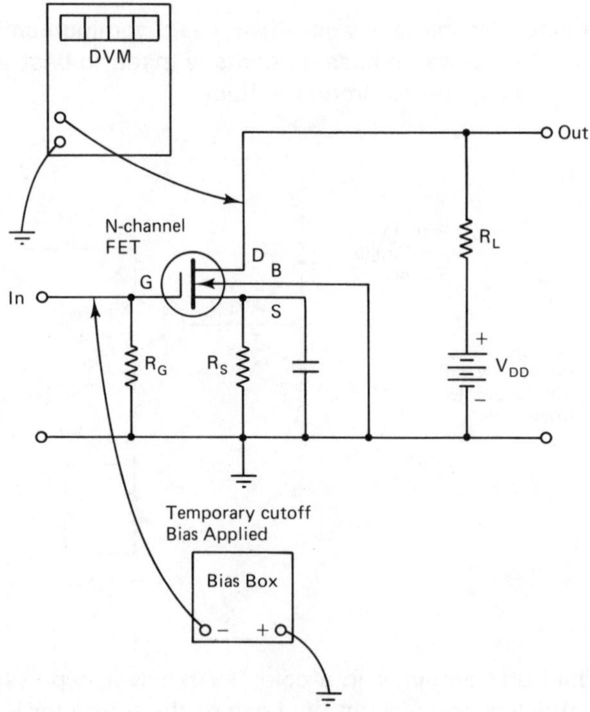

Note: Although the drain voltage will be practically the same as the V_{DD} voltage in a turn-off test, a small amount (such as 1 percent) will be dropped across R_L due to the small input current drawn by the DVM or TVM.

Figure 1–14 FET is biased to cutoff for turn-off in-circuit tests.

1. A DVM is connected between the drain terminal and ground of the FET.
2. A bias box is connected between the gate terminal and ground (in this example the bias box applies a negative voltage to the gate).
3. If the DVM indicates the V_{DD} voltage value when the output from the bias box is advanced sufficiently, the FET passes the turn-off test.
4. On the other hand, if the FET cannot be biased into cut-off, it is probably defective (although there is a lesser possibility of a fault in the associated circuitry).

CRITICAL VERSUS NONCRITICAL DC VOLTAGES

Much time can be saved in routine test procedures if the technician recognizes and distinguishes between critical and noncritical voltages. This is just another way of saying that some dc voltage values have relatively "loose" tolerances, whereas other dc voltage values have relatively "tight" tolerances. As a familiar example, the collector voltage and the emitter voltage in an amplifier stage is noncritical, whereas the base-emitter voltage is quite critical (see Fig. 1–15).

With a typical NPN silicon transistor in this circuit, the stage operation is optimum when R has a value of 200 kΩ; in turn, the base-emitter voltage will be 0.65 V. By way of contrast, when R has a value of 50 kΩ, the base-emitter voltage is 0.68 volt and the transistor is near saturation. Accordingly, the stage will then overload and distort the audio waveform at a low input voltage. In other words, the base-emitter bias voltage has a "tight" tolerance—a practical tolerance of 1 or 2 percent. If the base-emitter bias voltage varies 5 percent, the stage is virtually unworkable.

On the other hand, the collector voltage has a "loose" tolerance—it can vary as much as 20 percent without seriously affecting stage operation. Observe in the previous example that an emitter resistor is not included in the circuit. Consequently, the emitter voltage to ground is zero. If you include an emitter resistor (bypassed or unbypassed), the emitter voltage will then be more or less positive with respect to ground. In turn, the base-emitter bias voltage will not be 0.65 V and the value of R will need adjustment accordingly.

Note that if an emitter resistor is included in the previous circuit, that the base-emitter bias voltage will be somewhat less critical. For this (and other) reasons, audio amplifier circuits generally include emitter resistors. As an illustration, the current gain of a transistor (h_{fe}) increases rapidly as

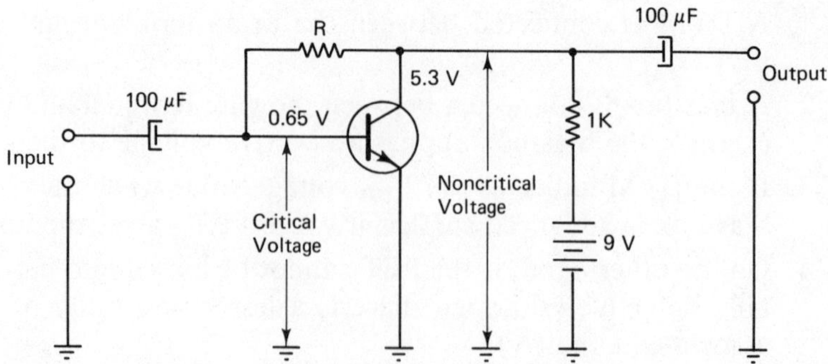

Experiment: Hobbyists, experimenters, apprentices, students, and technicians can rapidly learn how to save time in routine dc voltage measurements by constructing this basic audio amplifier configuration on an experimenter socket such as the Archer (Radio Shack) No. 276-174. A DVM should be used to measure the base-emitter voltage.

After you have completed the assembly and have verified the dc operating voltages, connect an ac voltmeter to the output terminals and apply a *small* 60-Hz input voltage (as from your ac power supply in your workbench test panel). Then re-measure the dc voltages in the circuit, and observe how they tend to change as the input voltage level is increased.

As previously noted, simple amplifier circuits are not completely linear, and partial rectification of the signal voltage will occur, particularly at higher input levels. When the signal voltage is partially rectified, a dc voltage is generated that affects the values of the resting (zero-signal) voltages.

Figure 1–15 Practical example of critical and noncritical dc voltages.

the ambient temperature rises. Unless the current gain is stabilized, a transistor is likely to develop "thermal runaway" and burn out. (See Fig. 1–16.) Therefore, an emitter resistor is generally utilized to provide improved current stability. Another advantage of including an emitter resistor is reduced stage distortion. As the value of the emitter resistor is increased, current stability improves, distortion is reduced, the input impedance increases, and the output impedance increases. Note that the stage gain will decrease unless the emitter resistor is bypassed. (Connect a large capacitor across the resistor in order to bypass it.)

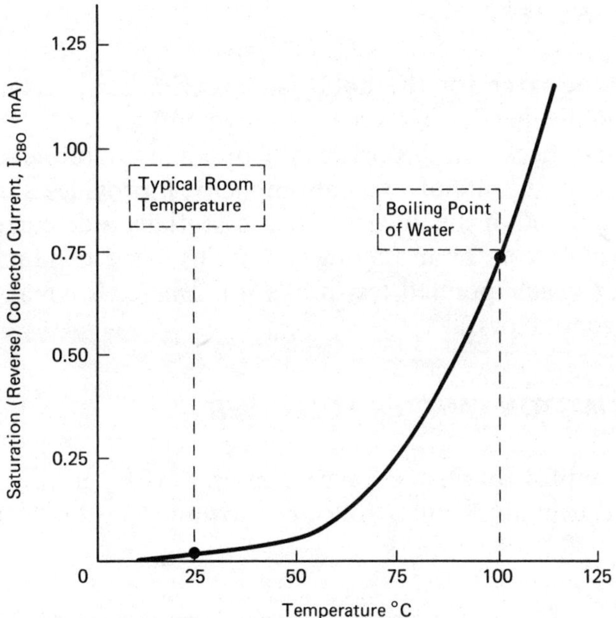

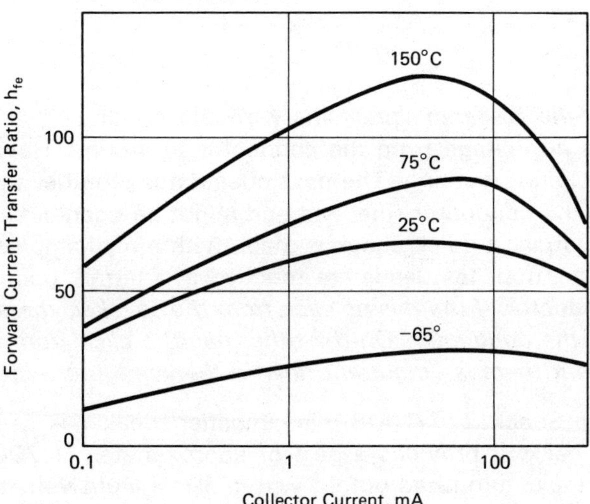

Figure 1–16 Transistor forward and reverse collector currents and h_{fe} variation versus temperature.

DC CURRENT "SNIFFER"

A real time-saver for the hobbyist, experimenter, technician, or troubleshooter is the dc current "sniffer" shown in Fig. 1–17. It is almost a "nothing" quick checker and consists only of a mini-amplifier/speaker and a pair of test leads. In spite of its total simplicity, it provides a sensitive dc current sniffer as noted in the diagram. It is essentially a dc current detector and a dc current tracer for use along any uninsulated conductor. (If you use a couple of needle-pointed test prods, the sniffer is equally useful along insulated conductors.)

DOUBLE-DARLINGTON EMITTER FOLLOWER

A very helpful source-resistance step-up device for your workbench is the double-Darlington emitter follower shown in Fig. 1–18. It consists of a bal-

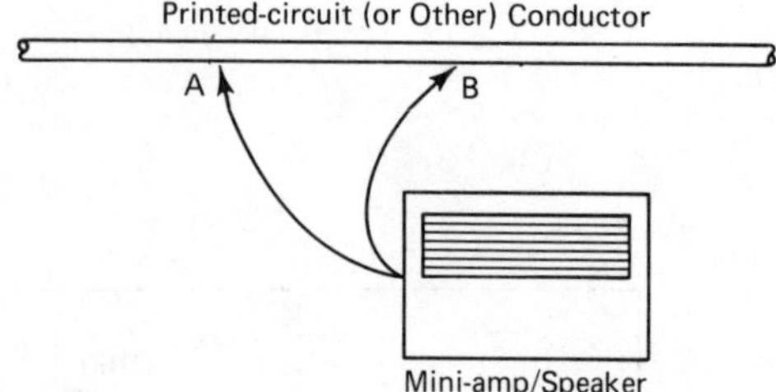

Trick of the Trade: In a preliminary troubleshooting checkout, we find that there is dc voltage from the conductor to ground. However, the circuit section is not working. The next question is whether there is dc current flow in the conductor (the load end might be open-circuited). To answer this question, a quick check is made with a mini-amp/speaker as shown here. The input test leads are touched at arbitrary points *A* and *B* along the conductor. *If there is no click from the speaker, there is no dc current flow in the conductor. On the other hand, a click from the speaker indicates that there is dc current flow in the conductor.*

A Radio Shack 277-1008 mini-amplifier/speaker is very useful for this quick check. It provides a gain of approximately 1,700 times, and develops maximum rated output with a 1-mV input voltage change. When the volume control is advanced to maximum, even a comparatively small dc current flow in the conductor will produce sufficient IR drop along the conductor to provide a clearly audible click from the speaker.

Figure 1–17 A useful dc checker-tracer.

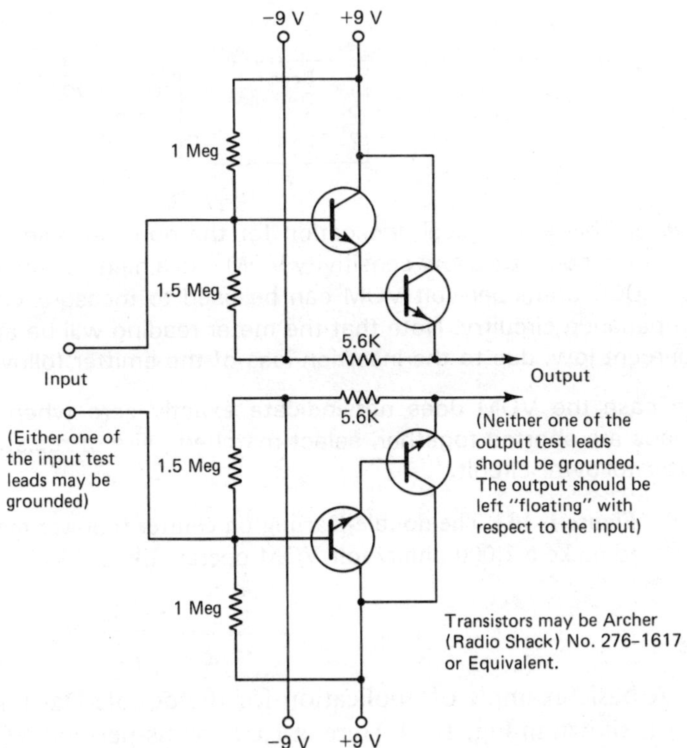

Note: This is a handy-dandy double-Darlington emitter follower for your workbench. It is a current amplifier that provides high input impedance and low-output impedance. Thus, you can use it for troubleshooting in high-impedance circuits; its output will drive almost any quick tester or servicing instrument. This current amplifier can be used in dc or ac tests. It is a balanced configuration, and there is normally 0 V across the output leads. When a voltage is applied to the input leads, it appears across the output leads. The input voltage is attenuated about 4 percent in passage through the current amplifier. Accordingly, if you are making dc or ac voltage measurements, your voltmeter will read approximately 4 percent low.

Figure 1–18 The double-Darlington emitter follower is a source-resistance step-up device for general electronic test work.

anced emitter-follower configuration that provides high-input resistance and low-output resistance for checking in high-resistance circuitry with comparatively low-resistance quick checkers, meters, and so on. Observe that the arrangement is direct-coupled, and that the transistors are biased to accommodate either positive-voltage or negative-voltage swings.

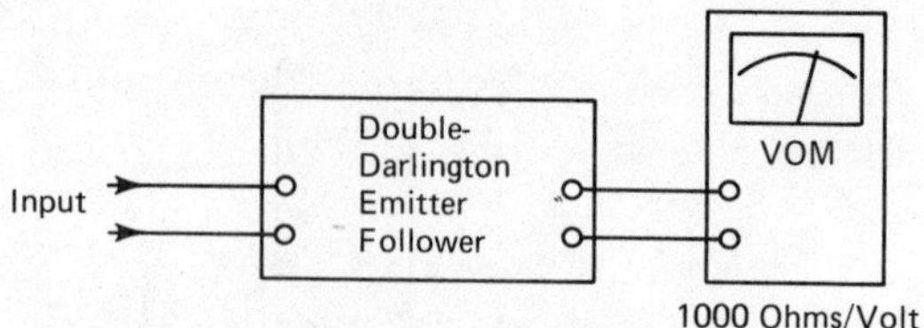

Note: This is a typical application for the double-Darlington emitter follower. It converts a low-sensitivity VOM into a high-sensitivity VOM. Thus, a 1,000 ohms-per-volt VOM can be used to measure voltages in high-impedance circuitry. Note that the meter reading will be approximately 4 percent low, due to the insertion loss of the emitter follower.

In case the VOM does not indicate exactly zero when the input test leads are shorted together, select matched pairs of resistors for the emitter-follower circuit.

Figure 1–19 The double-Darlington emitter follower may be used to make a 1,000 ohms/volt VOM operate like a TVM.

A basic example of application for the double-Darlington emitter follower is shown in Fig. 1–19. Here, a 1,000 ohms-per-volt VOM is being used to measure dc voltages in high-resistance circuitry. Again, this emitter follower may be used to drive a dc voltage monitor. Observe also that this emitter follower can be applied in ac circuitry to drive quick checkers or meters, as detailed subsequently.

TOLERANCES IN COMPARISON QUICK CHECKS

When making preliminary troubleshooting tests in a malfunctioning electronic unit, and a similar electronic unit in normal working condition is available, considerable time can often be saved by making comparative dc voltage measurements on the two units. We know that there are rated tolerances on electronic components and devices. Thus, many resistors commonly have ±20 percent tolerance; other resistors often have ±10 percent tolerance; occasional matched-pair resistors may have ±1 percent tolerance. Semiconductor devices ordinarily have rather wide tolerances.

The bottom line is that when we are making comparative dc voltage measurements on corresponding electronic units, we will seldom measure exactly the same voltage values at given test points. For example, let us consider comparative dc voltage values measured in the converter stages of similar AM radio receivers:

Collector Voltages

Radio I Radio II
8.99 V 8.84 V
Voltage tolerance: 1.7 percent

Base Voltages

Radio I Radio II
2.70 V 2.50 V
Voltage tolerance: 8 percent

Emitter Voltages

Radio I Radio II
2.32 V 2.10 V
Voltage tolerance: 10 percent

In evaluating the test data, we observe that the tolerances on dc voltage values in the converter comparison test do not exceed 10 percent. In other words, these dc voltage values are within normal tolerance. Therefore, we conclude that the malfunction is most likely to be in some part of the receiver other than the converter stage.

TAPE-RECORDER RECYCLING QUICK TEST

A tape recorder provides both recording and playback facilities, whereas a tape player lacks recording facilities. A tape deck lacks a built-in amplifier; it is operated with an external amplifier and speaker system. Tape decks may or may not include recording facilities. Tape recorders are designed as monophonic, stereophonic, or quadraphonic units. Audiophiles tend to prefer reel-to-reel machines over cartridge or cassette-type machines. Eight-track cartridge tape players, however, are popular because of their compactness and simplicity of operation. Most eight-track cartridge tape machines are designed as player decks. (A player deck lacks recording facilities.) All eight-track tape players provide stereo reproduction, and many qualify as high-fidelity units. (Note that cassettes dominate the market, however.)

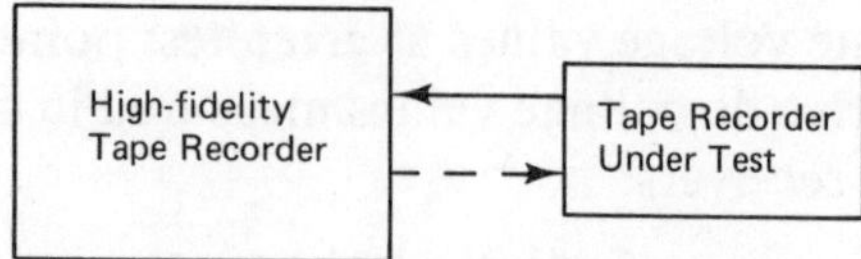

Note: In this quick test, either the recording channel, or the playback channel, or both, may be evaluated. For example, you can record a musical passage on the tape recorder under test from the high-fidelity tape recorder. Then you can rerecord the passage on the high-fidelity tape recorder. Next, rerecord this rerecorded passage on the tape recorder under test from the high-fidelity tape recorder. At each step, you may play back the recycled recording on the high-fidelity tape recorder, and evaluate the reproduction for distortion trouble symptoms.

Again, you can recycle a musical passage through the playback section of the recorder under test, rerecording it on the high-fidelity tape recorder. This mode of recycling provides data concerning distortion trouble symptoms related to the playback section of the tape recorder under test.

Figure 1–20 Tape-recorder recycling quick test.

Distortion problems can be a "headache" compared to most of the routine catastrophic failures in tape machines, due to difficulties in evaluating distortion trouble symptoms. When distorted output occurs, the audio technician usually checks the tape first. In other words, if a chromium-dioxide tape is used on a machine designed for ferromagnetic tape, both recording and reproduction will be impaired. Another basic type of distortion results from cross-talk between tape tracks. In this situation, the technician checks out the playback head height, inspects the tape for possible damage, and looks for foreign substances in the cartridge opening.

If distortion involves poor high-frequency response, the technician checks the playback head for oxide deposits, for excessive wear, or for misalignment. If no physical faults are found, he ordinarily proceeds to check out the amplifier for defects. If distorted output is accompanied by scraping, rattling, or buzzing sounds, the speaker falls under suspicion. Certain amplifier malfunctions can simulate a defective speaker, and a speaker substitution test is advisable in case of doubt. Loud popping sounds throw suspicion on the speaker voice-coil connections. If the trouble symptoms concern distortion on the recording function only, the technician generally checks out the ac bias section. However, similar trouble symptoms can be caused by a recording amplifier defect, a worn or fouled recording head, or even a faulty microphone.

A tape-recorder recycling quick test, as shown in Chart 1-3 is often helpful in troubleshooting distortion trouble symptoms. This quick test consists merely of rerecording and replaying the same musical passage several times via a high-fidelity (hi-fi) tape recorder. The number of times that the recording can be recycled without intolerable deterioration is a figure of merit for the tape recorder under test. Moreover, as sound-track deterioration progresses, the most serious distortion factor becomes greatly exaggerated. In turn, analysis of otherwise puzzling trouble symptoms can be facilitated.

Chart 1-3

POWER AMPLIFIER QUICK CHECKS

Most power amplifiers are configured in a complementary-symmetry arrangement, such as follows.

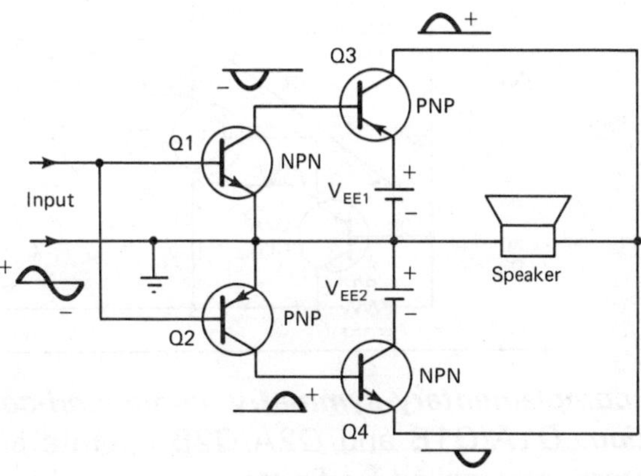

Common-emitter version of the direct-coupled complementary-symmetry amplifier. The transistors operate in Class B. Class AB operation may be employed to minimize crossover distortion.

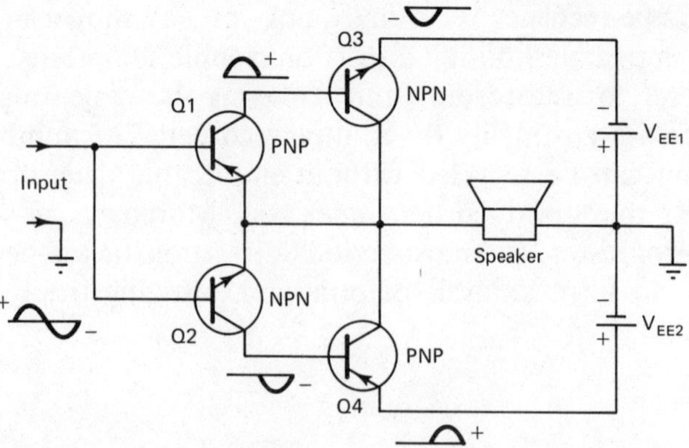

Common-collector version of the direct-coupled complementary symmetry amplifiers. The stage gain is less than in the common-emitter arrangement, but the input resistance is greater.

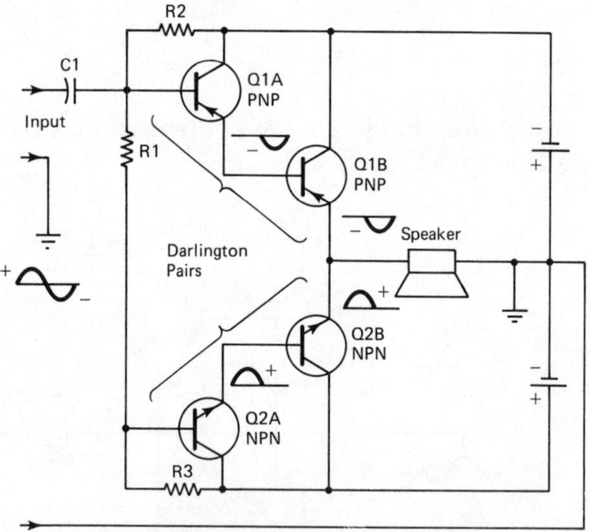

A complementary-symmetry compound-connected configuration. *Q*1A/*Q*1B and *Q*2A/*Q*2B operate as compound transistors, also called Darlington pairs, double emitter followers, or beta multipliers. Each of the compound transistors functions as an emitter follower.

These are symmetrically arranged push-pull amplifiers. In normal operation there is practically no dc current flow through the speaker.

QUICK CHECK

A basic quick check is to measure the dc voltage drop (if any) across the speaker. If there is an appreciable voltage drop, the two sections are unbalanced and the dc current distribution is incorrect. Collector-junction leakage in a transistor is a prime suspect. Otherwise, there will be a fault in a bias circuit.

Notice that the basic complementary-symmetry circuit is sometimes elaborated into a quasi-complementary-symmetry arrangement as shown in the following diagram. In turn, it is helpful to employ a quick check that immediately shows whether the current distribution is normal. As indicated in the diagram, the dc voltage at the emitter of Q3 is normally equal to one-half of V_{cc}.

If this voltage is incorrect, look for collector-junction leakage in a transistor or for a bias-circuit fault. On the other hand, if the voltage at the emitter of *Q*3 is correct, look for a capacitor fault. For example, *C*1 or *C*2 might be open.

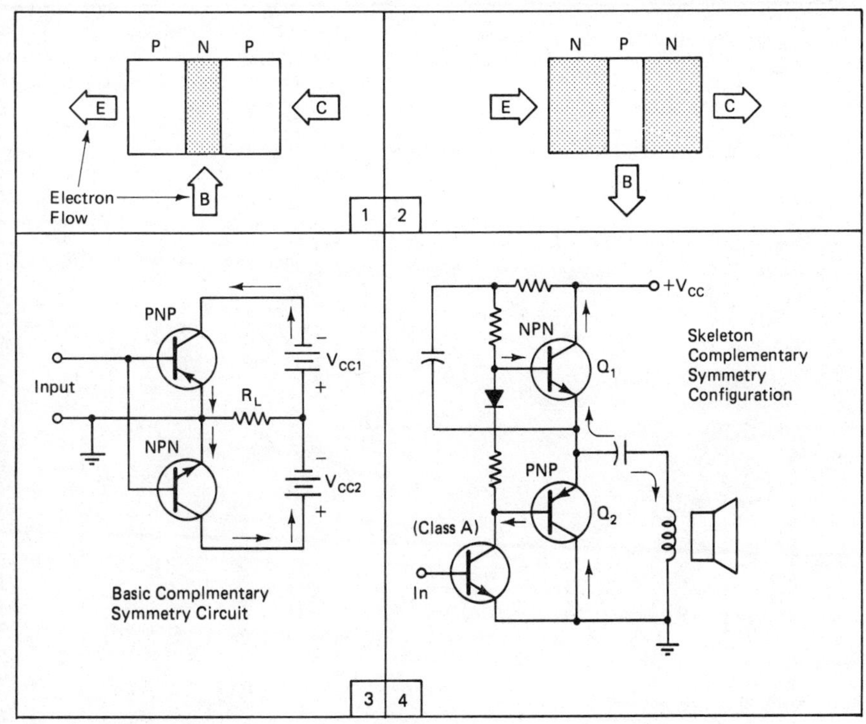

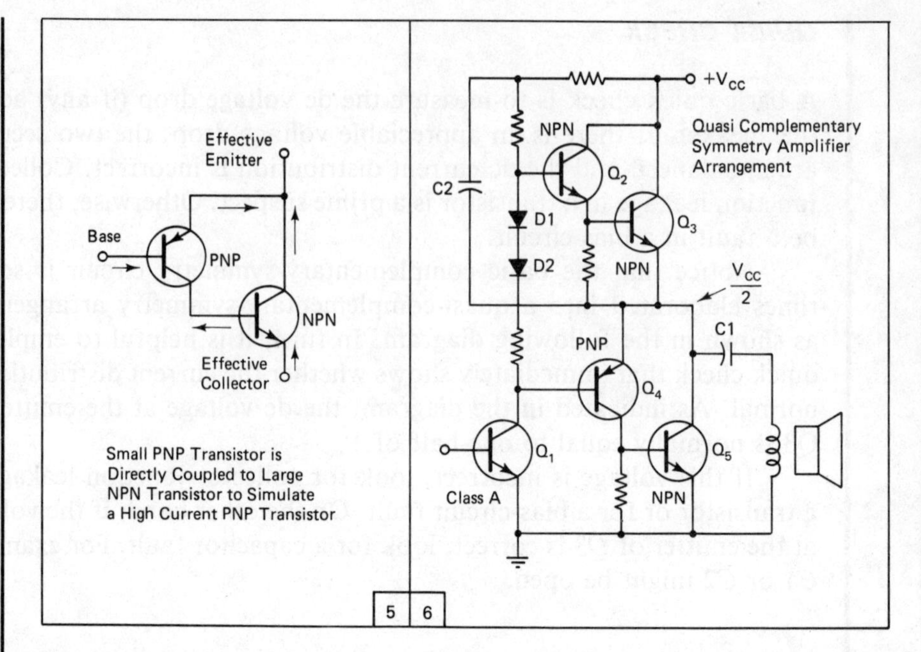
Effective
Emitter
Base
PNP
NPN
Effective
Collector
Small PNP Transistor is
Directly Coupled to Large
NPN Transistor to Simulate
a High Current PNP Transistor
+Vcc
Quasi Complementary
Symmetry Amplifier
Arrangement
NPN
Q2
C2
D1
Q3
D2
NPN
Vcc/2
C1
PNP
Q4
Q1
Q5
Class A
NPN
5
6

2

Resistance Checks and Measurements

OHMMETER OPERATION

The ohmmeter is second only to the dc voltmeter in its general-purpose usefulness. Although it is an outwardly simple instrument, much time can be wasted if its characteristics are not observed. For example, as shown in Fig. 2–1, the red test lead of a VOM may or may not be the positive lead when the meter is operated on its ohmmeter function.* The only way to be certain is to make the test depicted in the diagram. Ohmmeter test polarity is of no concern when checking a resistor out-of-circuit. On the other hand, ohmmeter test polarity can be the most important consideration when checking a resistor in-circuit, or when checking simiconductor devices out-of-circuit.

As shown in Fig. 2–2, a germanium diode normally has a barrier potential of approximately 0.2 V. This means that the voltage remains constant after the device reaches 0.2 V, and any changes in applied voltage result only in current fluctuations (unless, of course, the diode may be defective). If the applied voltage is greater than 0.2 V, and is applied in the forward direction,

*The red test lead of a VOM is always the positive lead for the dc voltage function, although it may be the negative lead for the ohmmeter function.

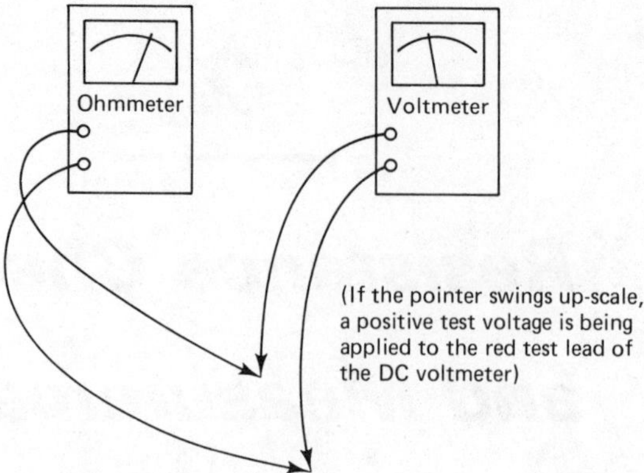
Ohmmeter
Voltmeter
(If the pointer swings up-scale, a positive test voltage is being applied to the red test lead of the DC voltmeter)

Note: When operated on its dc voltage function, a VOM has its test leads polarized red:positive, black:negative. This may also be true when a VOM is operated on its ohmmeter function—or it may not be true. In turn, the polarity of the ohmmeter test leads should be checked with a dc voltmeter. Most semiconductor tests with an ohmmeter require that the operator apply the test leads in specified polarity.

Observe that in the case of a DVM, its red lead is positive and its black lead is negative on both dc voltage and ohmmeter functions.

Figure 2–1 Use of a dc voltmeter to check the polarity of the ohmmeter test leads.

the germanium diode will conduct more or less current. Next, as shown in the diagram, a silicon diode normally has a barrier potential of approximately 0.5 V. This means that a silicon diode will not conduct any current unless the applied voltage is greater than 0.5 V (unless, of course, the diode is defective). If the applied voltage is greater than 0.5 V, and is applied in the forward direction, the silicon diode will conduct more or less current.

A silicon diode starts to conduct in the forward direction at approximately 0.5 V, but a germanium diode starts to conduct in the forward direction at about 0.2 V. It is most helpful to make the experiment shown in Fig. 2–3. It is evident that a germanium diode has lower forward resistance than a silicon diode. An ohmmeter measures this forward resistance in terms of an E/I ratio and the value of this ratio is indicated in ohm units on the ohmmeter scale.

A normal silicon diode has an extremely high reverse resistance and this resistance is practically unmeasurable with service-type ohmmeters. On

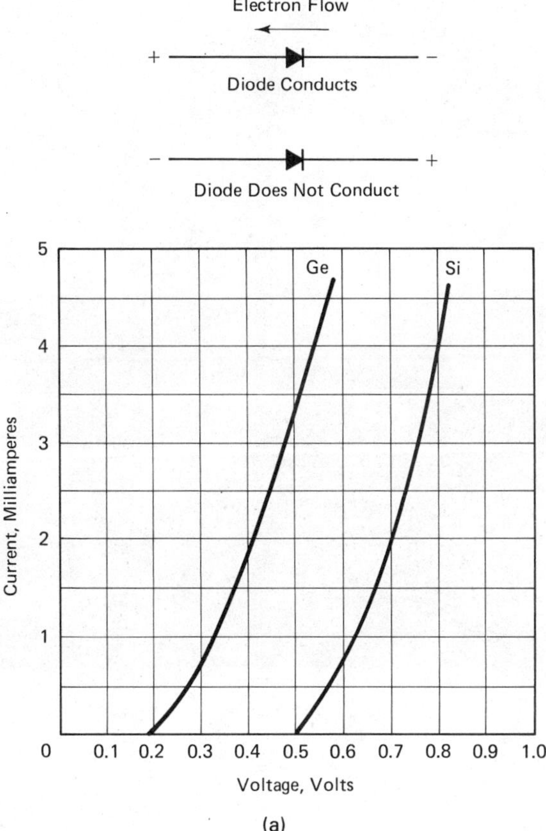

(a)

Note: If the Si voltage-current characteristic were extended several volts in the reverse direction, the current flow would be practically zero, until the Zener point was reached. When the Zener voltage is reached, the reverse current increases with great rapidity. This Zener current can confuse ohmmeter measurements of reverse resistance in silicon diode and transistor devices if the ohmmeter applies sufficient test voltage to reach the Zener point. For example, the Radio Shack 50,000 ohms/volt multitester applies a maximum of 9 V on its $R \times 10$ K range, and applies a maximum of 1.5 V on its $R \times 1$ K range. In turn, a silicon diode or silicon transistor may appear to have low reverse resistance when tested on the $R \times 10$ K range of the ohmmeter, whereas the same diode or transistor measures infinite reverse resistance when tested on the $R \times 1$ K range of the ohmmeter.

Figure 2–2 Time-saving facts to observe when checking diodes with an ohmmeter. (a) Diode characteristics. (b) Ohmmeter in use. (Courtesy Radio Shack, a Tandy Corporation Company.)

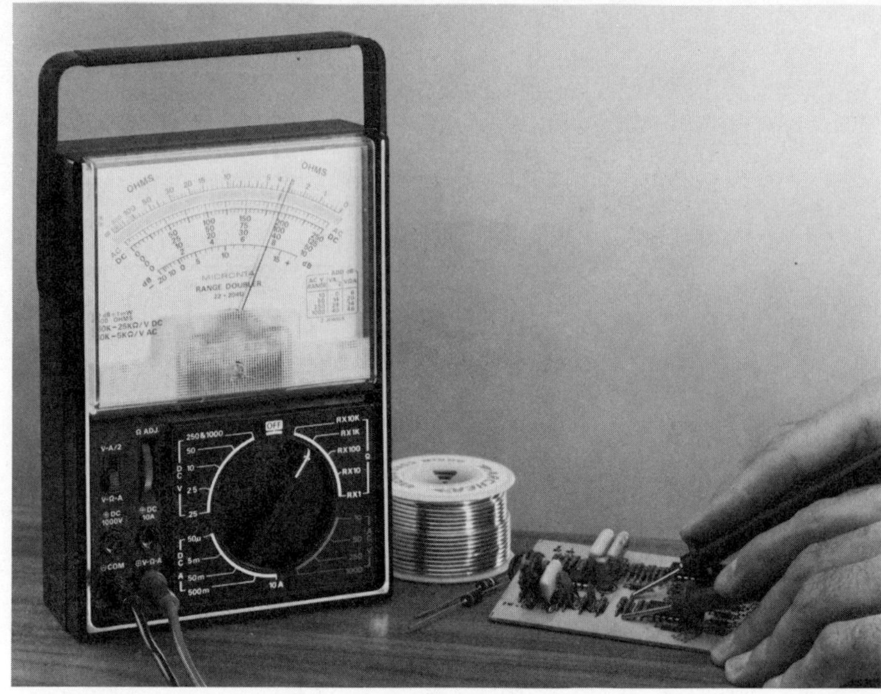

(b)

Figure 2–2 (*cont.*)

the other hand, a germanium diode in normal working condition has a reverse resistance that is detectable, even if you cannot accurately measure it with a 50,000 ohms/volt multimeter. *Notice that the reverse resistance of a silicon diode is easily measurable with a DVM that has megohm (or siemens) ranges*. Observe that ohms = 1/siemens, and that siemens = 1/ohms. The siemens unit of conductance was formerly called the mho.

It is helpful to observe the forward-resistance values for a typical silicon diode and for a typical germanium diode on various ranges of five different types of ohmmeters, as listed in Fig. 2–4. This example of test data shows the widely different resistance values that will be indicated on various ohmmeter ranges. *All of these values are correct for the particular experimental condition*. In other words, a semiconductor diode is a nonlinear resistance, and its resistance value is dependent upon the selected operating point (value of applied voltage).

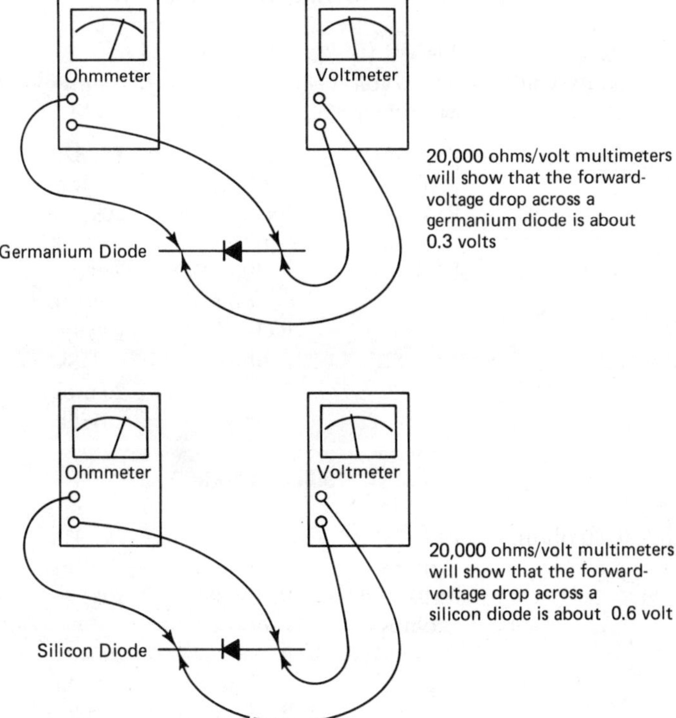

Note 1: These voltages are nominal. Germanium diodes drop between .2 and .3 V. Silicon diodes drop between .5 and .7 V.

Note 2: The ohmmeter applies a different test voltage to the diode depending upon the range to which the meter is set. Accordingly, the forward-voltage value that you measure will depend on the particular ranges provided by your ohmmeter, and the range to which the meter is set.

Figure 2–3 Experiment shows that the forward-voltage drop across a silicon diode is approximately twice the forward-voltage drop across a germanium diode.

CAUTIONARY NOTE

Electronics personnel should keep in mind the limited forward-current ratings of some mini and micro diode types. This is just another way of saying that if an ohmmeter produces an excessive forward-current flow through a

Germanium Diode

1000 ohms/volt Multimeter	*20k ohms/volt Multimeter*	*50k ohms/volt Multimeter*	*6-range DVM*	*Auto-ranging DVM*
450 ohms	Rx1: 18 ohms Rx10: 90 ohms Rx1k: 310 ohms	Rx1: 11 ohms Rx10: 47 ohms Rx100: 250 ohms Rx1k: 1600 ohms Rx10k: 2000 ohms	2k: 379 ohms 20k: 10.5k 200k: 10.5k 2000k: 115k 20 meg: 150k	10.5k

Silicon Diode

1000 ohms/volt Multimeter	*20k ohms/volt Multimeter*	*50k ohms/volt Multimeter*	*6-range DVM*	*Auto-ranging DVM*
1250 ohms	Rx1: 25 ohms Rx10: 200 ohms Rx1k: 1100 ohms	Rx1: 12 ohms Rx10: 88 ohms Rx100: 675 ohms Rx1k: 5000 ohms Rx10k: 6000 ohms	2k: 910 ohms 20k: — 200k: 50.8k 2000k: — 20 meg: 2.3 meg.	328k

Note: You need to know about the *range hold* function of an autoranging DVM when it is being used to check semiconductor devices. There is a range hold position that may be employed, if desired. When you switch your autoranging DVM to its range hold position, the instrument locks on the particular range that is being used at the time. In turn, if you measure the forward and reverse resistance values of a diode with your DVM set to its range hold position, you will measure quite different forward and reverse resistance values when your DVM is not set to its range hold position.

Figure 2–4 Forward-resistance values typically measured for silicon and germanium diodes with different types of ohmmeters operated on various ranges.

small diode, its junction will be burned out. As an illustration, a 1N34A germanium diode is rated for a steady forward-current flow of 50 mA. Like all commercial ratings, this is a conservative rating, and we can expect that the diode will withstand a moderate overload without damage. However, it would be very poor practice to test this diode with an old-style ohmmeter that could produce a forward-current flow of 100 mA, for example.

E/I RESISTANCE MEASUREMENTS

It can save time in some experimental and troubleshooting situations to remember that resistance is an *E/I* ratio, and if the situation is such that the workbench test panel is indicating the applied voltage and the resulting current flow, you need only to divide the current into the voltage to note the resistance. (See Chart 2–1.)

Experiment: A voltage-current graph for a 117-V 7-W light bulb is shown in Fig. 2–5. Any tungsten-filament lamp will have this type of nonlinear resistance characteristic. Using the ac power supply in your workbench test panel, plot a voltage/current curve for a chosen light bulb. Observe that the filament resistance is equal to the *E/I* ratio at any point along the curve. For example, we see in Fig. 2–5 that the filament resistance at 20 volts is equal to 20/.01, or 2,000 Ω. Again, the filament resistance at 60 V is equal to 60/.04, or 1,500 Ω.

It is helpful to note that a lamp filament has a positive resistance coefficient (its resistance increases as its temperature rises). On the other hand, a semiconductor diode has a negative resistance coefficient (its resistance decreases as its temperature rises).

Experiment: Connect your VOM test leads across a germanium diode and note its forward-resistance on the $R \times 1$ ohmmeter range. Then warm the diode between your fingers and watch the ohmmeter scale. Observe that the indicated resistance decreases as the diode warms up. This test illustrates the principle of a negative resistance coefficient. It has far-reaching consequences in electronic troubleshooting procedures.

MEASURED RESISTANCE OF STACKED DIODES

Two or more diodes may be connected in series (stacked) in various types of electronic equipment. Diodes may be stacked to attain higher peak inverse voltage (to prevent diode damage from substantial reverse-voltage values).

Chart 2–1

WORKING WITH RESISTORS

Resistors are generally marked with color bands or have the resistance value printed on the resistor body. The power rating of a resistor may be printed on its body if its power capability is 5 W or more. As a guideline, you can estimate the power rating of a resistor from its size as shown in the following:

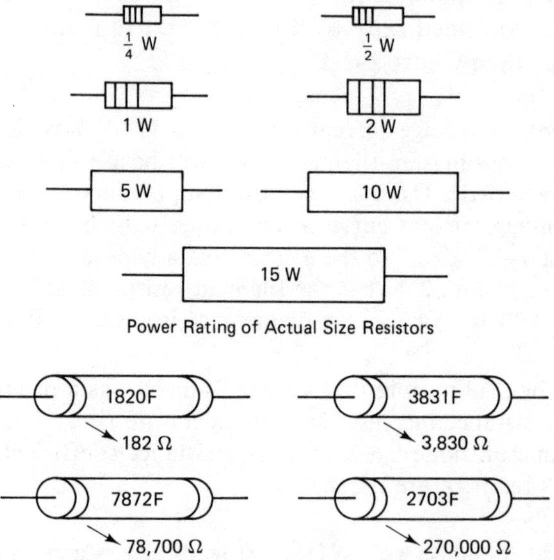

Power Rating of Actual Size Resistors

Identifying Unusual Precision Resistor Labels

You may encounter $\frac{1}{8}$-W resistors on some printed-circuit boards and in miniature electronic equipment. A $\frac{1}{8}$-W resistor is approximately half the size of a $\frac{1}{4}$-W resistor. Typical resistors have rated tolerances of ±20 percent, ±10 percent, or ±5 percent. Precision resistors have a rated tolerance of ±1 percent, or sometimes even less. Precision resistors may not have the same identification numbers as other resistors. For example, the precision resistors depicted in the diagram above illustrate how the last number on the right denotes the number of zeros to add, similar to the third color band on conventional color-coded resistors. The standard resistor color code is as follows:

Resistor Color Code				
Color	1st + 2nd Significant Figures	Multiplier	Tolerance	Failure Rate*
Black	0	1	–	–
Brown	1	10	± 1%	1.0
Red	2	100	± 2%	0.1
Orange	3	1000	± 3%	0.01
Yellow	4	10000	± 4%	0.001
Green	5	100000	–	–
Blue	6	1000000	–	–
Violet	7	10000000	–	–
Gray	8	100000000	–	–
White	9	–		Solderable*
Gold	–	0.1	± 5%	–
Silver	–	0.01	± 10%	–
No Color	–	–	± 20%	–

*When Used on Composition Resistors, Indicates Percent Failure per 1,000 Hours On Film Resistors, A White Fifth Band Indicates Solderable Terminal.

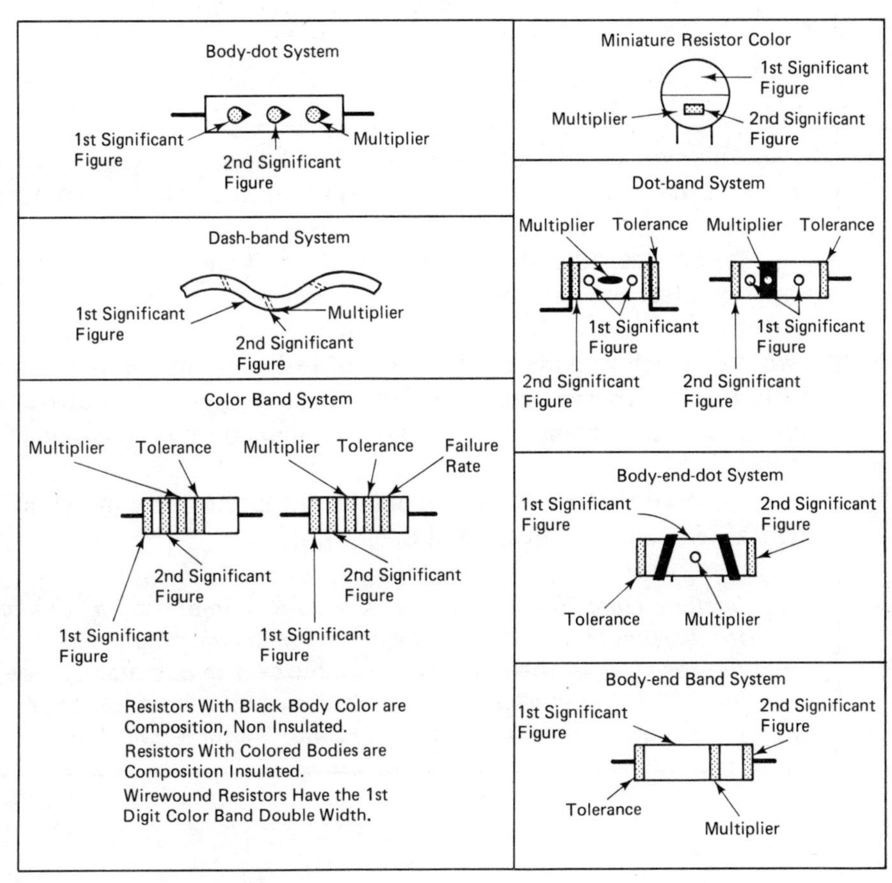

Potentiometers may have their resistance values marked on the back of their cases. However, if a potentiometer is unmarked, its resistance value can be measured with an ohmmeter. Potentiometer power ratings are related to their size; note, however, that a wirewound potentiometer dissipates a comparatively large amount of power for its size. Potentiometer resistance may increase uniformly as the shaft is turned; this is called a *linear taper*. An audio taper potentiometer has a resistance element such that its resistance increases nonuniformly as the shaft is turned (the potentiometer is said to have a logarithmic taper).

Experimenters and troubleshooters will save time and tempers by having one or more adjustable power resistors available. The most useful type is 100-Ω, 50 W adjustable resistor, as shown here.

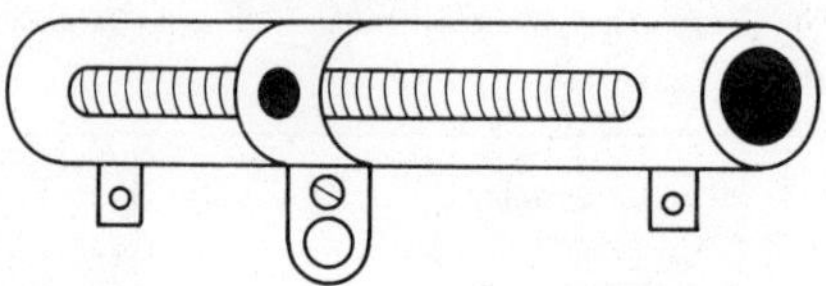

This type of resistor can be adjusted for any chosen value from 1 to 100 Ω. Its 50-W power rating ensures that it is not likely to burn out in routine experimental or troubleshooting procedures.

Resistance substitution boxes also find handy application during experimental and troubleshooting procedures. Commercial resistor substitution boxes are typically provided with three rotary control switches which provide a wide range of resistor combinations from 1 Ω to 10,000 Ω. Observe that the resistors in a substitution box are rated for typically ±10 percent tolerance and for maximum permissible current flow.

Observe that resistors in commercial electronic equipment have typical temperatures in normal operation.

Bottom Line: *You can frequently save prime time in preliminary troubleshooting procedures by quickly measuring the temperatures of the resistors in the unit under test. (Use a temperature probe with your DVM). For discussion, see the author's* Troubleshooting Electronic Equipment Without Service Data, *Prentice-Hall, 1984.*

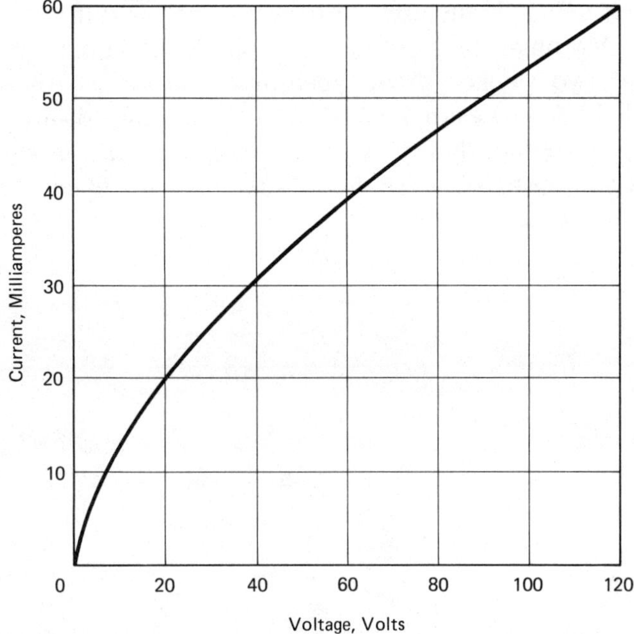

Note: In this example of nonlinear resistance, the tungsten filament has a ''cold'' resistance that is only about 0.1 of its ''hot'' resistance (at normal operating voltage). This is a steady-state plot of the voltage/current relation of the filament. It is evident that if the lamp is switched on from a ''cold'' condition, the filament will briefly draw about 10 times normal operating current. This is called the starting current surge. The excessive starting-current surge rapidly falls to the normal operating current value as the filament temperature quickly rises and the filament resistance attains its ''hot'' value.

Figure 2–5 A typical tungsten-filament voltage/current graph such as can easily be found with the ac power supply in the workbench test panel.

Diodes may also be stacked to obtain a dc level shift (to reduce a source voltage to a somewhat lower value that will remain essentially constant as current demand varies). In other words, if we connect two diodes in series and pass a forward current through the diodes, there will be two forward-voltage drops (about 1.2 V for silicon diodes) across the stack.

From the viewpoint of resistance measured with an ohmmeter, the forward resistance of a stack seems to increase faster than might be anticipated, based upon the measured resistance value for one diode. With reference to

Fig. 2-6, a VOM was used to measure the resistance of 1, 2, 3, 4, and 5 diodes. Although the resistance of a single diode measured 70 Ω, the resistance of two stacked diodes measured 180 Ω, and the resistance of five stacked diodes measured 1,200 Ω. This seeming disproportion is due, of course, to the fact that the forward resistance of a diode is nonlinear, and the test voltage across each diode is reduced as the number of diodes in the stack is increased.

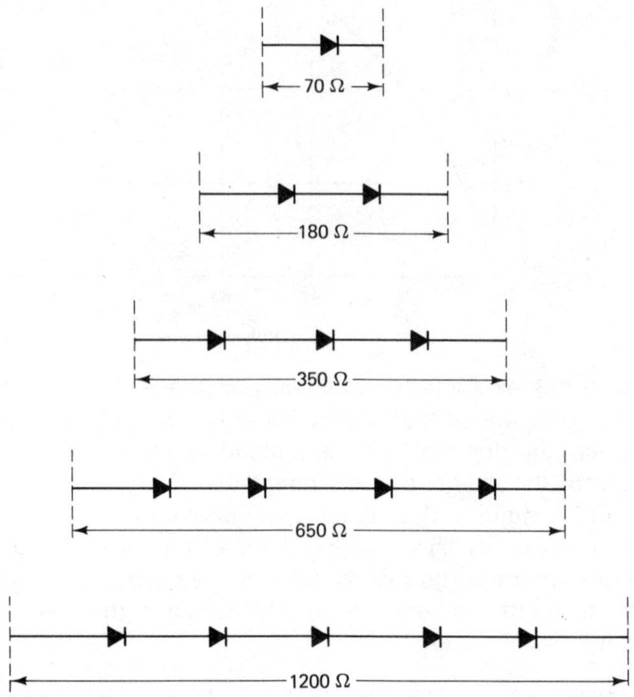

Note: Although a conventional ohmmeter indicates progressively higher forward-resistance values for series-connected diodes, this test is nevertheless useful, inasmuch as the technician can observe whether the measured resistance of a stack is "in the ballpark." For example, if a resistance of 70 Ω were measured for a 2-stack, the technician would conclude that one of the diodes is shorted. On the other hand, if a measured resistance of 1,500,000 Ω were obtained for a 2-stack, the technician would conclude that one (or both) diodes is practically open.

Figure 2-6 Example of measured values of forward resistance for stacked diodes.

LOW-POWER OHMMETERS

A low-power ohmmeter is a prime time-saver when measuring resistance values in semiconductor circuitry. We have seen that a germanium diode conducts zero current unless the applied forward voltage exceeds 0.1 V. Similarly, a silicon diode conducts zero current unless the applied forward voltage exceeds 0.5 V. (The same principle applies to bipolar transistors). The significance of this barrier potential in ohmmeter test procedures is that all semiconductor devices in a circuit will "look" open provided that the ohmmeter applies a test voltage that does not exceed the barrier potential. (See Fig. 2–7.)

Note in passing that a normal silicon diode conducts zero current unless the applied forward voltage exceeds 0.5 V. However, a faulty silicon diode may conduct a large value of current at 0.001 V. This is just another way of saying that if a diode is shorted, its resistance is zero regardless of the applied test-voltage value. Accordingly, time can be saved in this type of troubleshooting if the semiconductor devices are checked before in-circuit resistance measurements are made. Preliminary semiconductor device tests can be made in-circuit with modern transistor/diode checkers. Alternatively, in-circuit turn-off tests may be made.

It may be noted that some types of DVMs provide low-power ohmmeter facilities supplemented by a "diode check" function. This diode check function employs an ohmmeter source voltage of 1.5 V, so that front/back junction resistance tests can be made in the same manner as with a VOM. Other DVMs provide a set of low-power resistance ranges and a set of high-power resistance ranges. Still other DVMs provide alternate high-power and low-power resistance ranges. Some of the lowest priced DVMs have high-power resistance ranges only.

BIPOLAR TRANSISTOR CHECKOUT WITH OHMMETER

Although this one is not really new, it is of such practical value that it should be mentioned here. A tremendous amount of time can often be saved if the technician can verify the basing of a transistor and determine whether it is normal operating condition by means of a few ohmmeter measurements. This is an out-of-circuit test. Refer to Fig. 2–8 and proceed as follows:

1. Measure the resistance between each pair of transistor terminals.
2. The two lowest resistance values will be measured from base to emitter and from base to collector, thereby identifying the base terminal.

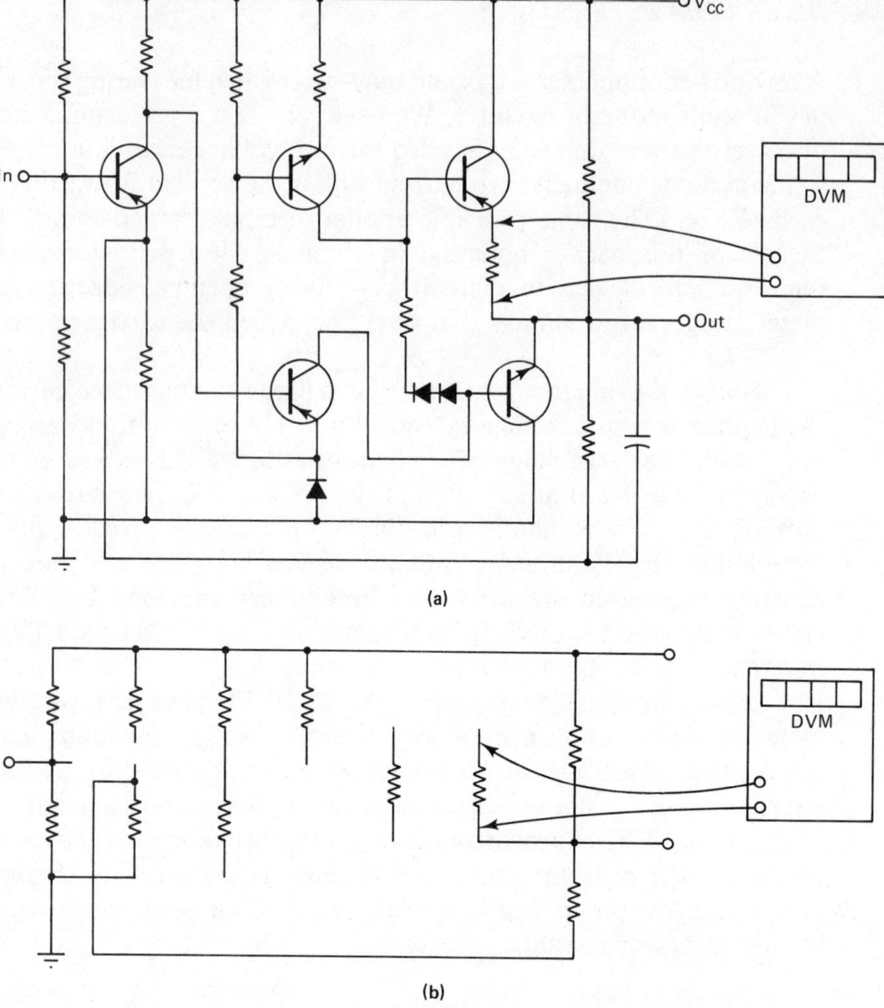

Note: Some low-power ohmmeters are designed to apply a maximum test voltage of less than 0.2 V to the circuit under test. If this type of low-power ohmmeter is used, in-circuit resistance measurements can be made in either germanium or silicon semiconductor circuitry. On the other hand, other low-power ohmmeters are designed to apply a maximum test voltage of less than 0.5 V to the circuit under test. If this type of low-power ohmmeter is used, in-circuit resistance measurements can be made only in silicon semiconductor circuitry.

Figure 2–7 Principle of in-circuit resistance measurements with a low-power ohmmeter. (a) Complete amplifier circuit. (b) Low-power ohmmeter "sees" this modified circuit.

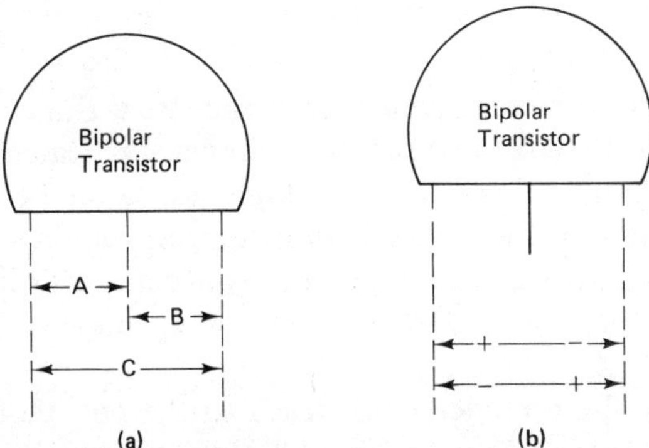

Note: First, "buzz out" the base.
Low forward resistance is measured in *A* and *B* tests. High forward resistance is measured in *C* test. The center terminal is identified as the base terminal.
A lower resistance is measured between the emitter and collector terminals when the test voltage is applied in normal operating polarity.

Note: Transistor basing requires attention. When a transistor is replaced, the basing may be different from that of the original transistor, although their appearances are identical.

Case History: An NPN replacement transistor checked out with its collector and emitter terminals reversed, as compared with the original transistor. Moreover, *the basing diagram on the replacement transistor packet was incorrect*. Therefore, the troubleshooter should not assume that the basing of a transistor is the same as would be expected. *Always verify the basing with ohmmeter tests.*

Figure 2–8 Transistor checkout with an ohmmeter. (a) The base terminal has a low forward resistance to each of the two other terminals. (b) A lower resistance is measured between the emitter and collector terminals when the ohmmeter applies a voltage that is polarized as in normal operation. (See also "Finger Test.")

3. Whether the transistor is a PNP or an NPN type is shown by the polarity of the ohmmeter test leads in measurement of forward resistance.
4. Whether the transistor is a silicon or germanium type is shown by the value of forward resistance (based on the technician's experience with the ohmmeter).
5. The collector and emitter terminals can be identified from the rule that a lower resistance is measured between these two terminals when the test voltage is applied in normal polarity.

FINGER TEST*

Unless the ohmmeter has megohm ranges, a finger test must be employed to carry out Step 5 in silicon-transistor tests. Stated otherwise, many ohmmeters cannot indicate the very high resistance between the collector and emitter terminals of a silicon transistor. In such a case, the finger test may be utilized to provide significant resistance indication, regardless of the ohmmeter that may be used. The finger test is made as follows:

1. Apply the ohmmeter test leads to the collector and emitter terminals of the transistor (which is collector and which is emitter is unknown at this time).
2. Pinch the base lead and one of the other leads between your thumb and forefinger to provide "bleeder resistance." Note the resistance reading, if any.
3. Pinch the base lead and the remaining other lead between your thumb and forefinger to provide "bleeder resistance." Note the resistance reading, if any.
4. Reverse the ohmmeter leads and repeat Steps 2 and 3.
5. The collector is the terminal that provides the lowest resistance reading when its test voltage is bled into the base terminal.

Trick of the Trade: If your skin is very dry, and you are using a comparatively insensitive ohmmeter, moisten your fingers slightly to bleed sufficient voltage into the base terminal so that a significant resistance reading is obtained.

IN-CIRCUIT RESISTANCE MEASUREMENT WITH CONVENTIONAL VOM

In some situations, it is feasible to use a conventional VOM for measuring resistance in-circuit, as exemplified in Fig. 2-9. Observe that an NPN transistor is present in this circuit. The V_{CC} supply voltage has been removed. In turn, the value of R_e can be measured if the positive test lead is applied at the emitter end of the resistor, and the negative test lead is applied at the

*This is an "out-of-circuit" test.

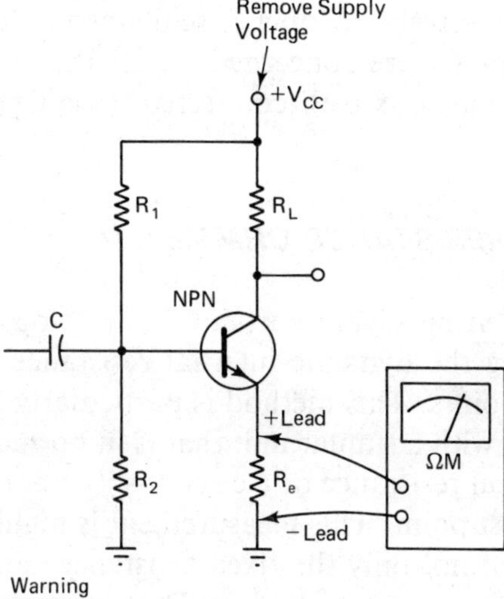

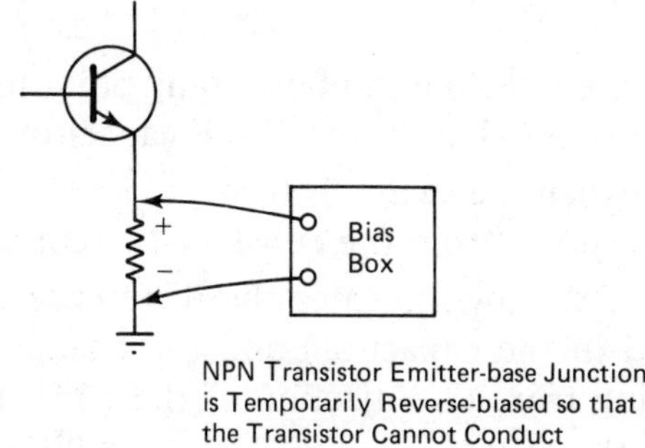

Figure 2–9 In-circuit resistance measurement with a conventional VOM. Reproduced by special permission of Reston Publishing Co. and Miles Ritter-Sanders from *Electronic Meters.*

ground end of the resistor. This is a feasible resistance measuring procedure because the ohmmeter reverse-biases the emitter-base junction whereby it is effectively open-circuited.

The same procedure can be used to measure the resistance of $R1$, $R2$, or R_L, provided that the technician first reverse-biases the emitter-base junction. As shown in Fig. 2–9, a bias box may be used to bias the emitter positive at a somewhat higher voltage than is applied by the ohmmeter. Then, the transistor is effectively an open circuit insofar as resistance measurements of $R1$, $R2$, or R_L are concerned. After the resistance measurements are completed, the bias box is disconnected. (See Chart 2–1.)

AUTOMATIC INTERNAL RESISTANCE OHMMETER

Much time can often be saved, particularly in "tough-dog" trouble situations, by measuring the dynamic internal resistance of a suspected circuit from various test points. This method is particularly helpful when made on a comparison basis with a similar unit that is in normal operating condition. The dynamic internal resistance of a circuit is its "hot" resistance to ground from any chosen test point. This measurement is highly informative because it takes into account not only the fixed resistance values, but also the junction resistances in the circuit under test. Dynamic internal resistance cannot be measured with an ordinary ohmmeter. However, it can easily be measured with the automatic internal resistance ohmmeter depicted in Fig. 2–10. It operates as follows:

1. When the test tip is applied at any point in a "live" circuit, the voltage at that point charges the 22-μF capacitor via the 100-kΩ resistor.
2. Then, when the switch is thrown, 1 mA of constant current flows into the test point from the constant-current source.
3. The DVM now indicates the difference between the original voltage (stored in the capacitor) and the change in voltage at the test point resulting from the injection of the 1 mA test current.
4. When the DVM is operated on its millivolt range, its readout is equal to the number of ohms of dynamic internal resistance between the test point and ground via the circuit under test.

A practical 1-mA constant-current source for use in checking typical semiconductor circuitry can be devised by connecting a 100,000-Ω resistor in series with the output from a 100-V power supply, as shown in the diagram. (Your workbench dc power supply is handy for this purpose, provided that it is capable of outputting 100 V.) This is a practical constant-current source because most bipolar transistor circuitry has an internal resistance considerably less than 100,000 Ω.

Observe that the arrangement shown in Fig. 2–10 is polarized for checking at a negative test point. If you wish to check at a positive test point,

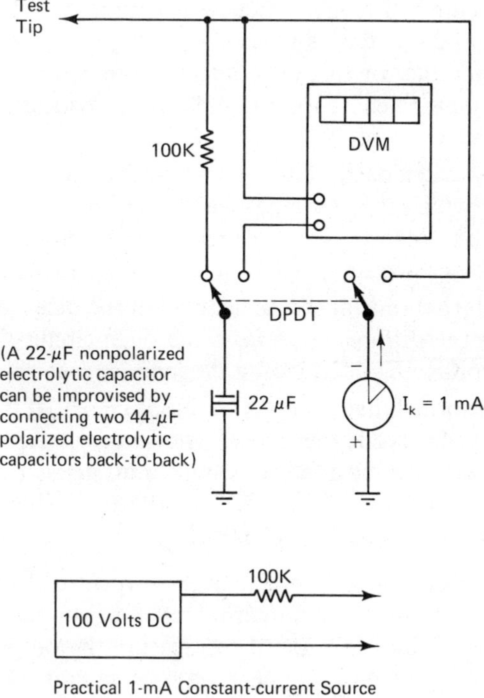

Note: The capacitor functions as a charge-storage device to hold the original voltage value which was present at the node under test. Then, when the switch is thrown, 1 mA of constant current is forced through the dynamic internal resistance. This produces a voltage increment of $I_k R_{in}$, or $0.001 R_{in}$ V. Thus, the DVM indicates the difference between the stored voltage value and its incremented value. Since $I_k = 1$ mA, the DVM mV readout equals the value of the dynamic internal resistance in ohms. (I_k may consist of a 100-kΩ resistor and 100-V power supply). The DPDT switch may be a miniature toggle type such as the Radio Shack 275-664.

Figure 2–10 Arrangement of a dynamic internal resistance ohmmeter.

the polarity of the constant-current source should be reversed. Note that the 22-μF capacitor will proceed to charge slowly from the constant-current source when the switch is thrown. This is just another way of saying that the initial readout on the DVM will gradually decay toward zero. In turn, the technician should observe the DVM readout as soon as the switch is thrown.

Experiment: This is a highly practical experiment that demonstrates the presence and value of a battery's internal resistance. A new or used 9-V transistor battery is suitable for this experiment. With reference to Fig. 2–11, a value of 1,000 Ω may be chosen for *R*; this will provide a reasonable current drain of about 9 mA.

With the switch open, note the voltmeter reading. This is the open-circuit voltage E_{oc}, which may be regarded as the EMF of the battery. Next, close the switch and note the voltmeter reading; also note the current reading.

The previous measurements provide necessary data for calculation of the battery's internal resistance. As denoted in the diagram, the internal resistance is equal to the difference between the open-circuit voltage E_{oc} and the closed-circuit voltage E_{cc}, divided by the load current value *I*.

It may be noted that a battery develops increased internal resistance as it ages. A battery also has higher internal resistance under heavy load than under light load. Keep in mind that consistent units must be used in calculations. For example, volts must be divided by amperes. Thus, if the current drawn is 5 mA, it will be calculated as 0.005 A.

Experiment: Another highly practical experiment is depicted in Fig. 2–12. It demonstrates the low value of internal resistance in a zener voltage regulated power supply. You can use a DVM to measure the output voltage from the power supply. This is typically 5.06 V on open circuit. Then, if you connect a resistance load in series with a milliammeter across the power-supply output, more or less current will be drawn, and the current value will be indicated by the milliammeter. Connect a resistive load of approximately 300 Ω to the power supply and note the current flow and also the reduction in power-supply output voltage that results. These measurements permit calculation of the internal resistance for the power supply. For example, if the load draws 15 mA, and the output voltage drops from 5.06 V to 5.02 V the internal resistance is equal to 0.04/0.015, or approximately 2.7 Ω.

WORKBENCH TEST EQUIPMENT REQUIREMENTS

A full complement of modern electronic test equipment can be a tremendous time-saver. However, not all technicians, experimenters, hobbyists, and students can justify the investment in a large number of high-performance instruments. Basic guidelines are as follows:

A good DVM is the keystone requirement; it can make highly accurate voltage measurements, and you can measure comparatively low voltage values, such as tenths of millivolts. A versatile DVM provides both low-power and high-power ohms ranges (plus a continuity beeper), and can measure very high resistance values in nanosiemens units. As previously noted, this

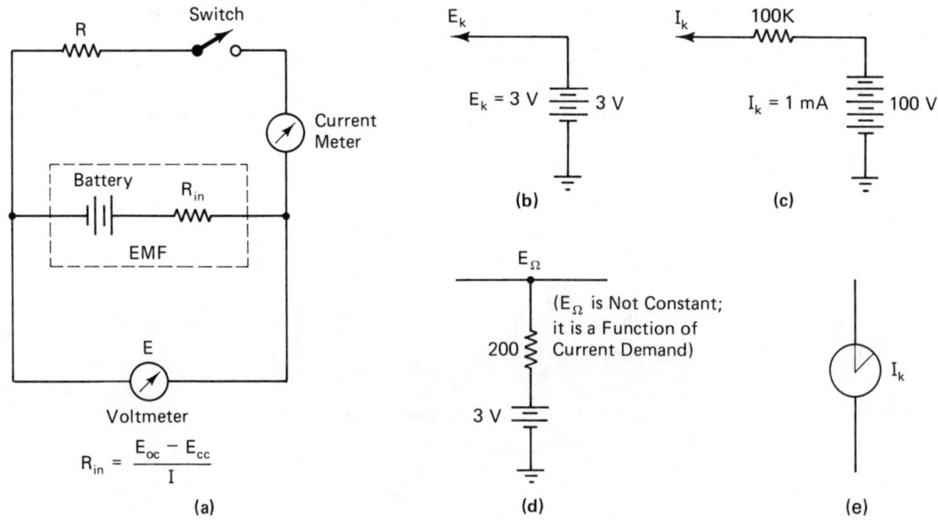

Note 1: The electromotive force (EMF) of a battery denotes the electric potential that is produced by chemical action. When this EMF drives current through the internal resistance of the battery, an IR drop results which subtracts from the EMF value. In other words, the terminal voltage of the battery under load is equal to the EMF minus the IR drop across the internal resistance. Again, if the battery is unloaded (except for the input resistance of a DVM) the terminal voltage is essentially equal to the EMF.

Note 2: Whenever you are making tests that require Ohm's-law calculations, you can save time by "punching out" the answers on your pocket calculator.

Figure 2–11 Practical arrangement for measurement of a battery's internal resistance. Electrical sources. (a) Battery is a practical constant-voltage source. (b) Battery and resistor form a practical constant-current source. (c) Battery and resistor form an ohmic electrical source; (d) Symbol for a constant-current source.

type of DVM also serves as a very sensitive null indicator and facilitates various test techniques that are impractical with less sophisticated meters.

Another time-saving feature of a good DVM is its ability to measure device and component temperatures accurately and thereby facilitate preliminary troubleshooting procedures. Moreover, a top-of-the-line DVM includes basic digital-logic probe test facilities. This feature can pinch-hit for a digital logic probe and also provides effective intermittent monitoring. An-

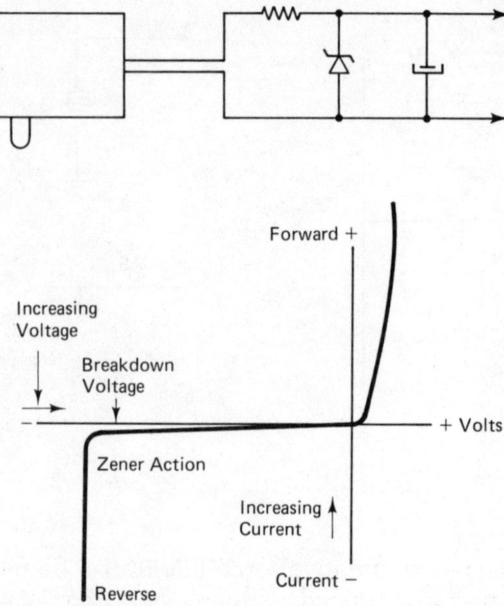

Note 1: This small voltage-regulated power uses a 9-V battery eliminator, such as the Archer (Radio Shack) 270-1552A, or equivalent with a 90-Ω $\frac{1}{4}$W resistor and a zener diode such as the Radio Shack 276.565, or equivalent, and a 500 μF 16-V electrolytic capacitor.

Observe that the 5.1-V zener diode operates as a shunt regulator. A zener diode is reverse-biased into its zener region. With the operating point of the diode placed in its zener region, a large current change corresponds to a very small voltage change. In other words, the internal resistance of the diode is very small within its zener region.

Note 2: If excessive current demand is imposed on the regulated power supply, the operating point will move out of the zener region. In turn, the output voltage will drop, and the internal resistance of the power supply will greatly increase. No regulation is provided unless the operating point is within the zener region of the diode characteristic.

Figure 2–12 The zener regulated power supply has a low internal resistance.

other helpful feature of a versatile DVM is its peak-hold function, which permits the troubleshooter to catch an over-voltage transient, for example.

This is not to say that you might as well go fishing if you do not have a top-of-the-line DVM. One of the factors that separates the old timers from the short timers is the ability of the latter to make outdated equipment perform well. For example, you can measure dc voltages in high-resistance circuits with an ordinary VOM if you use a charge-storage technique with a resistor and a large capacitor. Again, you can measure tenths of millivolts with an ordinary VOM if you use an op-amp preamplifier. And so on.

After the DVM, your next most important requirement is probably a really good oscilloscope. (Some technicians would say that a high-performance signal generator is the Number 2 requirement—but you will find that a good scope is an accurate frequency indicator for an inexpensive generator.) You will find it worthwhile to invest in a triggered-sweep dual-channel scope with a bandwidth of at least 15 MHz. Practical test work is often speeded up if the scope provides $A+B$ and $A-B$ displays in addition to conventional A & B displays.

The third most important test-instrument requirement (for most hobbyists and experimenters) is a good audio oscillator. It should have low distortion (good sine waveform) to facilitate high-fidelity tests. If the audio oscillator provides a choice of single-ended or double-ended output, you can make a comparatively wide variety of tests. Consideration is seldom given to whether an audio oscillator is line-powered or battery-powered. This is not necessarily a trivial point, inasmuch as certain types of test techniques are facilitated by battery operation (complete isolation of the audio oscillator from the power line).

Note, however, that you can often "get by" with an outdated audio oscillator if suitable expedients are employed. As an illustration, an uncertain output frequency can be precisely measured with a lab-type triggered-sweep scope. A distorted output can be "cleaned up" to some degree with RC filtering, and can be markedly improved with a sharply peaked tone amplifier. Again, a single-ended output can be converted into a usable double-ended (push-pull) output by passing the signal through a hi-fi audio transformer. Note that this expedient also provides considerable isolation from the power line (minimizes capacitive "sneak currents" to the power-line ground).

If your audio oscillator has extended high-frequency output, so much the better. If it also provides alternative square-wave output, that's still better! The basic requirement for good square-wave output is reasonably fast rise time, such as 20 ns. If you are seriously into digital troubleshooting, you will need a good logic probe, logic pulser, and current tracer. These basic testers are generally used to localize a trouble area, and follow-up tests with a scope or DVM will sometimes be in order.

A good wide-range AM signal generator is a "must" in any ambitious electronics shop. Basic generator requirements are frequency calibration accuracy, good waveform, and an accurate output level attenuator. An AM generator may also provide FM output. However, you will usually choose a separate FM generator with stereo signal output. Also, if you are into hi-fi, you will need a good harmonic-distortion meter.

Except in specialized situations, other instruments are less essential. For example, you may or may not opt for a versatile semiconductor tester. You may or may not opt for an FM/TV sweep-and-marker generator (although it can be a tremendous time-saver). Similarly, you can "get by" in color-TV service work without a color-bar generator, but it is a prime time-saver.

Finally, do not opt for any electronic test instrument unless you fully intend to use it on all appropriate jobs. The basic question is: "What is it going to do for me?"

3

AC Tests and Measurements

OVERVIEW

Hobbyists, experimenters, students, and apprentices are routinely concerned with various kinds of ac voltage and current. For example, an ac voltage may range in frequency from nearly zero to many MHz. It may be a pure sine wave, as in basic tests of hi-fi equipment, or it may be a mixture of a low-frequency sine wave and a high-frequency sine wave, as in intermodulation tests of audio amplifiers. Again, an ac voltage may be a square wave as used in tests of video-amplifier response. Or, it may be a pulse waveform as employed in digital-electronics test procedures.

An ac voltage may comprise a carrier (CW) component plus an amplitude-modulation component as in checking AM radio circuitry. Or, an ac voltage may consist of a constant-amplitude sine waveform which varies in frequency, as in tests of FM radio circuitry. As another example, an ac voltage may consist of a sine wave that is pulse-modulated, as in the color burst component of the composite color-television signal. We will find that much wasted time can be saved by utilizing various shortcuts and time-saving techniques, as explained in the following pages.

BASIC AC WAVEFORMS

Much wasted time can be avoided in ac voltage measurements if we have a practical understanding of the basic ac waveforms. With reference to Fig. 3–1, the two fundamental steady-state waveforms are the sine wave and the square wave. The square wave is an example of a complex wave (any complex wave can be built up from a combination of sine waves). Observe that the sine wave and the square wave depicted in Fig. 3–1 have the same peak-to-peak voltage. On the other hand, the sine wave has a root-mean-square (rms) voltage equal to 0.707 of peak, whereas the square wave has an rms voltage equal to its peak voltage.

Typical service-type ac voltmeters measure the rms voltage of a sine wave, but do not indicate the true rms voltage of a square wave. Note in passing that true-rms ac voltmeters are available, although they are comparatively expensive. It is helpful to recognize the relation of an rms ac voltage to an equivalent dc voltage. This relation states: The power produced by an rms ac voltage is the same as the power produced by an equal value of dc voltage. As an illustration, a lamp filament will glow with equal brightnesses if it is energized 117 V dc or ac.

Timbre: Observe that all three of the waveforms depicted in Fig. 3–1 have the same pitch. Their pitch is determined by their repetition rate. Note that an experienced troubleshooter can recognize some basic characteristics of a waveform from the sound that it produces through a signal tracer. The signal tracer should have uniform frequency response from 20 Hz to 20 kHz, and less than 1 percent distortion. (Otherwise, a critical evaluation of timbre cannot be made.) The speaker used with the signal tracer should also have hi-fi characteristics. Note that the best evaluation of timbre is made at moderately loud output level—low-level and very high-level outputs cannot be accurately evaluated.

It may also be noted that there are various interesting and sophisticated follow-up tests that can be made by the technician who is into timbre analysis. For example, the waveform can be passed through an RC differentiating or integrating circuit before it is applied to the signal tracer (such as a small hi-fi amplifier and speaker). The interesting point in this case is that when a true sine wave is differentiated or integrated, its timbre remains unchanged. On the other hand, when any complex wave is differentiated or

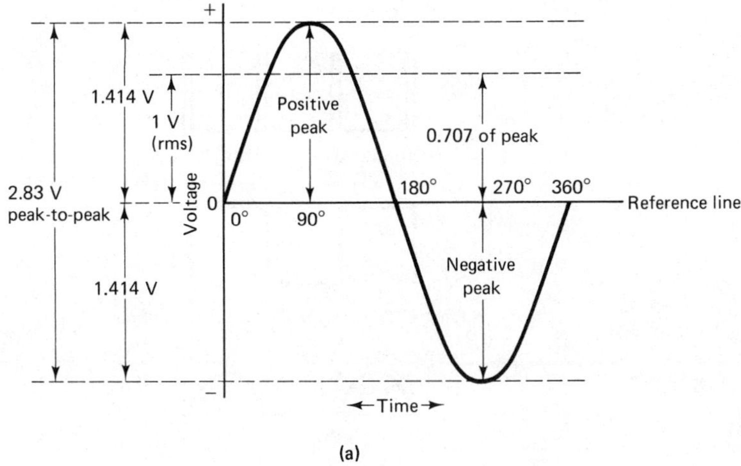

(a)

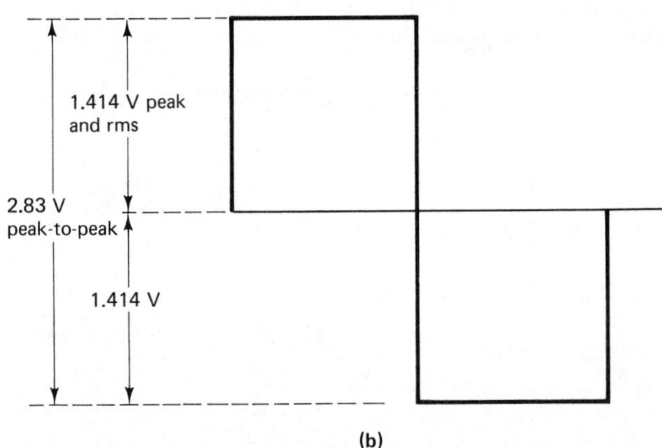

(b)

Note 1: An experienced hi-fi technician can detect very small percentages of distortion in an amplifier by listening to its reproduction of musical passages.

Note 2: When checking waveform timbre, an audio signal tracer consisting of a hi-fi amplifier such as the Realistic (Radio Shack) SA-150 may be used with a hi-fi speaker such as the Radio Shack No. 40-2043.

Figure 3–1 AC waveform parameters. (a) Sine wave. (b) Square wave.

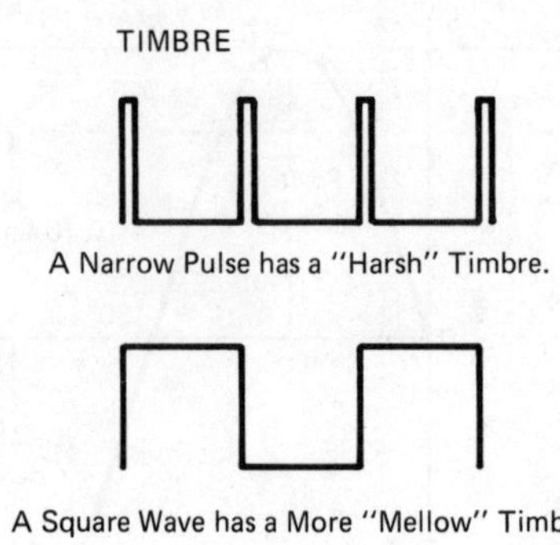
TIMBRE
A Narrow Pulse has a "Harsh" Timbre.
A Square Wave has a More "Mellow" Timbre.

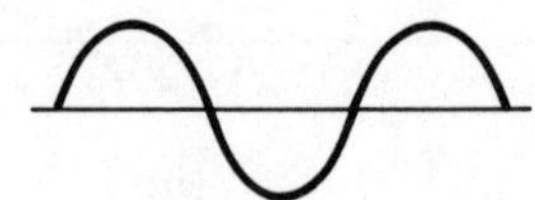

A Sine Wave has a "Thin" and "Soft" Timbre.

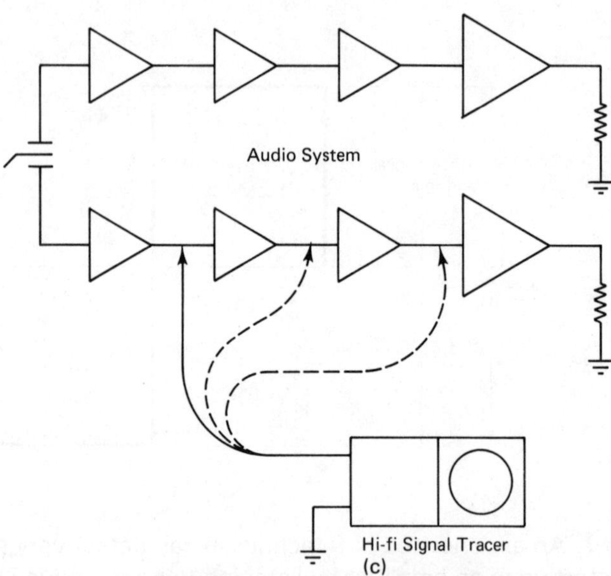
Audio System
Hi-fi Signal Tracer

(c)

Note 3: A hi-fi signal tracer can often save prime time in preliminary troubleshooting of hi-fi systems. If the troubleshooter "plays it by ear," he can often zero-in on an area that is causing distortion, hum, noise, or signal attenuation. When troubleshooting a stereo system, comparison tests can also be made to verify preliminary conclusions.

Figure 3–1 (*cont.*)

integrated, its timbre is changed in a manner that is a function of the complex waveshape.

COMPONENTS OF PULSATING DC WAVEFORMS

Electronic equipment often operates with pulsating dc voltages (see Fig. 3–2). One of the most common examples of pulsating dc-voltage operation is the ac voltmeter per se. In other words, most service-type ac voltmeters employ half-wave instrument rectifiers. Accordingly, the ac voltmeter responds to the average value of one-half cycle in the applied ac waveform (whereas the voltmeter scale is calibrated in rms values of a sine wave). As shown in Fig. 3–3, the average value of a sine wave half cycle is 0.318 of peak voltage.

As previously mentioned, the scale of a service-type ac voltmeter is usually calibrated in rms units to indicate the rms values of sine waveforms. It follows from Fig. 3–2 that a service-type ac voltmeter will not indicate the rms value of a square wave. (The rms value of a sine wave is equal to 0.707 of peak.) Inasmuch as the ac voltmeter responds to the half-cycle average value of the waveform (or 0.318 of peak), the scale on the ac voltmeter is calibrated to indicate 2.22 times the average value of the sine wave. *Now, if a square-wave input is applied to the ac voltmeter, the scale indication will be 1.11 times the peak voltage of the square wave.*

We have observed that in the case of a square wave, its rms voltage is equal to the half-cycle peak voltage, as illustrated in Fig. 3–3. The logical conclusion is that if the technician measures the rms voltage of a square wave with a service-type ac voltmeter that has a half-wave rectifier, he must multiply the scale reading by 0.9 to determine the rms voltage of the square wave.

APPARENT RMS VALUES OF BASIC PULSATING DC WAVEFORMS ON A HALF-WAVE AC VOLTMETER

A large amount of otherwise wasted time can be saved if we understand the apparent rms values of basic pulsating dc waveforms on a half-wave ac voltmeter. It follows from Fig. 3–2 that electronic technicians are concerned with both pulsating dc waveforms and with ac waveforms. Any pulsating dc waveform has an apparent rms value on a half-wave ac voltmeter, just as an ac waveform has an apparent rms value. As shown in Fig. 3–4, the meter will

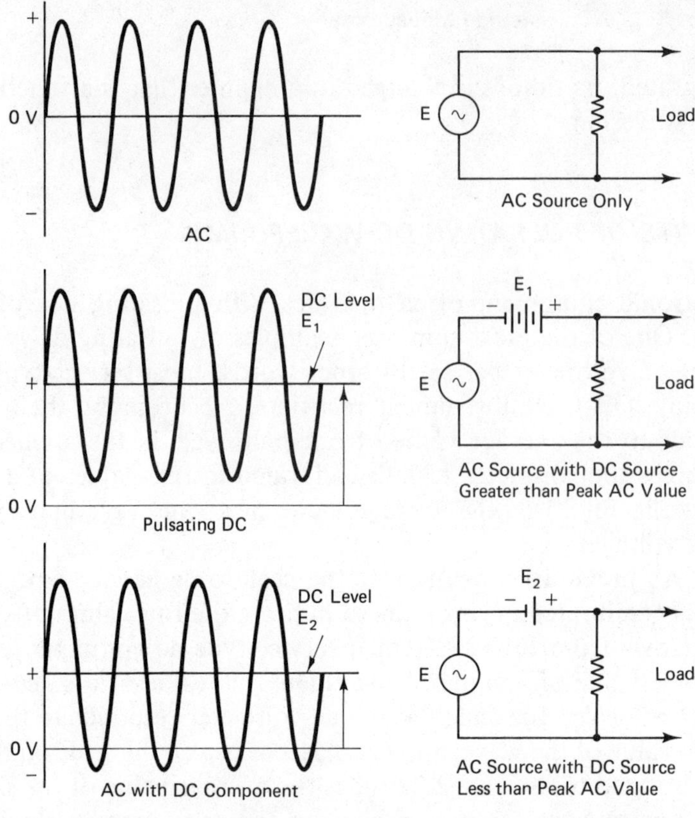

Note: When measuring component voltages in pulsating dc waveforms, or ac waveforms with dc components, the peak value of the waveform should not exceed the range limit of the voltmeter. Select a voltage range such that the peak value of the waveform is within the range. The frequency of the ac component should not exceed the frequency capability of the voltmeter.

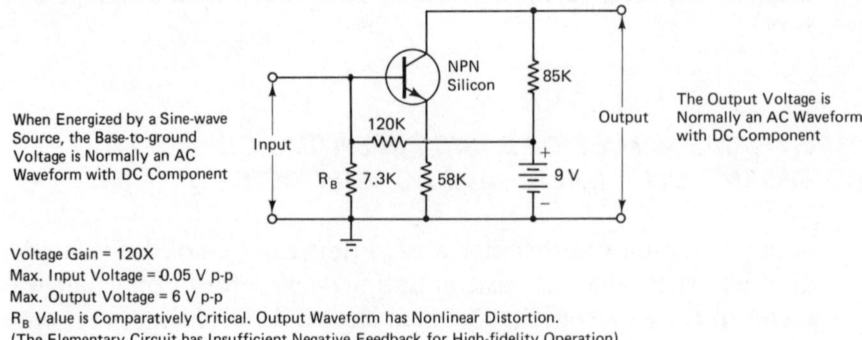

Voltage Gain = 120X
Max. Input Voltage = 0.05 V p-p
Max. Output Voltage = 6 V p-p
R_B Value is Comparatively Critical. Output Waveform has Nonlinear Distortion.
(The Elementary Circuit has Insufficient Negative Feedback for High-fidelity Operation)

Figure 3–2 Basic examples of ac waveforms with different dc components, and amplifier characteristics.

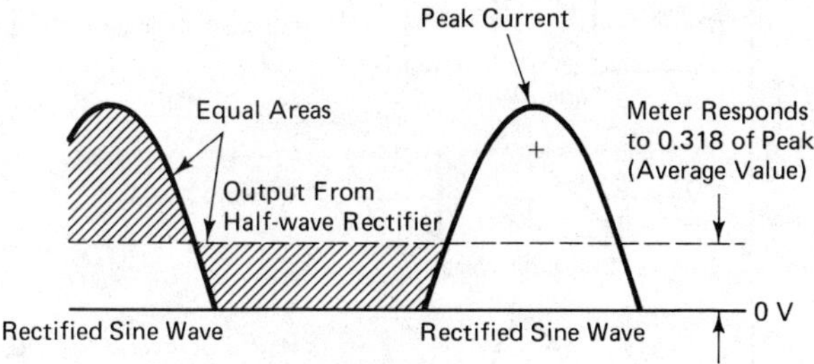

Pulsating DC Waveform

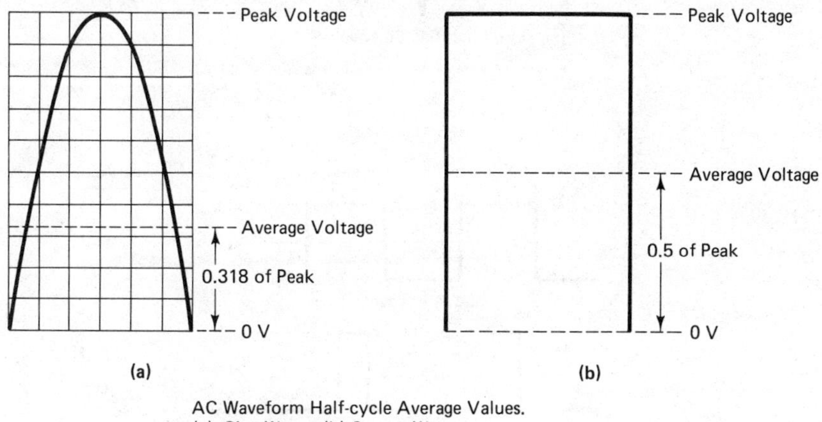

AC Waveform Half-cycle Average Values.
(a) Sine Wave; (b) Square Wave

Bottom Line: A DC Voltmeter always Indicates the Average Value of a Pulsating DC Waveform

Bottom Line: A dc voltmeter always indicates the average value of a pulsating dc waveform.

Figure 3–3 The average value of a half-rectified sine wave is 0.318 of peak. (a) Pulsating dc waveform. (b) Sine wave. (c) Square wave.

indicate that a square wave with 1-V peak has an rms voltage of 1.11 V, whereas its true rms voltage is equal to 1.

Also shown in Fig. 3–4 is the fact that the meter will indicate that a sawtooth wave with 1 V peak has an rms voltage of 0.555 V, whereas its true rms voltage is equal to 0.577 V. Again, the meter will indicate that a half-rectified sine waveform with 1 V peak has an rms voltage of 0.707 V, whereas its true rms voltage is 0.5 V. Or, the meter will indicate that a full-

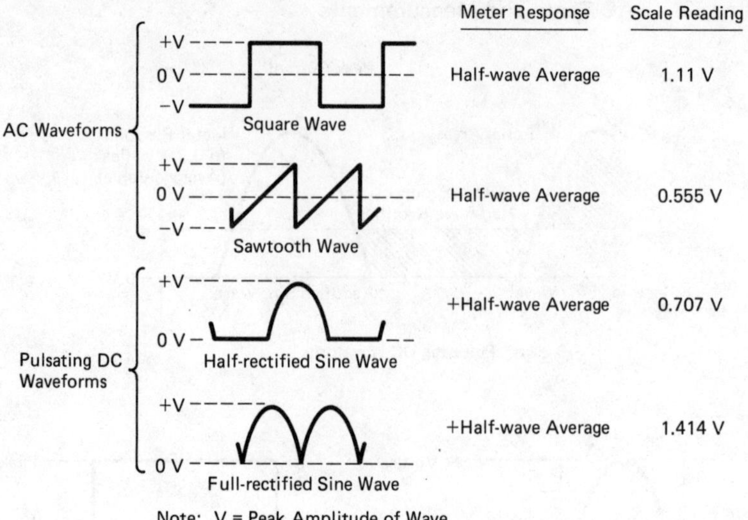

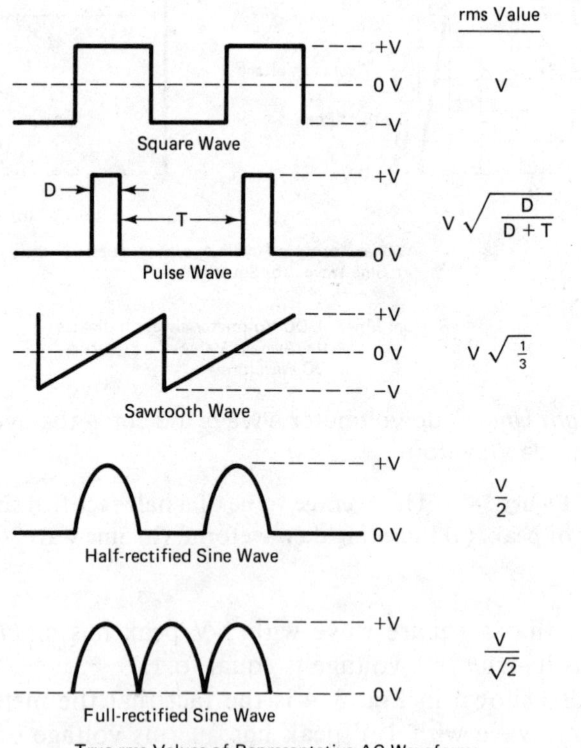

True rms Values of Representative AC Waveforms and Pulsating DC Waveforms.

Figure 3–4 Apparent rms values of basic ac and pulsating dc waveforms indicated by a half-wave ac voltmeter.

rectified sine waveform with 1 V peak has an rms voltage of 1.414 V, whereas its true rms voltage is 0.707 V.

IF YOUR AC VOLTMETER HAS A FULL-WAVE INSTRUMENT RECTIFIER . . .

Occasionally, the technician will be using an ac voltmeter that has a full-wave instrument rectifier (instead of a half-wave instrument rectifier). In such a case, the apparent rms readings in Fig. 3–4 will be the same for complex ac waveforms. On the other hand, the apparent rms readings will not be the same for pulsating dc waveforms. This is just another way of saying that if your ac voltmeter has a full-wave instrument rectifier, it will indicate an apparent rms value of 0.354 for the half-rectified sine wave, and will indicate an apparent rms value of 0.707 for the full-rectified sine wave.

Bottom Line: Save time, tempers, and misunderstandings by knowing your ac voltmeter, by knowing the basic waveform parameters, and by taking nothing for granted.

PEAK-RESPONSE AC VOLTMETERS

In addition to the familiar half-wave ac voltmeter, and the occasional full-wave ac voltmeter, you will sometimes encounter a peak-response ac voltmeter. All three types of ac voltmeters have scales calibrated in rms units. It follows from previous discussion that a peak-response ac voltmeter indicates 0.707 of peak for an ac square wave; it indicates 0.707 of peak for a sawtooth wave; it indicates 0.707 of peak for a half-rectified sine wave; it indicates 0.707 of peak for a full-rectified sine wave. In turn, the required scale factors follow from the true rms values with respect to peak listed in Fig. 3–4.

TRUE RMS AC VOLTMETER

Yes, Virginia, there is a true rms ac voltmeter, although it is comparatively costly. If the technician is using a true rms voltmeter, basic ac and pulsating dc waveforms will measure their true rms values, and no scale factors are employed. Technically speaking, ac voltmeters with instrument rectifiers respond to the average value of a half cycle, which is a linear function of area. On the other hand, true rms ac voltmeters respond to the square root of the sum of the squares in the waveform.

AC VOLTMETER RESPONSE TO PURE DC

Since a pulsating dc voltage consists of a dc level with a superimposed ac component, we may pause to consider the response of an ac voltmeter to a pulsating dc voltage with a zero ac component. It follows from previous discussion that ac voltmeter response to pure dc depends upon its instrument-rectifier circuitry. In other words, if an ac voltmeter with a half-wave instrument rectifier is connected to a 1.5 V dc source, the scale indication will be 3.33 V rms. However, if an ac voltmeter with a full-wave instrument rectifier is connected to a 1.5 V dc source, the scale indication will be 1.665 V rms. Of course, the true rms voltage in both of these examples is 1.5 V which brings us to Fig. 3–5.

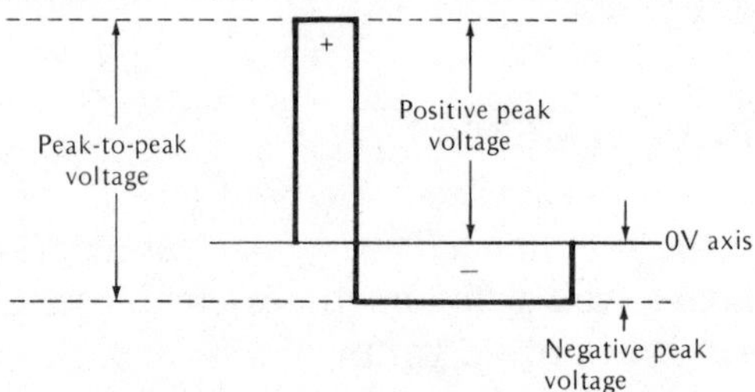

Note: The average value of the positive excursion is the same as the average value of the negative excursion.

Bottom Line: Since this is an ac waveform, there is just as much electric charge in its positive excursion as in its negative excursion, or, the average value of the ac waveform is zero. This means that the average value of the positive excursion is the same as the average value of the negative excursion. In conclusion, this waveform will *not* exhibit turnover when applied to an ac voltmeter with a half-wave instrument rectifier, or when applied to an ac voltmeter with a full-wave instrument rectifier.

Figure 3–5 A basic ac pulse waveform has unequal peak voltages, and turnover will be observed with peak-response voltmeters.

THE TURNOVER CHECK

I have found that turnover is not really understood by many technicians (and even some engineers!). Turnover is something that all of us had *better* understand, unless wasted time is no object. Turnover is often encountered in ac troubleshooting procedures. *Turnover occurs in a test situation if the meter reading changes when its test leads are reversed.* As a practical illustration, consider peak-voltage measurements of the pulse waveform shown in Fig. 3–5.

When a peak-reading ac voltmeter is applied, we might happen to measure the positive-peak voltage, for example. Then, when the meter test leads are reversed, we will measure the negative-peak voltage of the waveform. But since the positive-peak voltage in this waveform is greater than its negative-peak voltage, neither of the readings is *the* peak voltage of the waveform (it has two peak voltages). This, then, is an eminently practical example of turnover. By way of comparison, a basic sine wave has equal peak voltages, and turnover is not encountered in this situation.

As was noted in Fig. 3–5, an ac pulse waveform will not exhibit turnover when applied to an ac voltmeter with a half-wave or a full-wave instrument rectifier. Observe that if the same pulse waveform were not an ac waveform, but was a pulsating-dc waveform, then any ac voltmeter (except a true-rms meter) would exhibit turnover for the pulsating-dc waveform. To anticipate subsequent discussion, we will find that the oscilloscope simplifies complex-waveform measurements because we are then dealing with "pictures" instead of written rules.

DISTORTION CHECKS WITH AN AC VOLTMETER

An ac voltmeter can be used to make a practical test for serious waveform distortion, as shown in Fig. 3–6. Although the test is not quantitative, and although it can "miss" in the event that the waveform has both peaks equally clipped, it is nevertheless a practical time-saver in preliminary troubleshooting procedures. This basic method employs a peak-responding DVM (scale calibrated in peak volts). First, the positive-peak voltage of the output waveform is measured, and then the negative-peak voltage of the output waveform is measured. If peak clipping is occurring, these two values will be more or less unequal.

Note that a blocking capacitor is essential in series with the lead from the audio oscillator to the base of the transistor to avoid drain-off of base-bias voltage and introduction of distortion. Although a blocking capacitor is usually not required in series with the input lead to a peak-responding DVM, it is good practice to make a habit of always connecting a blocking capacitor in series with an ac voltmeter. (Most ac voltmeters will drain off collector dc voltage unless a series blocking capacitor is utilized.)

MEASUREMENT OF STAGE GAIN

An audio amplifier stage has a normal voltage gain and a normal current gain. It also has a normal power gain, which is the product of the voltage gain and the current gain. At this time, we will consider a practical method of

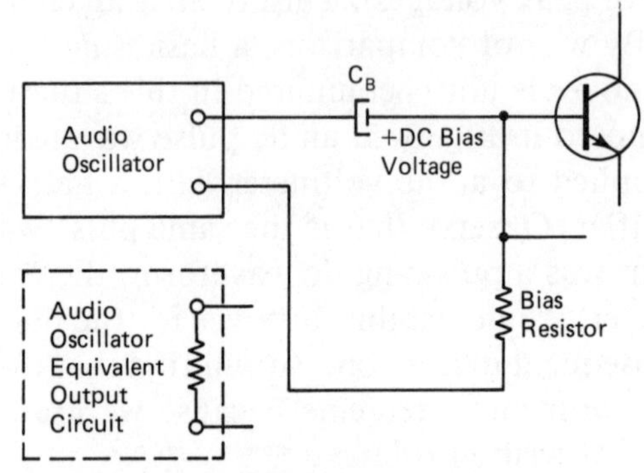

Blocking Capacitor C_B Prevents the Base-bias Voltage From Draining off Through the Audio Oscillator

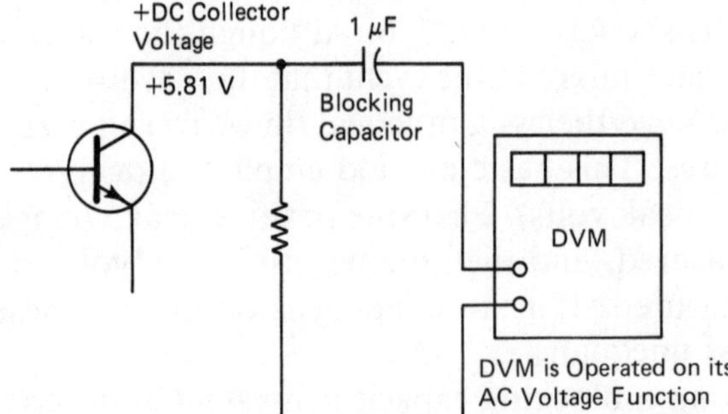

A Blocking Capacitor is Connected in Series with the AC Input Lead to the DVM.

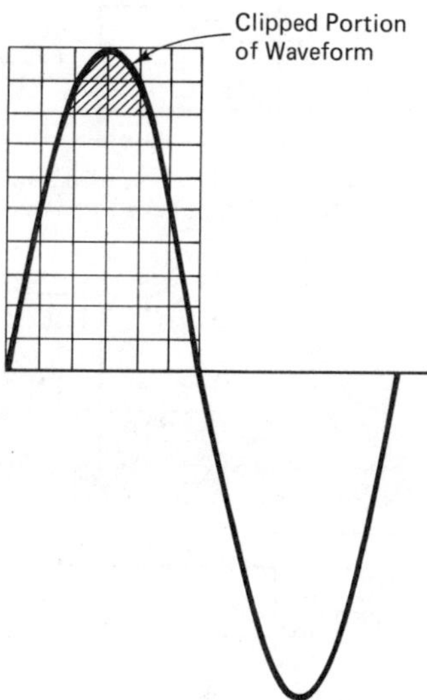

Note 1: When troubleshooting intermittent amplifier circuitry, the ac voltage can be monitored or recorded in the same general manner as a dc voltage level. This topic is detailed subsequently.

Note 2: When peak-clipping occurs, the form-factor of the clipped half-sine wave is changed, and this change, if substantial, can be determined by voltmeter tests. This is a qualitative test, not quantitative. To measure percentage harmonic distortion, a harmonic distortion meter must be used.

Figure 3–6 Peak-clipping distortion quick check with ac voltmeter.

measuring amplifier voltage gain. This measurement is customarily made at 1,000 Hz, and at maximum rated output for the amplifier. An audio oscillator is used as a source of test voltage, and a DVM is a practical necessity (to measure the input voltage).

Let us consider a practical situation in which the voltage gain is measured for the amplifier that was constructed earlier. Since a 9-V V_{CC} source is utilized, the maximum peak-to-peak output voltage that can be anticipated will be somewhat less than 9 V (transistor driven to the vicinity of cutoff, and to the vicinity of saturation). For the sake of illustration, let us assume that the voltage gain of the amplifier will be approximately 70 times. In such a case, the required input voltage would be about 8/70 V peak-to-peak, or 0.036 rms V.

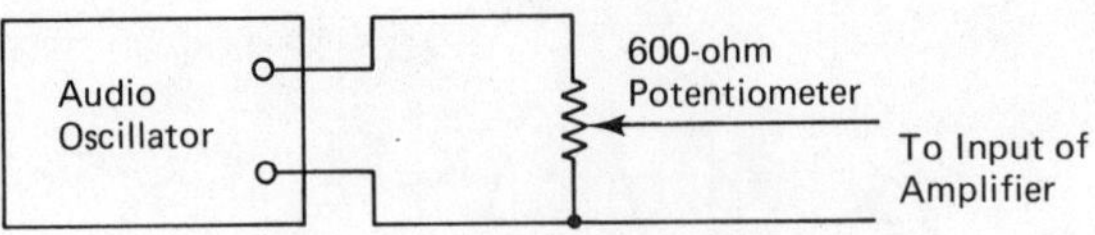

A Potentiometer can be used to Reduce the Minimum Output Level From an Audio Oscillator

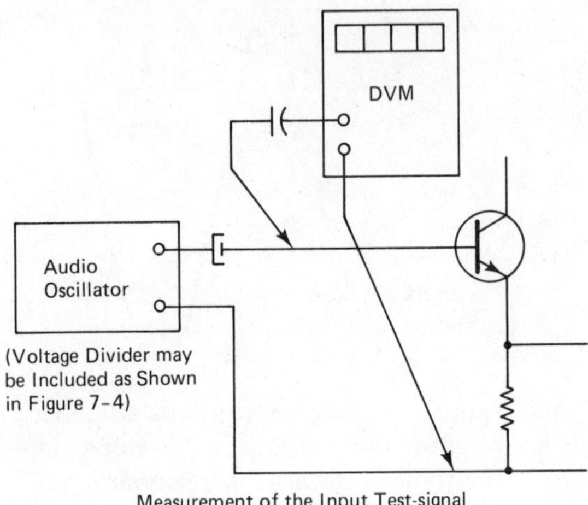

Measurement of the Input Test-signal Level to the Amplifier

Experiment: Measure the gain of your experimental amplifier at 1,000 Hz.

Repeat the measurement at 20 Hz, 500 Hz, 5,000 Hz, 10,000 Hz, and 20,000 Hz, using the same input voltage as before.

Determine the low-frequency value at which the voltage gain is 70.7 percent of the gain at 1,000 Hz.

Determine the high-frequency value at which the voltage gain is 70.7 percent of the gain at 1,000 Hz.

Bottom Line: The low-frequency cutoff point (70.7 percent output) is determined by the reactance values of the coupling capacitors in the amplifier.

The high-frequency cutoff point (70.7 percent output) is determined by the ratio of shunt capacitance to shunt resistance in the amplifier circuit.

Figure 3–7 Inputting and measuring an audio voltage for checking the gain of an amplifier.

Your audio oscillator may not be designed to provide a low-level output such as 36 mV. In turn, it becomes necessary to insert a voltage divider between the audio oscillator and the amplifier input terminals as depicted in Fig. 3–7. Before the test is started, it is good practice to set the potentiometer for zero output—in turn, the possibility of destructive overdrive to the transistor is avoided. Connect a DVM with a blocking capacitor from base to ground of the transistor, as shown in the diagram. Set the audio oscillator to 1,000 Hz and advance the potentiometer for an indication of, say, 0.035 rms mV. (See Chart 3–1.)

Next, disconnect the DVM from the amplifier input circuit and reconnect the DVM with the blocking capacitor across the amplifier output terminals. Note the reading on the DVM; this might be 2.5 rms V, for example. In turn, the voltage gain of the amplifier will be equal to 2.5/0.035 or 71 times, approximately. It is evident that if you repeat this voltage-gain measurement at various frequencies from 20 Hz to 20 kHz, you can obtain data for plotting the frequency response of your amplifier.

Chart 3–1

WORKING WITH CAPACITORS

We need to identify capacitors by their capacitance value, type, and working voltage (rated maximum voltage). Capacitors often have identification data stamped on their cases. Common types of capacitors are illustrated in the following:

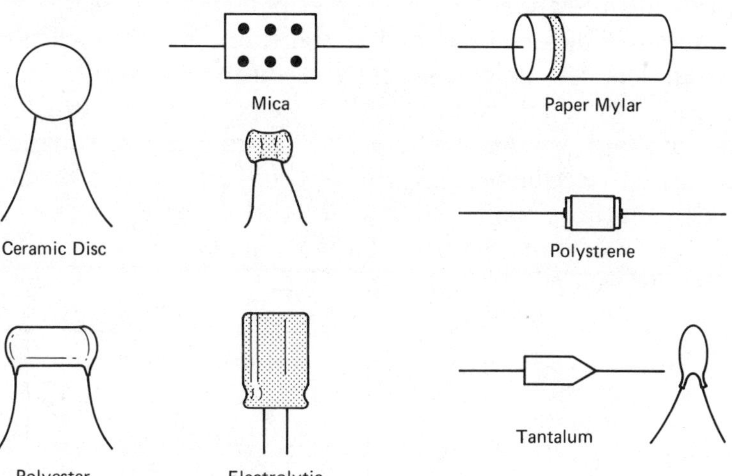

Observe that ceramic capacitors are frequently manufactured in disc form, but are also available in tubular and feedthrough types. A marking such as .01 denotes .01 μF, whereas a marking such as 270 denotes 270 pF (270μ-μF). A ceramic capacitor typically ranges in values from 5 pF to .05 μF. Working voltages up to 1 kV are common, and specialized capacitors have working voltages up to 25 kV.

Note that general-purpose ceramic capacitors can usually be substituted for mica, paper, or plastic-film capacitors in noncritical (moderate frequency) coupling, filtering, and bypass applications.

Mica capacitors are epoxy dipped and are ordinarily marked for easy identification. Mica capacitors can substitute for ceramic capacitors when temperature compensation is not of concern. Observe that some ceramic capacitors are temperature-compensated, with one of the following markings:

P 100 N 100 N P O

P denotes a positive temperature coefficient. In other words, for every degree of temperature rise between +25 and +85 degrees Centigrade, a P 100 capacitor will exhibit a capacitance increase of 100 parts per million.

N denotes a negative temperature coefficient. Stated otherwise, for every degree of temperature rise between +25 and +85 degrees Centigrade, an N 100 capacitor will exhibit a capacitance decrease of 100 parts per million.

N P O denotes no capacitance change between temperatures from +25 and +85 degrees Centigrade.

We will find that electrolytic capacitors are usually marked for ready identification, including polarity markings. Polarity markings are generally indicated as + and − signs, but may be spelled out in some cases. The case of a metal-cased electrolytic capacitor is generally its negative terminal. Note also that nonpolarized electrolytic capacitors are available, and must be used in certain circuit situations. A nonpolarized electrolytic capacitor is equivalent to two ordinary electrolytic capacitors connected in series back-to-back.

Two or more electrolytic capacitors may be enclosed in the same case, as exemplified in the following:

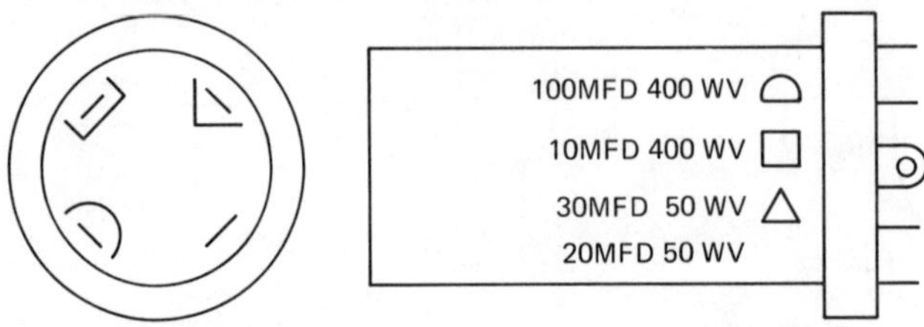

Ordinarily, the positive terminals of each capacitor are connected to terminals on the end of the case, and the negative terminals of all the capacitors are connected together and to the metal case. On the side of the case we will find stamped capacitance values, working voltage ratings, and the terminal symbols such as a square, a triangle, or a semicircle.

A capacitor substitution box is a handy item to keep beside you on the test bench. A capacitor box is outwardly similar to a resistor box. Rotary switches are provided for selection of any desired capacitance value from typically 100 pF to .1 μF or more in 100 pF steps.

Opinion among hobbyists and experimenters is divided concerning whether a commercial capacitor tester is "worth its salt." Some technicians prefer to keep a good supply of replacement capacitors on hand, and to make quick checks of suspected capacitors by means of the various methods described in this book. Others maintain that a commercial capacitor tester, with its precise indication of capacitance values, leakage resistance, and power factor is a basic test instrument for any serious shopowner. Circumstances alter cases, and it is up to the individual technician whether investment in a commercial capacitor tester can be justified.

AC VOLTAGE MONITOR/RECORDER ARRANGEMENT

Much time can be saved and procedural convenience provided by the availability of an ac voltage monitor-recorder arrangement, particularly when troubleshooting intermittent electronic equipment. With reference to Fig. 3–8, the addition of a peak-rectifier circuit to the dc voltage monitor/recorder described in the first chapter permits monitoring and/or recording of ac voltage levels in audio, radio, or television equipment. (The peak-rectifier circuit can be applied in an ac circuit with a very low frequency, such as 60 Hz, in an audio-frequency, radio-frequency, or VHF circuit.)

Observe in Fig. 3–8 that the optional DVM may be connected to the output of the peak probe in combination with the emitter follower. Thereby the operator can observe the peak voltage of the ac waveform that is being monitored and/or recorded. It should be noted that when peak-reading probes are applied in low-level circuits, a limitation is imposed on indication by the barrier potential of the diode in the probe, regardless of the DVM sensitivity.

This is just another way of saying that a germanium diode does not conduct until the signal level is at least 0.2 V; a silicon diode does not con-

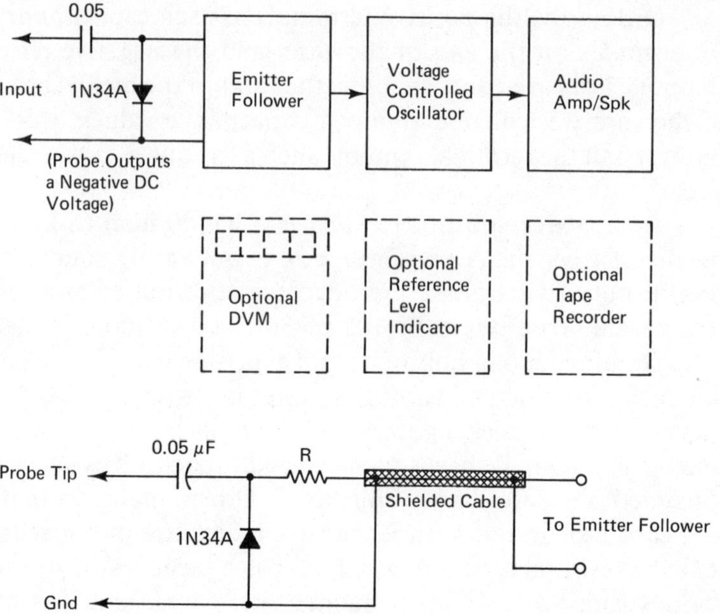

Note 1: This arrangement is similar to the dc voltage monitor/recorder described in the first chapter, except that a peak-rectifier circuit is provided for changing ac voltage into a dc control voltage. Since the emitter follower imposes slight loading on the peak-rectifier circuit, the input impedance to the arrangement is comparatively high. In turn, ac voltages can be monitored and/or recorded in relatively high-impedance circuits. If hum interference is encountered, construct the peak-rectifier circuit in a shielded housing and connect it to the emitter follower via shielded coaxial cable.

Note 2: When the ac voltage monitor/recorder is applied in high-frequency circuits, it is desirable to use a resistor *R* between the peak-rectifier circuit and the coaxial cable. This resistor may have a value of 100 kΩ. Its purpose is to contain the high-frequency currents in the probe housing, so that high-frequency energy does not enter the coaxial cable. In turn, standing waves of high-frequency voltage are avoided in the cable. Standing waves in high-frequency circuitry lead to falsely high or falsely low voltage indications depending upon the prevailing frequency.

Figure 3–8 An ac voltage monitor/recorder arrangement.

duct until the signal level is at least 0.4 V. Accordingly, the foregoing arrangement is not suited for checking very-low-level signals. Note also that this probe arrangement has a principal frequency range from 500 Hz to 10 MHz. At frequencies below 500 Hz, capacitor reactance progressively attenuates the signal level. At frequencies above 10 MHz, the input leads to the emitter follower may develop standing waves and cause erroneous level indication. However, this limitation can be avoided as shown in the diagram by employing a series isolating resistor *R* and a shielded input cable.

As a practical note, a simple peak-reading probe will objectionably load higher frequency circuits that have an internal impedance greater than 1,500 Ω, as explained in greater detail in the following pages. If you use a 1N34A germanium diode in the probe, the input voltage level should not exceed 25 rms V, to avoid possible damage to the diode. On the other hand, if you use a silicon diode in the probe, considerably higher input voltages may be applied. (The chief advantage of a germanium diode is its lower barrier potential.)

DIODE PROBE WITH ZERO BARRIER POTENTIAL

A helpful trick of the trade for monitoring voltages in low-level circuits is to forward-bias the diode in the probe, as shown in Fig. 3–9. If a small forward bias is applied to the diode in a peak-indicating probe, very low signal levels can be traced (although the full peak voltage of the low-level signal is not indicated by the DVM, due to the high internal resistance of the diode). A small forward bias may be applied by means of a bias box connected in series with the DVM, as shown in the diagram.

A germanium diode requires approximately 0.2 V of forward bias; a silicon diode requires about 0.4 V of forward bias. When the forward-bias voltage is adjusted so that the DVM is just beginning to indicate (for example, fluctuating between a zero reading and a 1 mV reading), the probe is then in its condition of maximum sensitivity for low-level signal tracing, and the DVM will respond to a very low level signal input to the probe. Note that a silicon epoxy diode has very high reverse resistance and will provide higher-precision measurements.

POSITIVE-PEAK/NEGATIVE-PEAK/PEAK-TO-PEAK HIGH-FREQUENCY AC PROBE

Here is another real time-saver. It is a combination diode probe arrangement that permits measurement of positive-peak/negative-peak/peak-to-peak voltages. It speeds up troubleshooting procedures because the three key voltages in a complex waveform can be measured with a single probe. (See Fig. 3–10.) The only disadvantage of this versatile probe is that it imposes twice

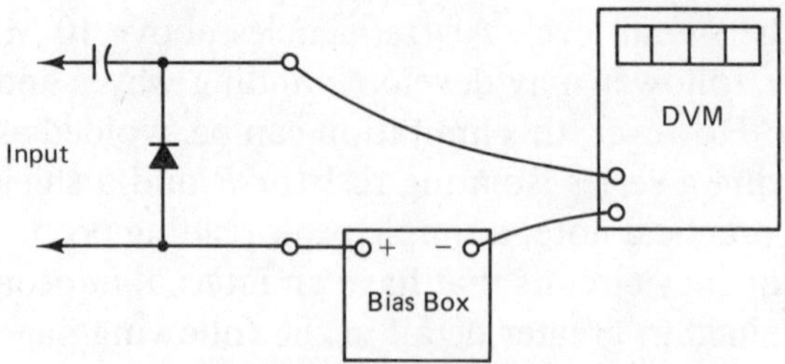

Note 1: This is a forward-biased positive-peak probe. If a negative-peak probe is used, the bias-box polarity is reversed.

The capacitor should be selected for very high insulation resistance, and the diode should be selected for very high reverse resistance.

Caution: When used with a TVM or DVM, a false readout may be obtained if the ac component of the output from the peak-reading probe exceeds the range being used on the dc voltmeter. (Make a cross-check, using the next higher range on the meter.)

Point-contact Diode: The old-style point-contact diode differs from the modern junction diode in that the former has zero barrier potential. As such it has practically the same small-signal sensitivity whether it is zero biased or forward biased. However the troubleshooter may find it difficult or impossible to locate a source for point-contact diodes. Therefore, small-signal tests must usually be made with forward-biased junction diodes.

Note 2: If the diode polarity is reversed, and the bias-box polarity is reversed, this arrangement becomes a forward-biased negative-peak probe.

This type of peak-indicating probe can be a major time-saver whenever the technician is working with low-level high-frequency circuits. It eliminates ''guesstimation'' in preliminary troubleshooting procedures.

Figure 3–9 A fraction of a volt applied as forward bias to a peak-reading probe provides maximum sensitivity to low-level signals.

the circuit loading that is imposed by a positive-peak or a negative-peak probe. In other words, a half-wave probe will objectionably load circuits that have an internal resistance greater than 1,500 Ω, whereas a peak-to-peak probe will objectionably load circuits that have an internal resistance greater than 700 Ω.

In order to avoid undue circuit loading in comparatively high-impedance circuits, this positive-peak/negative-peak/peak-to-peak probe may be utilized in combination with a probe preamp, as noted in Fig. 3–10. The

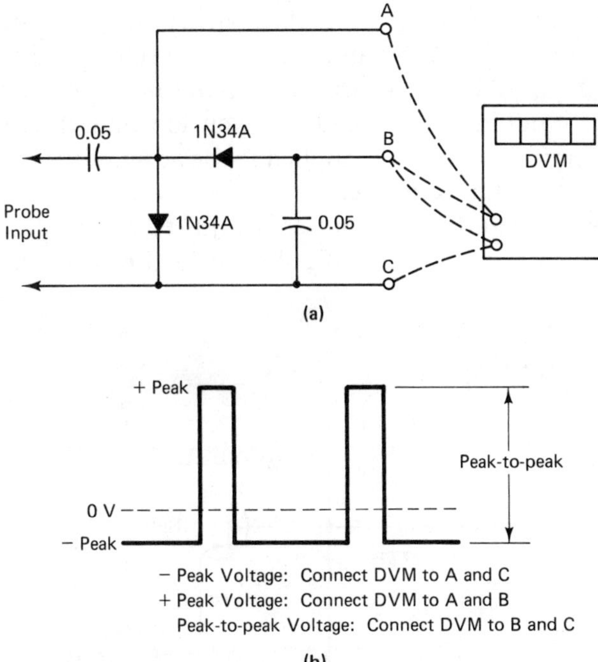

Note: The useful frequency range of this probe is from about 500 Hz to 10 MHz. This probe will objectionably load circuits that have a dynamic internal impedance greater than 700 Ω.

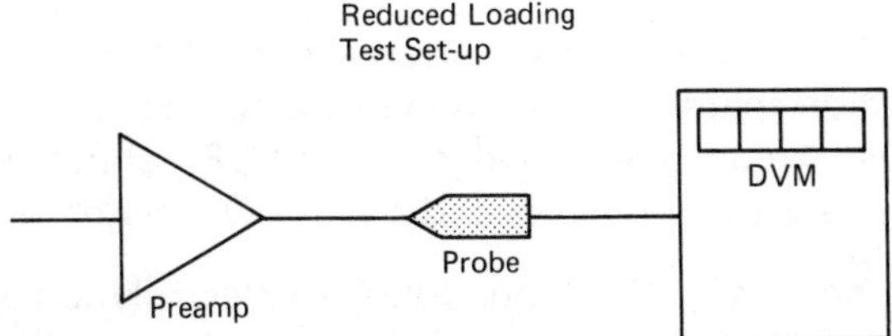

This elaboration of the probe arrangement provides high input impedance and also permits checking in low-level circuits. To obtain high input impedance, use a preamp that has high input impedance and which is adjusted for unity gain. To check in low-level circuits, use a preamp that has high input impedance and ×10 gain. In turn, divide the DVM readout by a factor of 10.

Figure 3–10 A time-saving positive-peak/negative-peak/peak-to-peak high-frequency ac probe arrangement. (a) Configuration. (b) Example of pulse-waveform measurements.

probe preamp may comprise a Darlington type emitter-follower, as depicted in Fig. 3–11, followed by an ac amplifier such as shown in Fig. 1–13(b) in Chapter 1. In other words, the emitter-follower illustrated in Fig. 3–11 provides high input impedance and relatively low output impedance. The output from the emitter-follower is applied to the ac amplifier, which, in turn, drives the high-frequency ac probe.

Observe that the ac amplifier is adjusted for the desired amount of gain by utilizing a suitable value of collector load resistor. Since the emitter-fol-

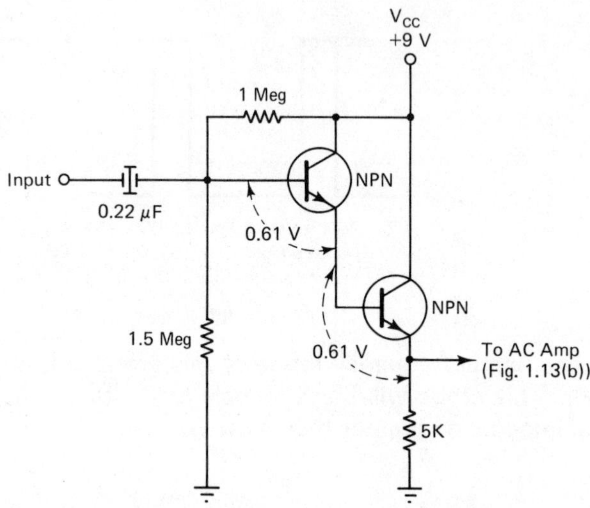

Note 1: This emitter-follower arrangement is similar to the direct-coupled configuration that was illustrated in the first chapter. However, this is an ac-coupled circuit that includes a 0.22 μF nonpolarized electrolytic capacitor, plus a 1-MΩ resistor and a 1.5-MΩ resistor for the bias circuit.

Observe that if a 0.22 μF nonpolarized electrolytic capacitor is not available, you can connect a pair of 0.44 μF electrolytic capacitors back-to-back. (A nonpolarized coupling capacitor is required because the circuit under test might have either a positive or a negative dc component.)

Note 2: Each transistor has a base-emitter bias voltage of 0.61 V. Since the transistors are connected in series, the total bias voltage from the base of the input transistor to the emitter of the output transistor is 1.2 V. We can think of the Darlington connection as a single transistor with a base-emitter bias voltage of 1.2 V.

Figure 3–11 An emitter-follower preamp for the high-frequency ac probe.

lower shown in Fig. 3–11 develops approximately 44 percent of the input voltage value, a system gain of unity is obtained by utilizing an amplifier collector load resistor that provides an amplifier gain of about 2.27 times. This is just another way of saying that 0.44 × 2.27 is, approximately, unity. Of course, the exact value of collector load resistance that you will require for precise ac voltage measurements must be determined by experiment due to commercial tolerances on devices and components.

Next, as also noted in Fig. 3–10, checks in low-level circuits should be made with the ac amplifier gain adjusted for a system gain value of 10 times. This is accomplished by utilizing a considerably higher value of collector load resistor in the amplifier. This expedient is employed for checks in low-level circuits in order to minimize the effect of barrier potential in the rectifier diodes. Stated otherwise, if the source signal level is amplified 10 times before application to the probe, barrier potential can be practically neglected. Of course, the DVM reading must be divided by 10 to obtain the true value of the signal voltage that is being measured.

Experimenters and hobbyists who are interested in amplifier parameters may note the three fundamental arrangements shown in Fig. 3–12. Three versions of common-emitter amplifiers are illustrated, each of which provides an input resistance of 20 kΩ. However, the first configuration obtains 20 kΩ of input resistance in an emitter-follower arrangement with an emitter resistor of 500 Ω. Next, the second configuration obtains 20 kΩ of input resistance in a common-emitter (CE) arrangement with an 18-kΩ series input resistor. Lastly, the third configuration obtains 20 kΩ of input resistance in a CE arrangement with unbypassed emitter resistor; the emitter resistor has a value of 500 Ω, and the collector load resistor also has a value of 500 Ω.

AUDIO CURRENT "SNIFFER"

One of the most useful time-savers that can be kept handy on the audio bench is the audio current "sniffer" shown in Fig. 3–13. It is the simplest possible arrangement that can be used to detect the presence of audio current, and to trace audio current through a wiring system. The "sniffer" is surprisingly sensitive, and will clearly indicate the presence of comparatively small ac currents. Its maximum effectiveness occurs at about 500 Hz, although it reproduces higher and lower audio frequencies also. This one is hard to beat.

Experiment: An instructive ac voltage experiment is shown in Fig. 3–14. This is an RC integrating circuit that develops an output voltage greater than its input

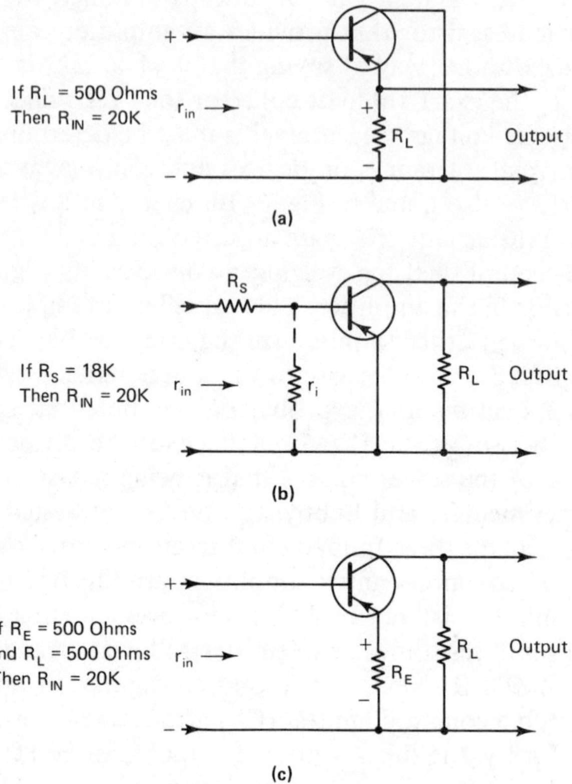

Disadvantages:

1. Small variations in current demand by the following stage cause large changes in the value of r_{in}.
2. Bias stability is poor if the bias is fed to the base via R_s. If bias is fed via a supplementary resistor r_i, the value of r_{in} is reduced accordingly. There is a small loss in current gain, in either case.

Advantages:

1. Low output resistance may be an advantage for impedance matching.
2. The total input resistance remains comparatively constant regardless of variations in current demand by the following stage.
3. Bias stability is comparatively good.

Figure 3–12 Preamp configurations for provision of 20 kΩ input resistance.

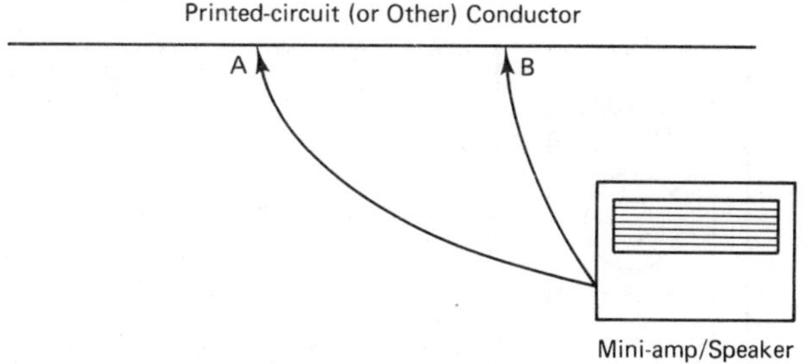

Trick of the Trade: In a preliminary troubleshooting checkout of an audio system, we often wish to quickly determine whether there is dc current flowing in a wire, and whether it may be accompanied by an ac audio current. It is easy to do this if we make a quick check with a mini-amp/speaker as shown above. The input teat leads are touched at arbitrary points *A* and *B* along the conductor. *If there is a click from the speaker, there is dc current flow in the conductor—but if there is no click there is very little or no dc current flowing in the conductor. If there is a tone from the speaker, there is ac audio current flow in the conductor—but if there is no tone there is very little or no ac audio current flowing in the conductor.* A Radio Shack 277-1008 mini-amp/speaker is very useful for this quick check. It provides a gain of about 1,700 times, and develops maximum rated output with a 1-mV input voltage change. When the volume control is advanced to maximum, even a comparatively small ac current flow in the conductor will produce sufficient IR drop along the conductor to provide a clearly audible tone from the speaker.

Figure 3–13 A useful audio current checker/tracer.

voltage at its resonant frequency. The voltage step-up that occurs is due to the phase shift in voltage from the 100-Ω to the 1,000-Ω to the 10-kΩ resistor. The diagram illustrates a basic arrangement. If you experiment with the RC values, particularly in the third section, you can obtain somewhat more voltage gain. Observe that this configuration can also be rearranged as an RC differentiating circuit, and a similar voltage bootstrap action will be obtained. (See also Chart 3–1.)

AC Current-Voltage Relations

Voltage-bootstrap action in Fig. 3–14 is based on ac current-voltage relations in the RC circuitry. In other words, the current is not in phase ("in step") with the applied voltage to an RC section. This is the reason why the sum of the ac voltage drops across the series resistors is greater than the ac volt-

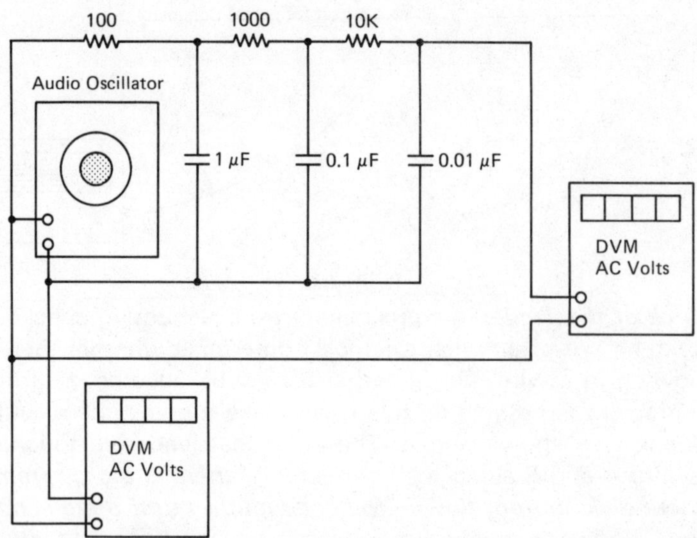

Note: This is a basic 3-section RC integrating circuit arranged for voltage-bootstrap action. It develops an output voltage that exceeds its input voltage at its resonant frequency. For example, if you set the audio oscillator to approximately 2 kHz with an output of 1 V, you will measure about 1.12 V at the output of the circuit. This basic circuit can be modified to provide a voltage gain of approximately 20 percent by experimenting with the resistance and capacitance values. The capacitance value in the output section may be varied, for example. The resonant frequency is affected in addition to the voltage output when component values are varied. Higher resistance values and smaller capacitance values in the second and third sections provide some advantage in gain. Observe that experimentally measured values depend on component tolerances, on waveform error in the audio oscillator, and on voltmeter accuracy.

Figure 3–14 An RC circuit arranged for voltage bootstrap action.

age applied to the circuit. Phase relations of current-voltage can often be disregarded in preliminary troubleshooting procedures. However, an understanding of ac circuit action must take phase relations into account.

IMPROVEMENT OF AUDIO-OSCILLATOR WAVEFORM

If you are using an economy-type oscillator, its output waveform may be noticeably distorted by harmonics. This can impair the accuracy of various ac tests and measurements. Accordingly, you can pass the output from the audio oscillator through one or more parallel-T RC filters to "clean up" the waveform, as exemplified in Fig. 3–15. The illustrated RC filter is arranged to trap out the second harmonic from a 1-kHz signal. Note that the second harmonic is usually the dominant harmonic in a distorted sine-wave signal.

In case that significant third-harmonic distortion might be present in the signal, you can add another parallel-T RC section to the section shown in Fig. 3–15. Observe that a third-harmonic filter uses the same resistive

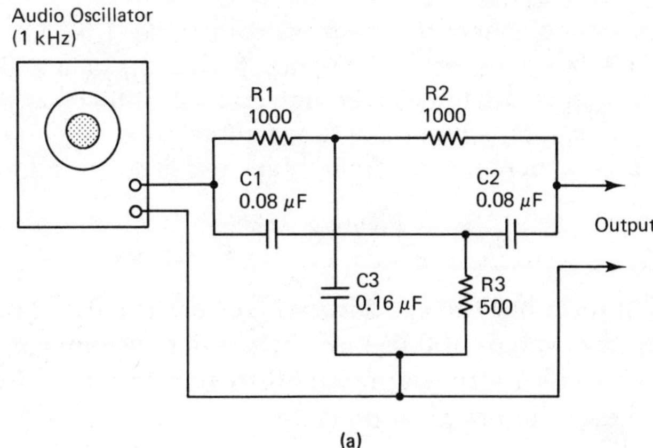

(a)

Note 1: This is a simple and effective filter for removing second-harmonic contamination from the 1-kHz audio generator output. It is a parallel-T RC filter with component values for trapping out any 2-kHz voltage that may be present in the generator output. Observe that the component values are somewhat critical for optimum trapping action. Accordingly, you may find it helpful to utilize potentiometers instead of fixed resistors.

To trim the filter, set the audio oscillator to 2 kHz and adjust the potentiometers for minimum output from the filter as indicated by an ac DVM connected across the filter output. Then, return the audio oscillator to 1 kHz, and you are ready to make tests with a purified sine-wave signal.

Note 2: $R1 = R2 = 2R3$

$C1 = C2 = 0.5C3$

f_o (trap frequency) $= 1/(6.2832 \times R1 \times C1)$

Figure 3–15 Parallel-T RC filter "cleans up" a poor waveform from an audio oscillator. (a) Configuration for a 1-kHz parallel-T RC filter. (b) Filter is also applied to advantage in timbre analysis.

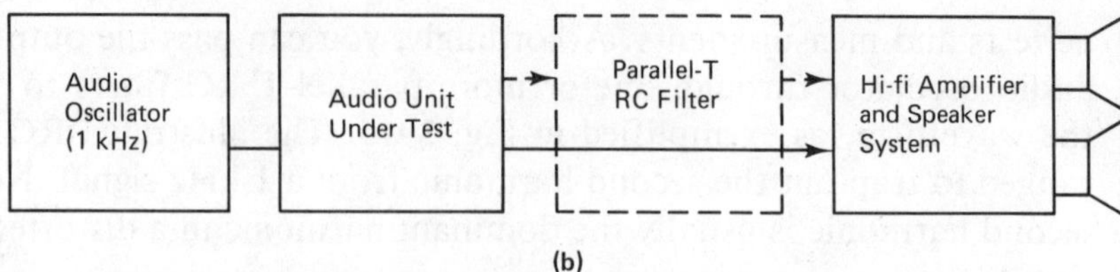

(b)

Note: This is a critical timbre-analysis arrangement for evaluation of an audio unit for small amounts of distortion. The procedure is to first observe the sound output from the hi-fi amp/speaker without the parallel-T RC filter in the signal path. Then again observe the sound output from the hi-fi amp/speaker with the parallel-T RC filter switched into the signal path.

This is a critical timbre test because the parallel-T RC filter removes most of the 1-kHz fundamental frequency from the signal and leaves the distortion components to pass through into the hi-fi amp/speaker. If you advance the gain to bring up the fundamental tone to its previous level, the distortion components will then ''stick out like a sore thumb.''

Figure 3–15 *(cont.)*

values as before, but the capacitor values are smaller. Thus, the two series capacitors have values of 0.053 μF each, and the shunt capacitor has a value of 0.318 μF. Any fourth-harmonic distortion is ordinarily very weak, and can be neglected in practical procedures.

AC VOLTMETER OPERATION WITH DOUBLE-DARLINGTON EMITTER-FOLLOWER

As shown in Fig. 3–16, a 1,000 ohms/volt VOM can be used in combination with a double-Darlington emitter-follower for checking ac voltage levels in high-impedance circuitry. The emitter-follower makes the VOM operate like a TVM. In other words, although the VOM has low input impedance, the emitter-follower effectively provides high input impedance in order to minimize circuit loading. (This type of emitter-follower was explained in Chapter 1.)

MODIFIED EMITTER-FOLLOWER WITH ZERO INSERTION LOSS

As previously noted, there is a small insertion loss imposed by a conventional emitter-follower, even when the Darlington connection is used. In other words, the output voltage is slightly less than the input voltage due to

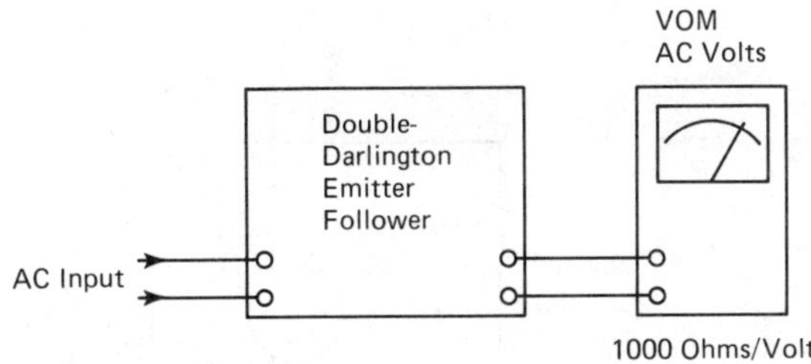

Note: This is a typical application for the double-Darlington emitter-follower. It converts a low-sensitivity VOM into a high-sensitivity VOM. Thus, a 1,000 ohms-per-volt VOM can be used to measure voltages in high-impedance circuitry. Note that the meter reading will be approximately 4 percent low, due to the insertion loss of the emitter-follower.

In case that the VOM does not indicate exactly zero when the input test leads are shorted together, select matched pairs of resistors for the emitter-follower circuit.

Figure 3–16 An ac voltmeter operation with a double-Darlington emitter-follower.

insertion loss in signal passage through the emitter-follower. If you want to employ the source-impedance step-up advantage of an emitter-follower, without incurring any voltage insertion loss, a modified emitter-follower configuration can be utilized, as shown in Fig. 3–17.

Observe that in this modified arrangement, a resistor R_T is included in series with the V_{CC} return lead. In turn, the arrangement operates as a combined emitter-follower and common-emitter amplifier (paraphase amplifier). If the value of the output trimmer resistor R_T is suitably chosen, the signal-voltage drop across R_T will equal the signal-voltage insertion loss of the emitter-follower. Accordingly, if we take the output from across the emitter resistor and the trimmer resistor, the output voltage will be exactly equal to the input voltage.

Although a comparatively large value of resistance is used for the emitter resistor, a relatively small value of resistance is used for the output trimmer resistor R_T. The precise value required for R_T depends on the current-amplification factor of the transistors. Thus, R_T might require a value of 100 Ω, or a somewhat lower value, or a somewhat higher value. For quick and easy adjustment of the configuration for unity voltage gain, a 200-Ω potentiometer may be utilized for R_T.

As a practical note, observe that both ends of the output circuit in Fig. 3–17 are above ac ground potential. This is just another way of saying that

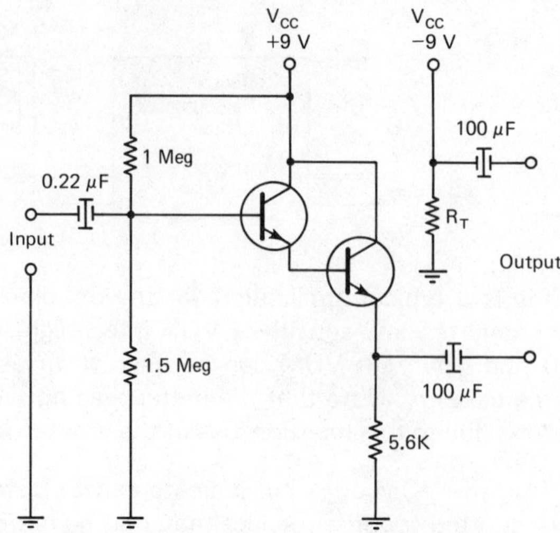

Note: This is a Darlington emitter-follower arrangement for ac voltage operation with an output trimmer resistor R_T. When a precise value is selected for R_T, the output voltage will be exactly equal to the input voltage. (There will be no insertion loss imposed by the emitter-follower.) You may prefer to use a 200-Ω potentiometer for R_T, so that the trimming action is easily adjusted. Observe that V_{CC} may be +9 V, or higher. High values of V_{CC} permit larger input voltage swings without overdriving the transistors.

Figure 3–17 An emitter-follower configuration with provision for insertion-loss cancellation.

a common ground connection cannot be employed between the modified emitter-follower and the load that is connected to the output. On the other hand, a common ground connection must be employed between the modified emitter-follower and the source that is connected to the input.

SHORTCUTS IN CB RADIO TROUBLESHOOTING

A citizen's-band (CB) radio is somewhat more elaborate than a conventional AM radio in that it comprises both a receiver and a low-power transmitter. A CB radio operates in the 27-MHz band, and employs quartz-crystal controlled oscillators. Useful quick checks in preliminary troubleshooting procedures are:

1. *Transmission OK, no reception.* When this difficulty occurs, check out each channel for possible output. Turn up the volume control and turn the squelch control to minimum. Any output, no matter how weak, noisy, or distorted, can provide helpful clues in trouble localization. Observe that even if there is no incoming signal, noise output will be audible in normal operation. In case that there is no noise output, the technician concludes that the channel is "stone dead."
2. *No transmission, no reception.* In case there is no noise output on any channel, it is still possible that the transmitter circuitry is basically functional. Accordingly, check the modulation indicator when the transmitting control is pressed. If the modulation indicator bulb varies in brightness as you speak into the microphone, it can be concluded that the transmitter circuitry is basically functional, even if the antenna radiation (if any) cannot be picked up by another CB radio.
3. If another CB radio cannot pick up antenna radiation from the CB radio under test, although the transmitter circuitry is basically functional, off-frequency transmission is a logical possibility. Off-frequency transmission is most likely to be caused by a defective transmitter quartz crystal. Note that a marginally defective crystal may intermittently jump from one frequency to another after it has been operating normally for a period of time.
4. Next, if the transmitter operates without a crystal inserted in the oscillator circuit (with off-frequency output), it is probable that the oscillator, driver, or RF power amplifier is self-oscillatory. Look for an open decoupling or bypass capacitor in the transmitter network.
5. In the event that the transmitter output is weak, it is advisable to make a systematic checkout of the transmitter circuitry. *If transmission has been attempted without the antenna fully extended, the RF power-output transistor is likely to become damaged or burned out.* However, if the RF power-output transistor is ok, look for leaky or open capacitors in the transmitter RF circuitry.
6. In case that there is weak or no modulation of the RF carrier, or if there is ample modulation with distorted signal output, look for a malfunction in the modulator section. The trouble might even be found in a defective microphone.
7. Note that if there is no noise output on any receiving channel, but the transmitter operates normally, the trouble will be found in the receiver section between the antenna and the detector.
8. When there is one dead receiver channel (no noise output), there is likely to be a defective quartz crystal in the associated circuit. (Some-

times the trouble will be tracked down to a defective channel-selector switch.)

9. If there is only one operative receiver channel (noise output), look for a defect in the channel-selector switch.
10. An inoperative squelch control throws suspicion on the automatic volume-control circuitry. Leaky capacitors are common culprits.
11. Weak reception can be caused by antenna defects. Otherwise, look for a defective stage in the signal channel, using signal-injection or signal-tracing tests.
12. Receiver overloading and distortion on strong signals, with reception ok on weak signals, points to off-value AVC voltage. Here again, a leaky capacitor is a common culprit.
13. In case of intermittent reception, the controls should be inspected for excessive wear. Defective contacts in the channel-selector switch can cause intermittent reception. A quartz crystal sometimes develops an intermittent defect. Any device or component can become intermittent. A recording voltage monitor can often simplify and speed up the localization of an intermittent fault.

4

Impedance Measurements and Quick Checkers

GENERAL CONSIDERATIONS

There is greater potential for time-saving shortcuts in impedance measurement and checking than in other categories of electronic troubleshooting, chiefly because impedance is not fully understood by many technicians. From the most basic viewpoint, impedance is equal to ac voltage divided by ac current. Impedance has a resistive component and a reactive component; the reactive component may be capacitive, or it may be inductive. *One of the most common errors among apprentices is an assumption that the impedance of a coil is equal to its resistance as indicated by an ohmmeter.*

This is not so. However, suitable quick checkers that are little less complex than ohmmeters can be easily constructed and used to speed up routine checkouts of ac circuit impedances. It is not necessary to wrestle with complex algebraic calculations in order to efficiently troubleshoot ac circuitry—instead, a few "target" quick checkers and practical guidelines for working with impedances will often come up with the right answer before the math whiz gets his formulas organized.

As an example of practical guidelines that we need when working with impedance, let's make the following comparisons of resistance and impedance:

1. Resistance is measured in ohms; it is equal to dc volts divided by dc amperes. *It is also equal to ac volts divided by ac amperes.*
2. Resistance remains the same at any frequency.

3. Impedance is measured in ohms; it is equal to ac volts divided by ac amperes. *Impedance cannot be measured with an ohmmeter*—however, we can measure impedance with an ac voltmeter and an ac ammeter.
4. Impedance changes with frequency. A capacitive impedance has fewer ohms as the frequency increases; an inductive impedance has more ohms as the frequency increases.
5. Let's observe the fact a tuned circuit has both capacitive impedance and inductive impedance, depending upon the frequency—*but at resonance, its capacitive action cancels out its inductive action with the result that only the resistance of the tuned circuit will be measured.*

CHARACTERISTICS OF SERIES- AND PARALLEL-RESONANT CIRCUITS

Quantity	*Series circuit*	*Parallel circuit*
At resonance: Reactance $(X_L - X_C)$	Zero; because $X_L = X_C$	Zero; because non-energy currents are equal
Resonant frequency	$\frac{1}{2\pi\sqrt{LC}}$	$\frac{1}{2\pi\sqrt{LC}}$
Impedance	Minimum; $Z = R$	Maximum; $Z = \frac{L}{CR}$, approx.
I_{line}	Maximum	Minimum value
I_L	I_{line}	$Q \times I_{line}$
I_C	I_{line}	$Q \times 1_{line}$
E_L	$Q \times E_{line}$	E_{line}
Phase angle between E_{line} and I_{line}	0°	0°
Angle between E_L and E_C	180°	0°
Angle between I_L and I_C	0°	180°
Desired value of Q	10 or more	10 or more
Desired value of R	Low	Low
Highest selectivity	High Q, low R, high $\frac{L}{C}$	High Q, low R,
When f is greater than f_o: Reactance	Inductive	Capacitive
Phase angle between I_{line} and E_{line}	Lagging current	Leading current
When f is less than f_o: Reactance	Capacitive	Inductive
Phase angle between I_{line} and E_{line}	Leading current	Lagging current

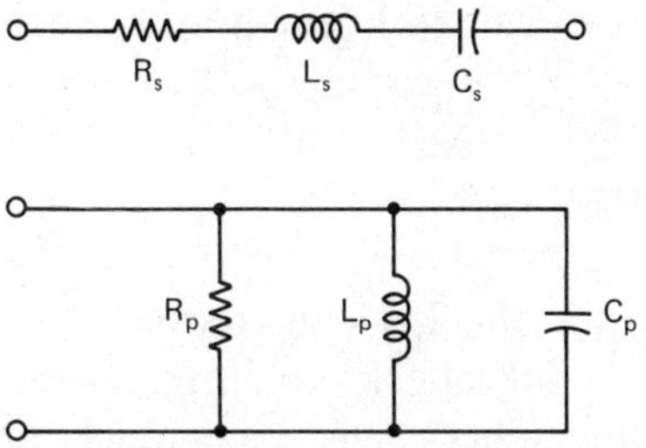

R_s L_s C_s

R_p L_p C_p

L_s, C_s, and R_s in Series Circuit
L_p, C_p, and R_p in Series Circuit

$$Q = \frac{\omega L_s}{R_s} = \frac{1}{\omega C_s R_s} = \frac{R_p}{\omega L_p} = R_p \omega C_p = \frac{\sqrt{L_s/C_s}}{R_s} = \frac{R_p}{\sqrt{L_p/C_p}}$$

General Formulas	*Formulas for Q Greater than 10*	*Formulas for Q Less than 0.1*
$R_s = \frac{R_p}{1 + Q^2}$	$R_s \simeq \frac{R_p}{Q_2}$	$R_s \simeq R_p$
$X_s = X_p \frac{Q^2}{1 + Q^2}$	X_s / X_p	$X_s \simeq X_p Q^2$
$L_s = L_p \frac{Q^2}{1 + Q^2}$	$L_s \simeq L_p$	$L_s = L_p Q^2$
$C_s = C_p \frac{1 + Q^2}{Q^2}$	$C_s \simeq C_p$	$C_s \simeq \frac{C_p}{Q^2}$
$R_p = R_s (1 + Q^2)$	$R_p \simeq R_s Q^2$	R_p / R_s
$X_p = X_s \frac{1 + Q^2}{Q^2}$	$X_p \simeq X_s$	$X_p \simeq \frac{X_s}{Q^2}$
$L_p = L_s \frac{1 + Q^2}{Q^2}$	$L_p \simeq L_s$	$L_p \simeq \frac{L_s}{Q^2}$
$C_p = C_s \frac{Q^2}{1 + Q^2}$	$C_p \simeq C_s$	$C_p \simeq C_s Q^2$
$B_L = \frac{1}{X_L}$	$B_L = \frac{1}{X_L}$	$B_L = \frac{1}{X_L}$
$B_c = \frac{1}{X_c}$	$B_L = \frac{1}{X_L}$	$B_L = \frac{1}{X_L}$
$Y = \sqrt{G^2 + B^2}$	$Y = \sqrt{G^2 + B^2}$	$Y = \sqrt{G^2 + B^2}$

Reproduced by special permission of Reston Publishing Company and Frank Weller from *Handbook of Electronic Systems Design.*

That's really all we need to know about impedance!

AUDIO IMPEDANCE CHECKER

An audio impedance checker can frequently be an important time-saver because it provides quick checks of impedance relations in "dead" circuitry. An audio impedance checker is particularly helpful in tough-dog situations when a normally operating unit is available for comparison tests. Let's start with the simple example shown in Fig. 4–1. In this case there was no sound output from the amplifier, although the V_{CC} supply voltage was normal. This was one section of a stereo amplifier, so that a normally operating unit was present. As shown in the diagram, an abnormally high impedance was found at point X. This could have resulted from either an open speaker or a defective coupling capacitor. In this case history, the coupling capacitor had lost practically all of its capacitance.

Impedance checks in audio circuitry may be made at any frequency from 20 Hz to 20 kHz. Preliminary tests (such as exemplified in Fig. 4–1) may be made at 1,000 Hz. Follow-up tests are sometimes desirable at a low frequency and at a high frequency. This is just another way of saying that a marginal electrolytic capacitor will clearly appear defective in a low-frequency test, whereas it would "look ok" in a high-frequency test. On the other hand, faulty inductive circuits such as in cross-over networks may "show up" clearly defective in a high-frequency test, but "look ok" in a low-frequency test.

IMPEDANCE AT BATTERY CLIP TERMINALS

Another important time-saver (particularly in comparison tests) is an impedance check at the battery-clip terminals. With reference to Fig. 4–2, "tough-dog" troubleshooting situations are frequently caused by open capacitors (or capacitors with a poor power factor) that are associated with the V_{CC} line. As exemplified in the diagram, various decoupling capacitors are employed to prevent feedback and interaction of subsections in the electronic equipment. If any one of the decoupling capacitors becomes open-circuited, short-circuited, or develops a poor power factor, the fault becomes evident in an impedance comparison test.

This comparison impedance check is made by disconnecting the battery from the electronic unit, and connecting the test leads from the impedance checker to the V_{CC} terminals. It is advisable to operate the audio generator

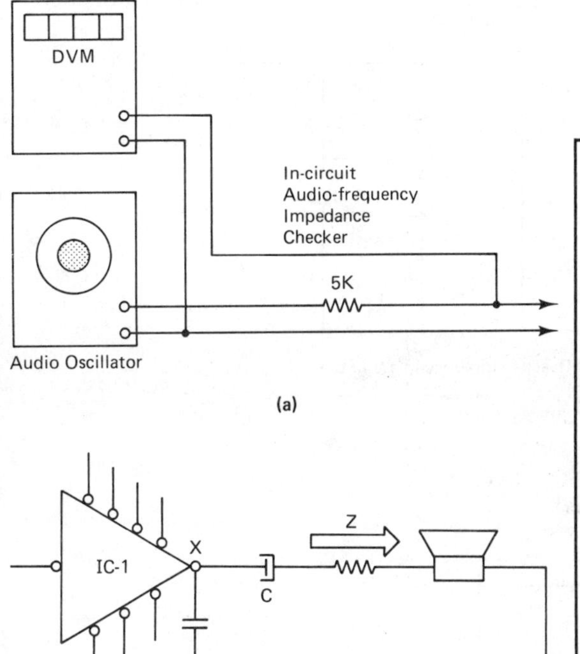

Trouble Symptom: No Sound Output

(b)

The DVM is operated on its AC voltage function. The audio oscillator is set to 1 kHz with an output level of 2 or 3 volts, as indicated when the output leads are open-circuited.

Practical Note: The peak ac voltage from the *test point to ground* should not exceed 500 mV, in order to avoid junction turn-on and possible confusion of test results.

This is an example of an AF integrated circuit driving a speaker through an electrolytic capacitor and a resistor.

An audio impedance checker is also valuable for testing other types of electronic circuitry.

In this example, there was no sound output from the speaker. All of the dc voltages at the IC terminals ''made sense'' on a comparative basis. However, when the audio impedance checker was applied at point *X,* the DVM indicated that the impedance was much higher in the ''bad'' amplifier than in the ''good'' amplifier:

''Bad'' Amplifier—56 mV
''Good'' Amplifier—5 mV
(DVM) indicates a 2.75 V on open circuit)

The comparatively high ac voltage reading in the ''bad'' amplifier indicated that the impedance to ground from point *X* was abnormally high. In this type of circuitry, *electrolytic capacitors are ready suspects.* In turn, capacitor *C* fell under suspicion. When a signal voltage was next applied to the input of IC-1, an ac voltmeter showed signal present on the left-hand end of *C,* but zero signal on the right-hand end of *C.*

Therefore, the troubleshooter concluded that *C* was open-circuited. When a test capacitor was ''bridged'' across *C,* the speaker resumed sound output.

Figure 4–1 Audio impedance checker. (a) Test setup. (b) Example of application.

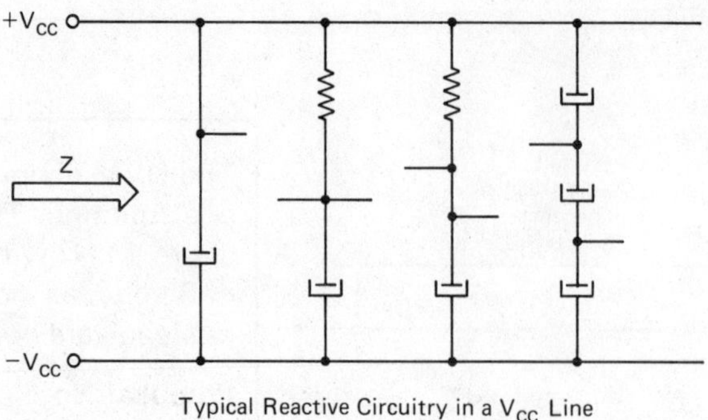

Typical Reactive Circuitry in a V_{CC} Line

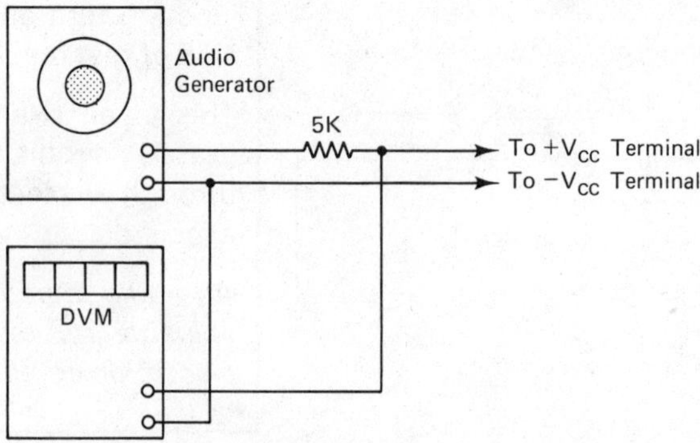

Note: A capacitor with a poor power factor is equivalent to a normal capacitor connected in series with substantial resistance. Accordingly, the resistance/reactance proportions of the network are upset by the poor power factor, and the total impedance is changed.

As previously noted, the DVM reading should be maintained at less than 500 mV from the V_{CC} terminal to ground to ensure that semiconductor junctions in the unit under test will not be turned on and give a misleading reading.

Bottom Line: An impedance quick checker can show immediately whether a network subsection should be checked out in greater detail, or passed as ok.

Figure 4–2 Impedance quick check at battery clip terminals.

at a comparatively low frequency, such as 60 Hz. Advance the generator output voltage to obtain an adequate ac-voltage indication on the DVM. Then, repeat the test on the comparison unit to determine whether there may be a significant difference in the readings.

As a practical note, observe that the DVM might indicate 2 V while the impedance checker is disconnected. Then, when the impedance checker is connected to the V_{CC} terminals in the unit under test, the DVM reading might drop to 0.3 V. In other words, the V_{CC} network is drawing ac current, and the resulting IR drop across the 5-kΩ resistor would be equal to 2 V minus 0.3 V. If one of the capacitors were short-circuited, the impedance would be lower and the DVM would indicate less voltage. On the other hand, if one of the capacitors were open-circuited, the impedance would be higher and the DVM would indicate more voltage.

Impedance tends to seem complicated when we first start to work with it. However, as we proceed through some basic experiments, impedance checks and measurements become almost as easy as resistance checks and measurements. Then, when we look back, we wonder how we could have believed that impedance was difficult. (See also Chart 4–1.)

Chart 4–1

DVM CAPABILITIES IN IMPEDANCE MEASUREMENTS

A DVM does not measure impedance directly (unless supplemented by additional devices and components). In other words, a DVM must measure impedance indirectly; it measures an ac voltage and it measures the corresponding ac current flow. Then, the technician divides the voltage reading by the current reading to calculate the impedance in ohms.

A typical top-of-the-line DVM measures ac voltage up to 750 rms V, and measures ac current up to 2 rms A. Accordingly, a wide range of impedance values can be determined. Since impedance increases as the frequency increases, the frequency capability of a DVM on its ac voltage and ac current ranges is of interest to the operator. Manufacturers' ratings should be observed in this regard.

Practical examples of impedance measurements made by the experimenter and hobbyist are as follows:

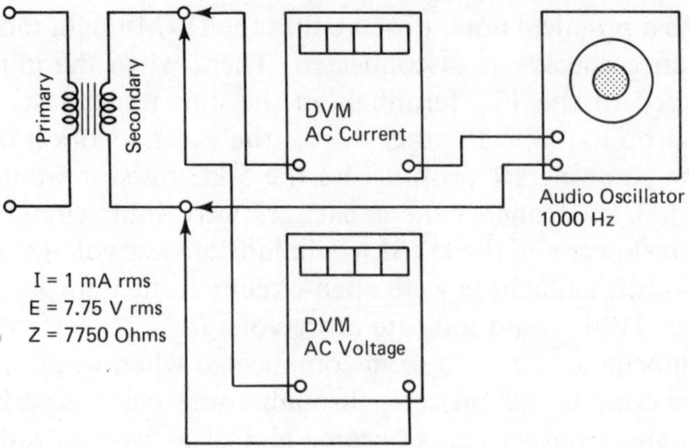

With the Primary Open, the Secondary has an Impedance of 7750 Ohms at 1000 Hz

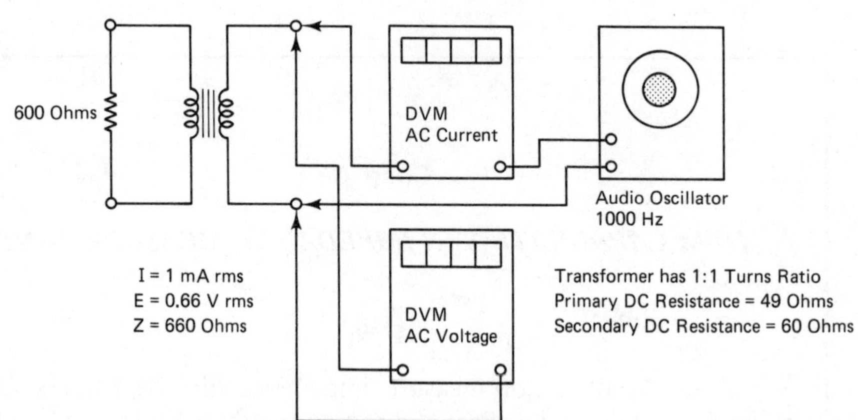

With the Primary Connected to a 600 Ohm Load, the Secondary has an Impedance of 660 Ohms at 1000 Hz

We Say That the Primary Circuit Impedance (600 Ohms) is Reflected into the Secondary Circuit

Note that the secondary impedance measured 660 Ω instead of 600 Ω. This difference is due to tolerances on components, a small waveform error, and tolerances on instrument accuracy.

More elaborate service-type DVMs can be used with ac current values of up to 2 A rms. However, many DVMs do not provide an ac current function. Accordingly, the technician must measure ac current flow with the aid of a shunt resistor connected in series with the circuit. The voltage drop across the shunt resistor is measured, and the current value is calculated from Ohm's Law. For example, if a shunt has a resistance of 1,000 Ω, a current of 1 mA will produce a voltage drop of 1 V. With reference to the diagram on the previous page, the shunt-DVM combination would replace the DVM for ac current measurement.

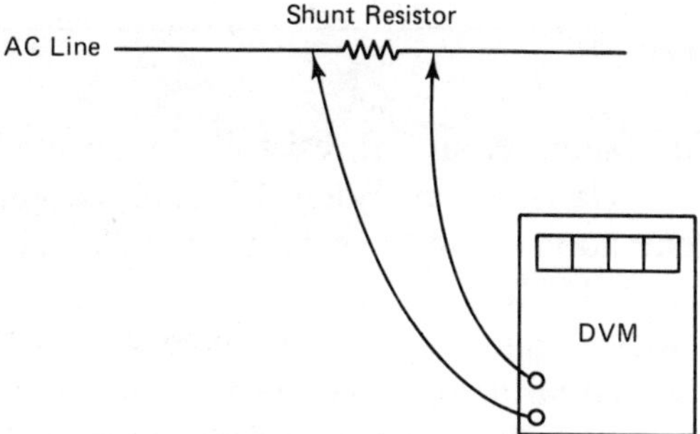

Note: DVM is Operated on its AC Voltage Function.

Measurement of Transformer Turns Ratio

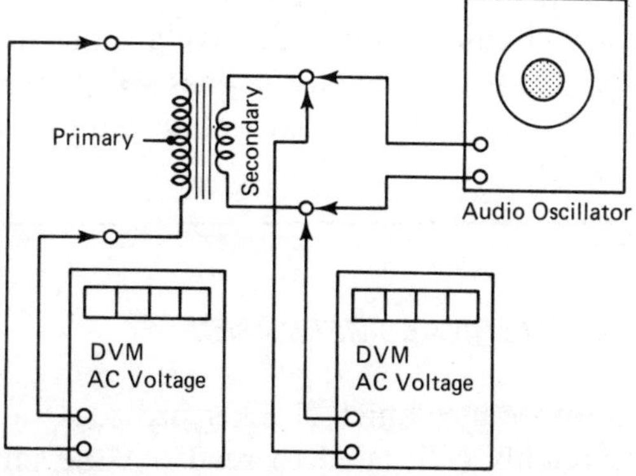

Transformer Turns Ratio is Measured with an Audio Oscillator nad a Pair of DVMs (or a Single DVM). The AC Voltage Ratio of the Primary and Secondary is Equal to the Turns Ratio.

As an example, an Archer (Radio Shack) miniature audio transformer (No. 273-1380) has a turns ratio of approximately 11-to-1. In other words, if 1 V rms is applied to the secondary, about 11 V rms will be measured across the primary.

Measurement of Reflected Impedance from Secondary to Primary of an 11-to-1 Transformer

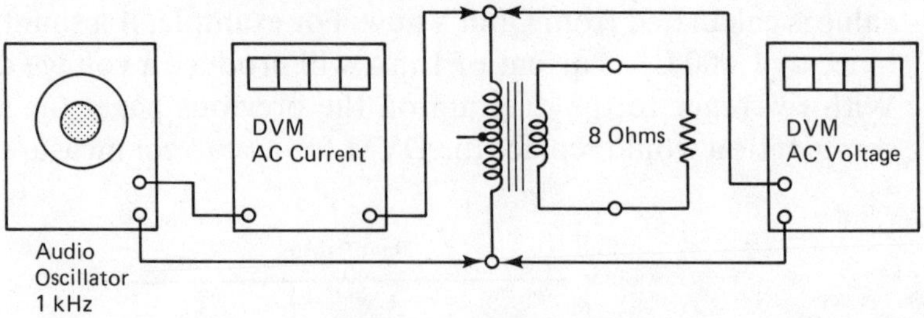

In this example, an 8-Ω resistor is connected across the secondary of an 11-to-1 transformer. When 1 V rms is applied to the primary, the current demand is 0.9 mA. In turn, the primary impedance is equal to 1/0.0009, or 1,111 Ω.

Note that the measured impedance ratio is 138-to-1 at 1 kHz. This is essentially equal to the square of the measured turns ratio (121-to-1). The factory rating on the impedance ratio is 113-to-1. These discrepancies are due to tolerance and a small waveform error.

We say that an 11-to-1 transformer will match an impedance of 1111 ohms to 8 Ω. This means that maximum power will be transferred from the primary to the secondary of the transformer if the primary-to-secondary circuit impedances have a ratio of 138-to-1.

Moderate impedance mismatches do not result in a serious loss of efficiency. However, large mismatches result in a serious loss of audio power transfer.

MEASUREMENT OF INTERNAL IMPEDANCE

Although an impedance quick checker provides valuable comparative data, it cannot be readily calibrated to read out the impedance value in ohms. However, we can employ a different approach and measure the internal impedance of a circuit in ohms, as shown in Fig. 4-3. The procedure is simple, and utilizes a blocking capacitor and a resistor with an ac DVM. The blocking capacitor may have a value of 0.1 μF. The electronic unit under test should be energized by a generator that is set at a normal operating frequency for the unit (such as 1 kHz for audio circuitry). Then proceed:

1. With the blocking capacitor connected in series with the DVM, measure the ac voltage from the test point to ground.

2. Using a resistor of suitable value which will reduce the test-point voltage at least 10 percent, but not more than 20 percent, shunt the resistor across the DVM as shown in the diagram, and note the reduced reading.

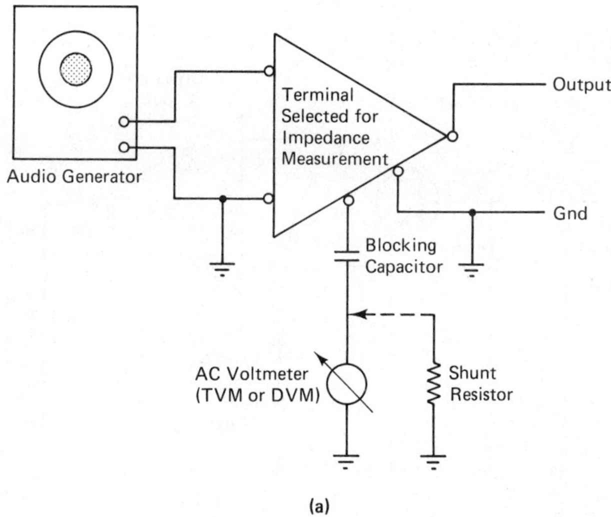

Note 1: This internal impedance meter cannot be applied in "dead" circuits because it operates with respect to the signal voltage in the circuit under test. An internal impedance measurement is more informative than an internal resistance measurement inasmuch as the impedance measurement takes the capacitive reactances into account along the path of signal flow. Observe also that there is no simple relation between the internal resistance of a circuit and its internal impedance.

Note 2: A measurement of the internal impedance from the point under test to ground may be made at more than one frequency. As an illustration, a basic test is made at 1 kHz. In most situations, faults in the vicinity of the test point will "show up" in a 1-kHz test. On the other hand, if the test result is inconclusive in a "tough-dog" situation, the impedance measurement may be repeated at a frequency of 20 Hz, and at a frequency of 20 kHz, for example. In other words, faults in reactive components will sometimes "show up" clearly only at a very low test frequency, or at a very high test frequency.

Figure 4-3 (a) Test setup for measuring the internal impedance of a system from a selected test point to ground in "live" circuitry. (b) Test setup for measuring the internal impedance of a network from a selected test point to ground in a low-level circuit.

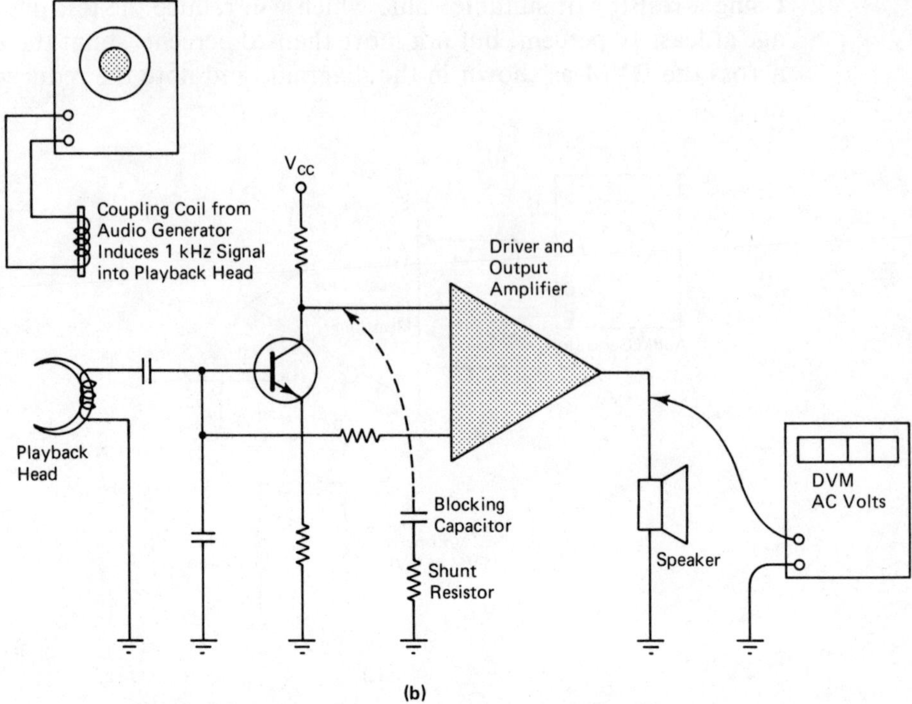

(b)

Note 3: Since a service-type DVM cannot measure low-level ac voltages such as those in preamplifier circuits, the test signal must be amplified before it is measured. This example shows how this may be done in a tape-recorder preamp network. Proceed as follows:

1. Apply a 1-kHz test signal, as from a prerecorded test tape, or induce a 1-kHz test signal into the play head from a coupling coil.
2. Connect an ac DVM at the output of the following amplifier section.
3. Disable the automatic level control (ALC) circuit, if present, and adjust the input signal level for normal operation.
4. Connect a shunt resistor with a blocking capacitor as shown in the diagram. This shunt resistor should have a value that reduces the output voltage by 10 or 15 percent, but not more than 20 percent.
5. Ohm's law is then applied to calculate the value of the internal impedance at the point under test.

Figure 4–3 *(cont.)*

3. Apply Ohm's law to calculate the internal impedance. For example: If you measure 0.4 rms V initially, and then measure 0.32 rms V when a 4-kΩ resistor is shunted across the DVM, it follows that the internal impedance of the circuit is 1,000 Ω from the test point to ground. (The 4,000-Ω resistor draws 80 μA at 0.32 V, and the internal impedance is accordingly equal to 0.08 divided by 80×10^{-6}, or 1,000 Ω.)

To speed up Ohm's-law calculations when making this test, reach for your pocket calculator and "punch out" the answer. Measurement of internal impedance, particularly on a comparative basis, can be a major time-saver in tough-dog troubleshooting situations.

Low-Level Circuit Impedance

Observe that the internal impedance of low-level circuitry can be measured as shown in Fig. 4–3(b).

MEASUREMENT OF AMPLIFIER INPUT IMPEDANCE

Audio amplifier input impedance can be easily measured as shown in Fig. 4–4. A moderate level of input signal is applied (do not overdrive the amplifier). A DVM is utilized to measure the input voltage, and another DVM is used to measure the input current. Note that the same DVM may be employed in two steps, if desired. As a practical example of test data observe the following measured values:

Input voltage	0.033 V rms
Input current	0.059 mA rms
Voltage/current	560 Ω, approx.
Input impedance	560 Ω

It is customary to make amplifier impedance measurements at a frequency of 1 kHz. If you are in a mood for experimenting, you may wish to measure the input impedance of this amplifier at a very low frequency and at a very high frequency, such as 20 Hz and 20 kHz.

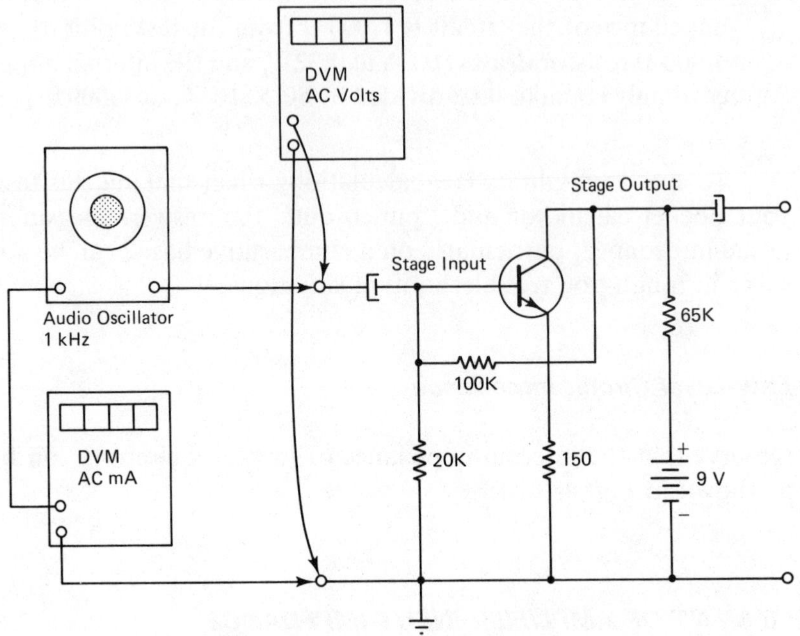

Note: This is the experimental amplifier that was previously constructed. Its measured input impedance is in the vicinity of 560 Ω. This is a moderately low value of input impedance that results from the employment of negative voltage feedback (100-kΩ resistor) and only a small amount of current feedback (150-Ω resistor).

Good Practices Note: It is sometimes claimed that the quickest way to measure amplifier input impedance is to connect a resistor in series or in shunt with the input circuit while observing the output voltage from the amplifier. Then, it is said, if the output voltage is reduced to one-half, the input impedance is equal to the value of the test resistor. However, it may be observed that this method is inaccurate to the degree that doubling or halving the normal input impedance disturbs the input parameters and changes the negative-feedback action, with resulting change in amplifier gain. Therefore, it is good practice to employ the method shown in the above diagram.

Figure 4-4 Measurement of amplifier input impedance.

MEASUREMENT OF AMPLIFIER OUTPUT IMPEDANCE

Audio amplifier output impedance can be readily measured as depicted in Fig. 4–5. This method employs the Ohm's-Law relation between output voltage and output current demand. The amplifier should be operated at about half of rated maximum output power, and at a test frequency of 1 kHz. The

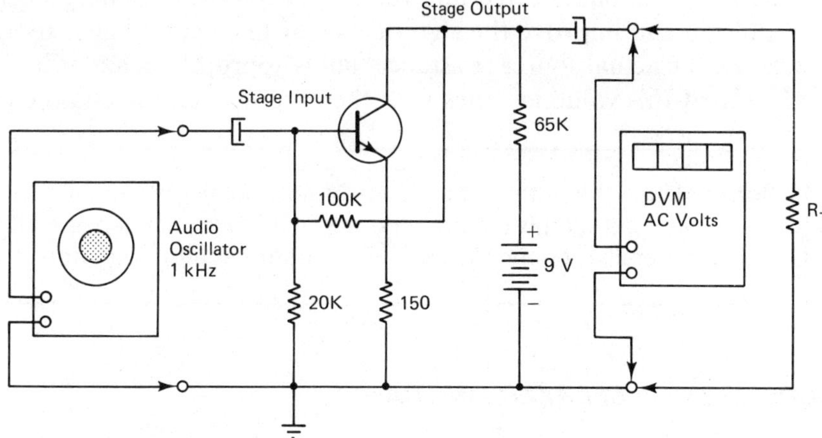

Note: This method of measuring the output impedance of an amplifier is based on the relation between the output voltage and the output current demand. To make the test, choose a value for R_T that reduces the output voltage at least 10 percent, but not more than 20 percent of its initial value. Note the initial voltage and the loaded voltage values. Apply Ohm's Law and calculate the current flow through R_T. Then divide the difference between the initial voltage and the loaded voltage by the current value; the answer is in ohms of output impedance. (This procedure is essentially the same as the one we use to measure the internal impedance of an audio system at any selected test point.)

Good Practices Note: It is sometimes claimed that the quickest way to measure amplifier output impedance is to shunt a test resistor across the output terminals, using a resistance value that reduces the output voltage to one-half of its initial value. Then, it is said, the value of the output impedance is equal to the value of the test resistor. However, this method is inaccurate to the degree that it changes the output circuit parameters and modifies negative-feedback action, thereby changing the gain of the amplifier. Therefore, it is good practice to avoid reduction of the output voltage more than 20 percent.

Figure 4–5 Measurement of amplifier output impedance.

technician may also repeat the measurement at a low frequency such as 20 Hz, and at a high frequency such as 20 kHz. In general, both of these impedance values will differ more or less from the 1-kHz impedance value.

Observe that this measurement procedure is made with respect to a source-resistance value of 50 Ω (most audio oscillators have an output resistance of 50 Ω). In practice, the source might have considerably higher resistance, such as 1,000 or even 5,000 Ω. In such a case, the input circuit conditions will be reflected into the output impedance value. Therefore, the technician can improve the significance of his measurement by taking into account the actual source resistance that is appropriate, and inserting a series resistor of this value in series with the output from the audio oscillator.

Bottom Line: The output impedance of an amplifier is generally dependent upon the input circuit parameters, and the input impedance of the amplifier is generally dependent upon the output circuit parameters.

MEASUREMENT OF SPEAKER IMPEDANCE

Although we often speak of the dc resistance of a speaker voice coil as if it were the speaker impedance, this is not so. It is easy to measure the speaker impedance at any desired frequency as shown in Fig. 4–6. Observe that the speaker impedance approaches the dc resistance at very low frequencies, and becomes comparatively high at high frequencies. In this example, the speaker impedance more than doubled over the frequency range from 100 Hz to 10 kHz.

MEASUREMENT OF MICROPHONE OUTPUT IMPEDANCE

Microphones are available with various output impedances. The most common rated impedance is 600 Ω. Some microphones are provided with a switch whereby the user can switch to either high-impedance output or low-impedance output. An example of measured output impedance for a 600-Ω unidirectional microphone is shown in Fig. 4–7. Like a speaker, a microphone has an impedance that tends to vary with frequency. In this example, the microphone output impedance more than doubles over the frequency range from 10 kHz to 100 Hz. However, the average impedance is approximately 600 Ω.

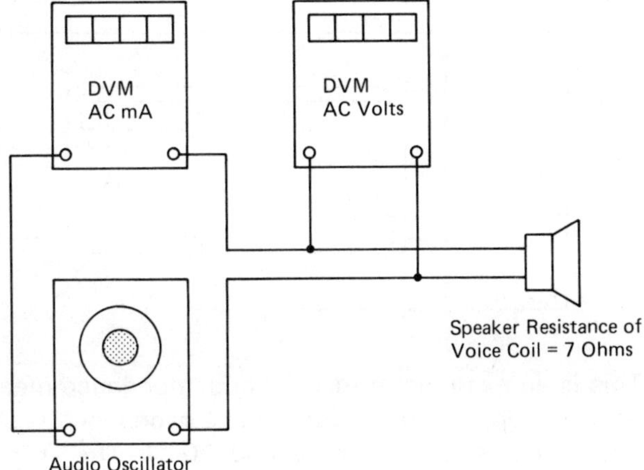

Note: In this example, impedance of a 12-inch woofer was measured. The test data were as follows:

At 100 Hz
I = 10 mA rms
E = 0.083 V rms
Z = 8.3 Ω

At 1,000 Hz
I = 10 mA rms
E = 0.133 V rms
Z = 13.3 Ω

At 10 kHz
I = 10 mA rms
E = 0.19 V rms
Z = 19 Ω

Observe that the impedance at 100 Hz is somewhat greater than the dc resistance of the voice coil. At 10 kHz, the impedance is more than twice as great than at 100 Hz.

Figure 4–6 Measurement of speaker impedance.

MEASUREMENT OF IMPEDANCE AT RADIO BATTERY-CLIP TERMINALS

As previously noted, an impedance check at the battery-clip terminals is an important preliminary test when troubleshooting a malfunctioning radio. The quick-checker that was noted provides informative comparison data,

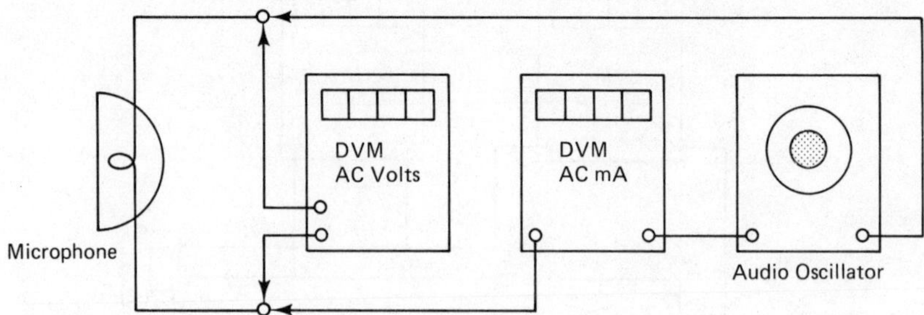

Note: This is an example of microphone impedance measurement. A unidirectional microphone rated for 600 Ω impedance was used in this experiment. Its dc resistance measured 50 Ω. The ac test data were as follows:

At 100 Hz
I = 1 mA rms
E = 0.719 V rms
Z = 719 Ω

At 1,000 Hz
I = 1 mA rms
E = 0.603 V rms
Z = 603 Ω

At 5,000 Hz
I = 1 mA rms
E = 0.422 V rms
Z = 422 Ω

At 10 kHz
I = 1 mA rms
E = 0.355 V rms
Z = 355 Ω

This type of microphone contains a line matching transformer between the electrodynamic unit and the output terminals. In other words, the measured impedance is a reflected value, and the dc resistance measured at the output terminals is the resistance of the secondary winding on the matching transformer. The matching transformer can be connected for either 600-Ω or 10-kΩ output impedance.

Figure 4–7 Measurement of microphone output impedance.

but does not measure the impedance value in ohms. With reference to Fig. 4–8, the impedance value can be easily measured with a pair of DVMs and an audio oscillator. If the troubleshooter measures a substantial impedance value at a test frequency of 1 kHz, he concludes that there is a fault in the V_{CC} decoupling network. In normal operation, the impedance will not exceed a few ohms.

Experiment: In this experiment, we measure the impedance of a wirewound resistor over the audio-frequency range. As shown in Fig. 4–9, an audio oscillator and two DVMs are connected to the terminals of a wirewound resistor, and the impedance of the resistor is measured at several different frequencies from 60 Hz to 10 kHz. In general, it will be found that a wirewound resistor has an impedance that is approximately equal to its dc resistance at low frequencies. On the other hand, its impedance at higher audio frequencies is likely to be quite different from its dc resistance.

Experiment: Next, we measure the impedance of a wirewound resistor with an 8-Ω dc resistance value over the audio-frequency range. This is a resistor such

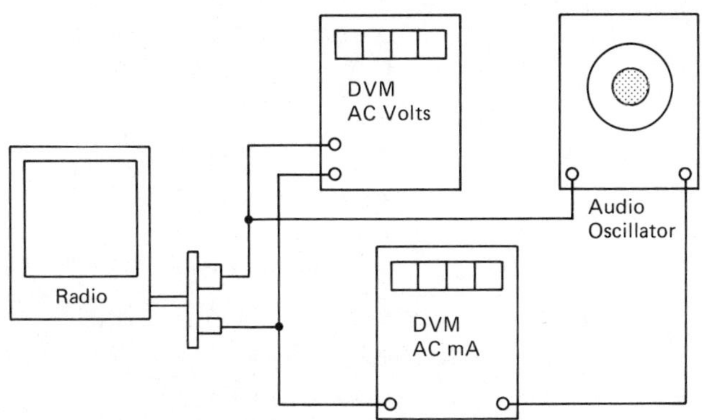

Note: This is an informative quick check that can sometimes "finger" the cause of elusive trouble symptoms. As an illustration of typical test data, an impedance of 5 Ω was measured for a small AM/FM radio, at a test frequency of 1 kHz. As would be expected, the impedance at a higher test frequency was less—2 ohms at 5 kHz. Note that the test voltage should not exceed 0.5 V peak (0.707 V rms) in order to avoid turn-on of semiconductors in the radio circuitry and possible confusion of test results. The power switch should be turned on while the test is being made. Otherwise the V_{CC} line will be open-circuited and an extremely high impedance value will be measured.

Figure 4–8 Measurement of impedance at battery-clip terminals.

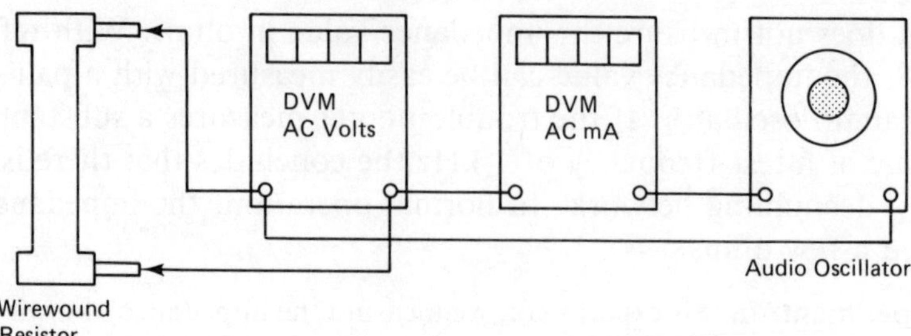

Note: Some wirewound resistors have significant inductance. For example, a 12-W 1,250-Ω wirewound resistor exhibited the following ac impedance values:

At 60 Hz
I = 1 mA rms
E = 1.25 V rms
Z = 1,250 Ω

At 500 Hz
I = 1 mA rms
E = 1.29 V rms
Z = 1,290 Ω

At 1,000 Hz
I = 1 mA rms
E = 1.35 V rms
Z = 1,350 Ω

At 5,000 Hz
I = 1 mA rms
E = 2.05 V rms
Z = 2,050 Ω

At 10 Hz
I = 1 mA rms
E = 4.74 V rms
Z = 4,740 Ω

Thus the wirewound resistor had an impedance equal to its dc resistance value at 60 Hz, but exhibited increasing impedance at higher frequencies. At 10 kHz, its impedance was more than 3 times its dc resistance value.

Figure 4–9 Measurement of impedance of wirewound resistor.

as used as an output load for an audio amplifier during troubleshooting procedures. As shown in Fig. 4–10, the impedance of the resistor is significantly different from its dc resistance value at higher audio frequencies. Observe that the change in resistance is opposite from the change which was measured in the previous experiment.

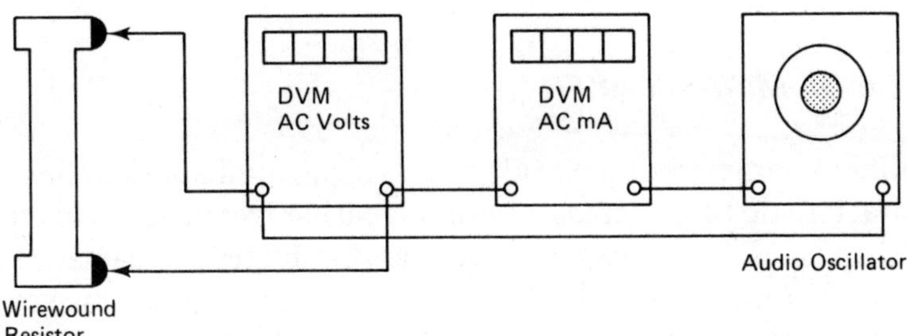

Note: Some wirewound resistors have distributed capacitance that outweighs their inductance. For example, a 5-W 8-Ω wirewound resistor exhibited the following ac impedance values:

At 60 Hz
I = 1 mA rms
E = 0.008 V rms
Z = 8 Ω

At 1,000 Hz
I = 1 mA rms
E = 0.007 V rms
Z = 7 Ω

At 5,000 Hz
I = 1 mA rms
E = 0.005 V rms
Z = 5 Ω

At 10 Hz
I = 1 mA rms
E = 0.005 V rms
Z = 5 Ω

Thus, the wirewound resistor had an impedance equal to its dc resistance value at 60 Hz, but exhibited decreasing impedance at higher frequencies. At 10 kHz, its impedance was 63 percent of its dc resistance value.

Figure 4–10 Another example of wirewound resistor impedance.

Bottom Line: A wirewound resistor may exhibit increasing impedance with increasing frequency, or vice versa. This depends on whether the resistor's inductive reactance or its capacitive reactance dominates the impedance characteristic.

IMPEDANCE OF MINI-SPEAKER

Unlike a large woofer speaker, a typical mini-speaker has an impedance characteristic that decreases with increasing frequency. This results from the mini-speaker's stray capacitance that dominates its reactance characteristic. In the example of Fig. 4–11, the mini-speaker's impedance exhibited a minimum value at approximately 5 kHz, and then started to rise at higher frequencies. This "dip" is frequently encountered in small cone-type speakers.

CROSSOVER OPERATION

When we use a woofer and a tweeter in a speaker system, a suitable crossover circuit is required inasmuch as the tweeter is not rated to withstand the power for which the woofer is rated at low frequencies. In other words, the tweeter must be connected in series with a high-pass filter of some kind. The woofer may be connected in series with a low-pass filter, although this is not absolutely necessary because a woofer can dissipate the sum of the low-frequency audio power and the high-frequency audio power.

If ultimate high-fidelity characteristics are not required, we can save time in practical situations by merely connecting an appropriate capacitor in series with the tweeter, as shown in Fig. 4–12. Observe that at some frequency, the impedance (reactance) of the capacitor will be equal to the impedance of the tweeter. This equal-impedance frequency is called the crossover frequency. Or, the capacitor functions as a crossover. Next, we need to know the required capacitor value with respect to the tweeter impedance.

For example, a 4-μF capacitor connected in series with a tweeter that has an impedance of 8 Ω will provide a crossover frequency of approximately 5,000 Hz. Again, if we use an 8-μF capacitor, the crossover frequency will be approximately 2,500 Hz. As a practical note, nonpolarized electrolytic capacitors are utilized as crossovers. Observe in Fig. 4–12 that at the cross-

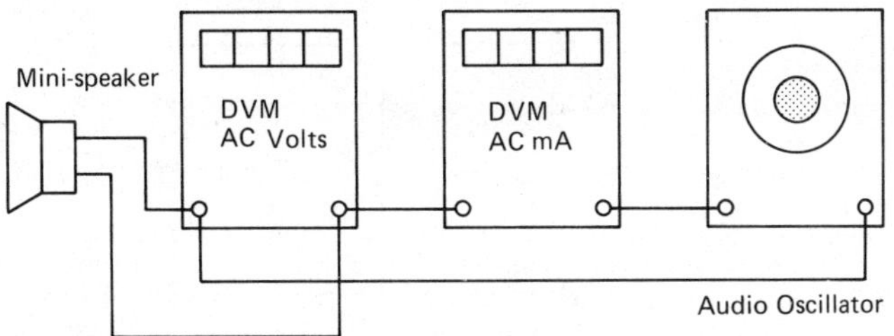

Note: A typical mini-speaker has distributed and stray capacitance that outweights its inductance. In turn, its impedance exhibits a decrease as the frequency increases. For example, a mini-speaker with 8 Ω dc resistance showed the following impedances versus frequency:

At 60 Hz
I = 1 mA rms
E = 0.008 V rms
Z = 8 Ω

At 1,000 Hz
I = 1 mA rms
E = 0.007 V rms
Z = 7 Ω

At 5,000 Hz
I = 1 mA rms
E = 0.004 V rms
Z = 4 Ω

At 10 Hz
I = 1 mA rms
E = 0.005 V rms
Z = 5 Ω

Thus, in this example, the impedance decreased to one-half of the speaker's dc resistance value at 5,000 Hz. Then, the impedance started to rise again at higher frequencies.

Figure 4–11 Measurement of mini-speaker impedance.

over frequency, the power demand of the tweeter will be one-half of its high-frequency power demand.

This is just another way of saying that the series capacitor prevents low audio frequencies from appreciably energizing the tweeter. Conversely, the series capacitor permits high audio frequencies to energize the tweeter. Al-

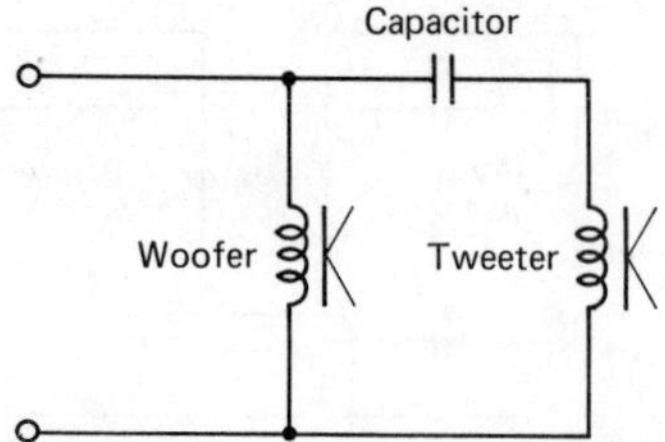

Note: The impedance (reactance) of a capacitor at a given frequency is equal to $1/(6.283 \times f \times C)\ \Omega$, where f is in Hz and C is in Farads. As previously noted, a microfarad is one-millionth of a Farad.

To save prime time, use your pocket calculator to ''punch out'' capacitive circuit problems.

Shortcut: Keep in mind that 0.159 μF has a reactance of 1,000 Ω at 1,000 Hz. With this basic relation in mind, you can solve many capacitance problems ''in your head.''

Figure 4–12 The simplest crossover arrangement.

though the capacitor does not prevent high frequencies from energizing the woofer. This may or may not be a significant consideration in regard to fidelity. As an illustration, if we are using a woofer that is capable of reproducing frequencies above the crossover point, then both of the speakers will reproduce the mid-range frequencies, and this may result in overemphasis of the mid-range interval.

To cope with overemphasis of the mid-range frequencies, we can either choose an optimum crossover frequency (optimum capacitor value) with respect to the woofer frequency response, or, we can connect a coil in series with the woofer, as shown in Fig. 4–13. As we know, a coil reacts oppositely with respect to a capacitor as the frequency increases; the reactance of the coil increases, whereas the reactance of the capacitor decreases. Observe that to obtain an impedance of 8 Ω at 5,000 Hz, our coil will require an inductance value of about 1.25 millihenry (0.25 mH). Again, to obtain a crossover frequency of 2,500 Hz, our coil will require an inductance of about 0.5 mH.

With these facts in mind, we can proceed to consider the total impedance that a basic speaker system with a crossover presents to the preceding amplifier. If we are using 8-Ω speakers as in Fig. 4–13, we will find that the continually changing impedance versus frequency of the coil branch and the capacitor branch will hold the total (input) impedance of the system at approximately 8 Ω. In fact, if the speakers were a purely resistive load (which

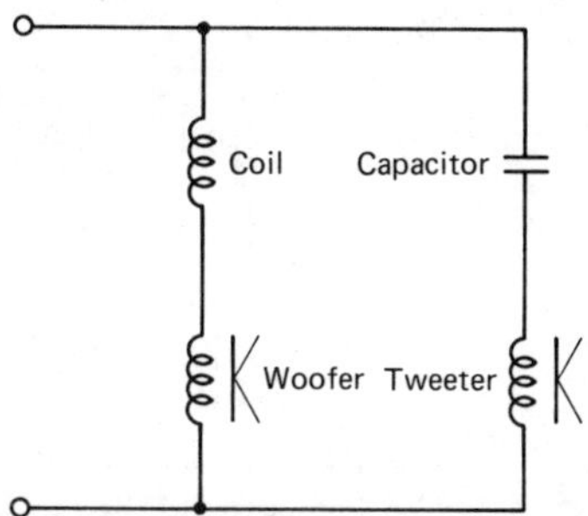

Note 1: Coils used in crossover networks are air-core type, and must be wound with sufficiently large wire so that they do not dissipate an objectionable amount of low-frequency power. Electrolytic capacitors may be used in the tweeter branch, provided that the nonpolarized type of capacitor is utilized.

Note 2: The reactance of a coil with respect to frequency is equal to 6.283 $\times f \times L$ where f is in Hz and L is in henrys. A millihenry is equal to 0.001 of a henry, and a microhenry is equal to one-millionth of a henry. Use your pocket calculator and save time by "punching out" inductive circuit problems.

Shortcut: Keep in mind that 0.159 H has a reactance of 1,000 Ω at 1,000 Hz. This basic relation will permit you to figure out many inductance problems "in your head."

Figure 4–13 Crossover network comprising a coil and a capacitor.

we know they are really not), the input impedance of the system would not vary by more than 1 percent.

With reference to Fig. 4–14, the foregoing system impedance remains 8 Ω at 1/10 and at 10 times the crossover frequency. However, we know that speakers have some inductance in addition to their resistance. This is the reason that a woofer has a higher impedance at higher frequencies. In turn, speaker inductance affects the ideal value shown in Fig. 4–14 and the woofer impedance value is somewhat higher (except at very low frequencies). It also follows from previous discussion that since a tweeter generally has decreasing impedance at higher frequencies, an efficient tweeter can be chosen that will largely compensate for the tendency of the woofer to increase in impedance as the frequency increases.

We could also consider finer points, such as connecting a variable resistor in series with the tweeter to adjust the tonal balance of the crossover

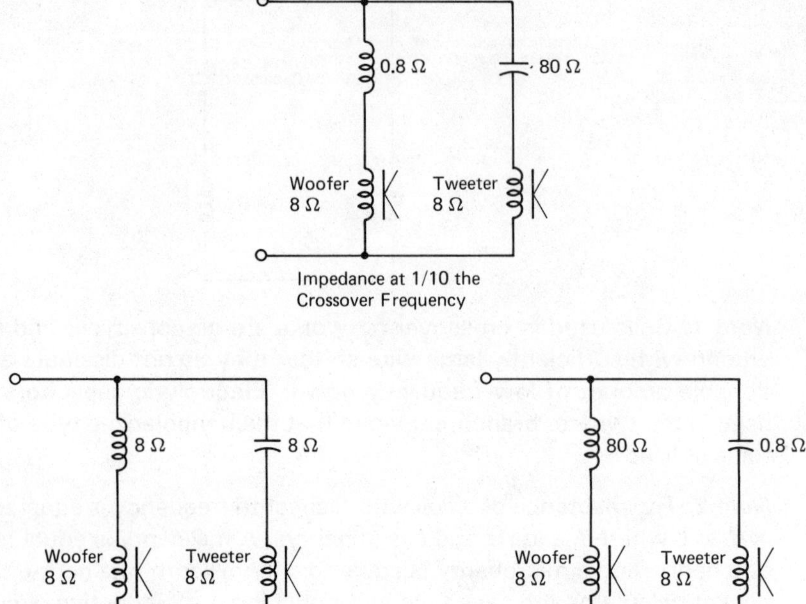

Note: Commercial crossovers may be employed, such as the Radio Shack No. 40-1296, or equivalent. Dyed-in-the-wool experimenters and audio buffs may wish to work with the more elaborate three-way crossover such as the Radio Shack No. 40-1299.

Figure 4–14 Examples of crossover input impedance at three different frequencies. Reproduced by special permission of Reston Publishing Company and Christopher Robin from *How To Build Your Own Stereo Speakers.*

system. However, these finer points are more in the province of the audio buff who may be referred to specialized audio technology books.

MEASURING INDUCTANCE WITH A DVM

Hobbyists, experimenters, technicians, troubleshooters, and just about everyone who is into electronics occasionally want to know the inductance of a transformer winding, or the inductance of a choke. Inductance can be easily and quickly measured as illustrated in Fig. 4–15. This method employs a DVM, a small step-down transformer, and an assortment of resistors with values from 50 to 10,000 Ω (or a resistor decade box). We will first observe a shortcut technique, and then consider a graphical technique that provides more precise answers.

As seen in Fig. 4–15, the inductor to be checked is connected in series with the resistor decade box. The RL series combination is then connected to the secondary winding of a 6.3 V step-down transformer operated from the 117-V power line. We measure the voltage across the inductor and the voltage across the decade box with an ac DVM. We compare these readings and change the resistance value until the two voltages are the same.

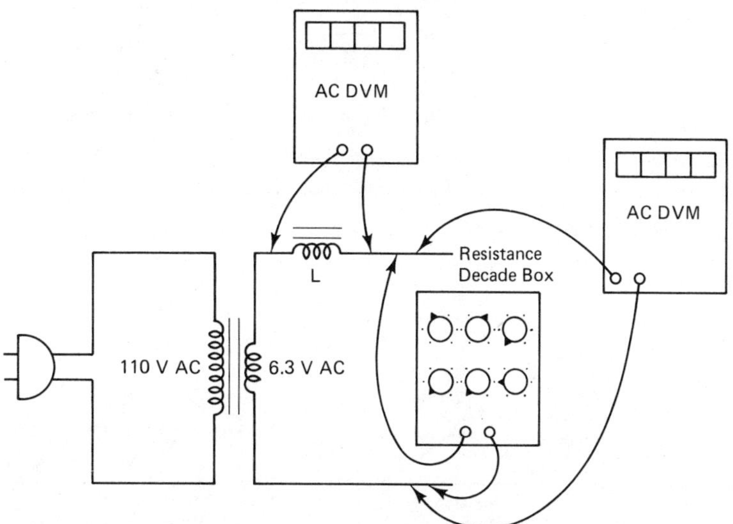

Note: This method of inductance measurement is based on adjusting the resistance decade box so that the ac voltage across *L* is the same as the ac voltage across the box. In other words, the inductive-reactance value is then equal to the resistance value. Accordingly, the inductance in henrys is equal to the resistance value in ohms divided by 376.8 as explained in the text.

Observe in the diagram that *L* is assumed to be purely inductive (without any winding resistance). This assumption is ordinarily justifiable in a shortcut check.

Figure 4–15 Shortcut method of inductance measurement with a DVM.

For example, if the inductor and resistor voltages are equal when we adjust the decade box to 5,000 Ω, we recognize that the inductive reactance of the inductor is also 5,000 Ω. Since the inductive reactance $X_L = 6.2832 \times f \times L$, it follows that $L = 5{,}000/376.8$ or $L = 13.2$ H. As another illustration, if the voltages across the inductor and the decade box are equal when the resistance is 1,200 Ω, then $L = 1{,}200/(6.283 \times 60) = 1{,}200/376.8 = 3.2$ H. This is a shortcut method, and it is sufficiently accurate for most practical procedures.

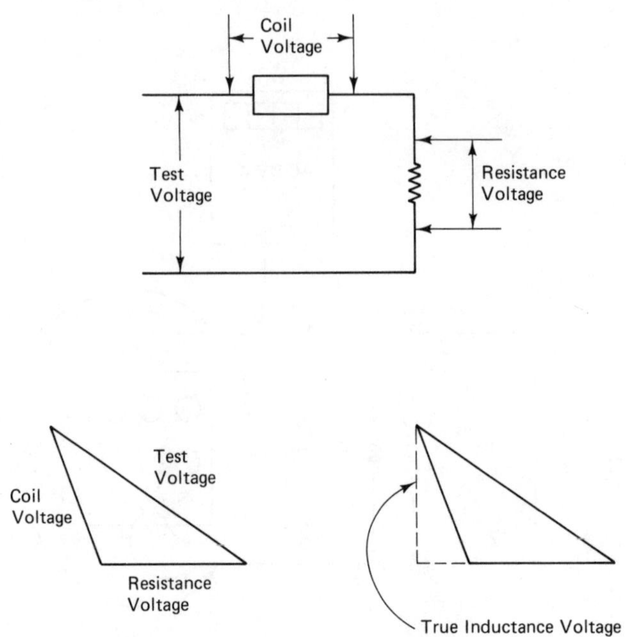

Note: We measure the three ac voltages in the test circuit and draw them in triangular form as shown above. The length of each line is equal to the voltage value. Then, we complete a right triangle as shown by the dotted lines above. Finally, the true inductance voltage has been found; this is somewhat less than the coil voltage. The true inductance value is equal to the length of the vertical line divided by 376.8 and this true inductance value is somewhat less than the shortcut value.

Figure 4–16 Precise measurement of inductance with a DVM.

Next, a more precise method of inductance measurement is shown in Fig. 4–16. It requires three measurements and we must draw a diagram to find the inductance value. The advantage of the precise method is that we can be sure that our answer is accurate and free from assumptions. Dyed-in-the-wool experimenters can appreciate the higher confidence level provided by the precise method of inductance measurement. What we are basically doing in Fig. 4–16 is taking the winding resistance of the inductor into consideration.

5

Signal Tracers and Analyzers

OVERVIEW

Signal tracers and analyzers comprise a wide spectrum of test equipment that is designed to track signal voltage or current flow through circuitry in order to identify its presence or absence, and to indicate various signal characteristics such as magnitude, frequency, distortion, noise components, interference components, modulation, and stability. We have previously noted a specialized type of audio signal tracer consisting of a hi-fi amplifier and speaker supplemented by RC differentiating or integrating circuits. This ''timbre'' analyzer is basically a quick checker for estimation of sine-wave purity, and for preliminary analysis of complex waveforms.

The following discussion is primarily concerned with those types of signal tracers and analyzers that fall into the time-saver and shortcut categories. As an illustration, an ac DVM can serve as a signal tracer, inasmuch as it indicates the presence or absence of signal voltage. On the other hand, the DVM does not function as a distortion quick checker—the DVM cannot distinguish between a pure sine wave and a slightly distorted sine wave. Observe that a ''timbre'' analyzer functions as a distortion quick checker because the ear critically discriminates between a waveform before and after passage through an RC differentiating circuit, in the event that the waveform consists of a sine wave with more or less harmonic content.

AUDIO SIGNAL-TRACER/RESIDUE ANALYZER

An alternative type of "timbre" analyzer employs a parallel-T RC trap, as shown in Fig. 5-1. This is a residue-analyzer type of distortion tracer and quick checker. If distortion is present in the equipment under test, this distortion will produce residual harmonic voltages. After the fundamental frequency (1 kHz) is filtered out by the parallel-R RC network, the residue passes into the hi-fi amplifier and speaker and produces a corresponding sound output. The operator can easily determine by ear whether there is significant distortion present and can form a rough judgment of its nature. (See also Chart 5-1.)

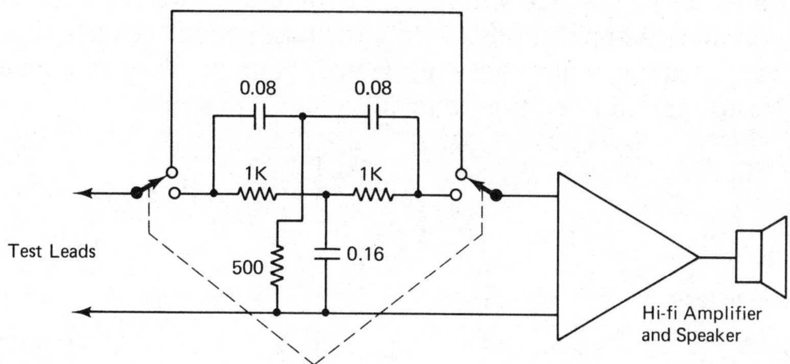

Note: This is an effective audio distortion tracer and quick checker that employs a parallel-T RC filter (trap). In application, a 1-kHz test signal is used, and the switch is first set to apply the voltage from the test point directly to the hi-fi amplifier and speaker. This provides the reference tone from the speaker. Then the switch is thrown to place the RC network in series with the voltage from the test point and the hi-fi amplifier and speaker. This provides the comparison tone (if any) from the speaker. In other words, if there is no distortion to the equipment under test, the RC network will trap out practically all of the 1-kHz voltage and there will be little or no sound output from the speaker. On the other hand, if there is distortion present, there will be more or less sound output from the speaker for the comparison tone. This comparison tone will lack the 1-kHz fundamental frequency component, but will contain all of the distortion products, such as a second harmonic, a third harmonic, and so on.

Figure 5-1 Distortion tracer and quick-checker utilizing a parallel-T RC filter.

Chart 5–1

WORKING WITH SEMICONDUCTORS

Experiments, hobbyists, technicians and troubleshooters sometimes encounter difficulty in checking out semiconductor devices. In turn, various time-saving methods and shortcuts such as in-circuit tests are of basic interest.

Some of the more fundamental quick-check techniques were noted in previous chapters. For example, procedures have been explained for in-circuit turn-on and turn-off tests of bipolar and MOSFET transistors. Front-to-back ratio checks for diodes were noted. Low-power ohmmeter applications for in-circuit resistance measurements of transistor circuitry have been illustrated. Note the three fundamental transistor amplifier configurations:

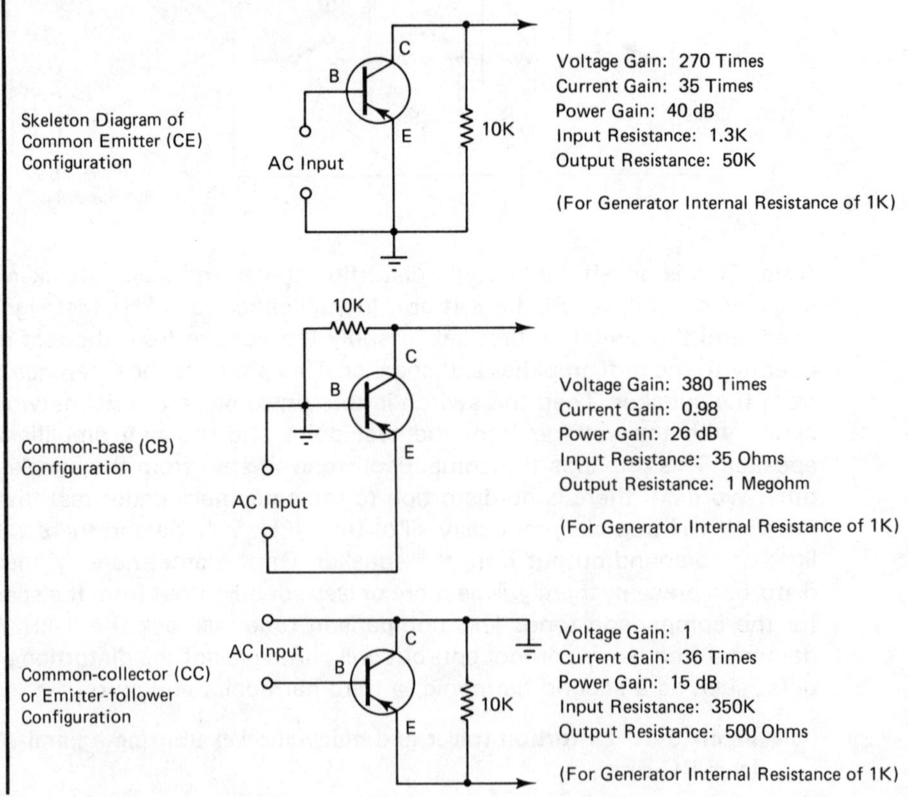

The basic MOSFET amplifier configurations and characteristics are as follows:

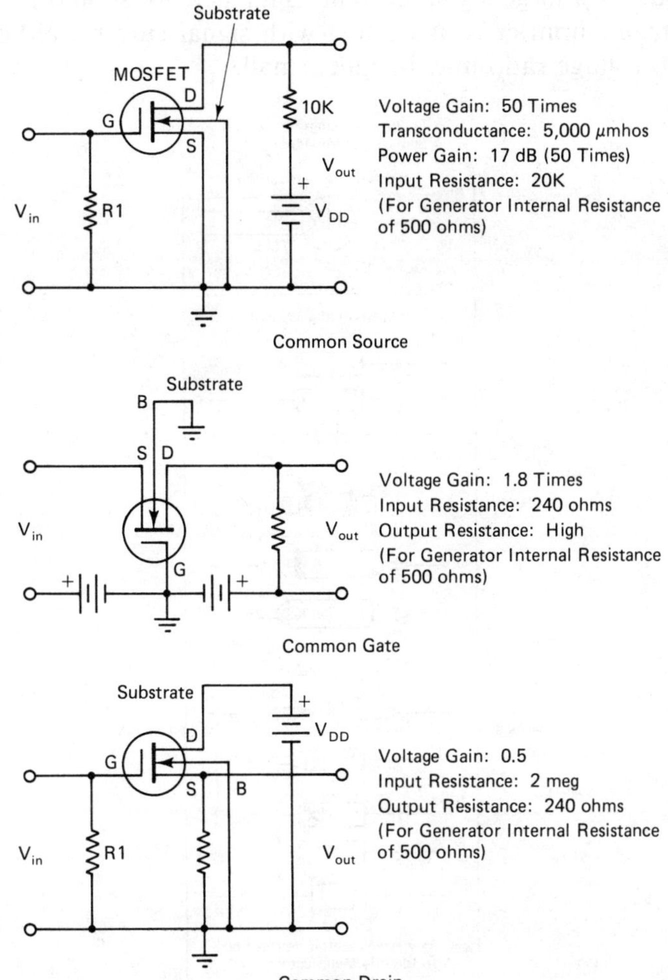

It is evident from these examples that MOSFET amplifiers have less gain than corresponding bipolar transistor amplifiers. On the other hand, MOSFET amplifiers have considerably higher input resistance, which is often desirable in practical applications.

Because bipolar transistors have a comparatively low input resistance, we sometimes say that they are current-operated devices that are used in current amplifiers. Conversely, because MOSFETs have a comparatively high input resistance, we sometimes say that they are voltage-operated devices that are used in voltage amplifiers.

We should recognize, however, that a current amplifier is associated with signal voltage, although the voltage may be quite small that produces a large signal current. Similarly, we should recognize that a voltage amplifier is associated with signal current, although the current/voltage ratio may be quite small.

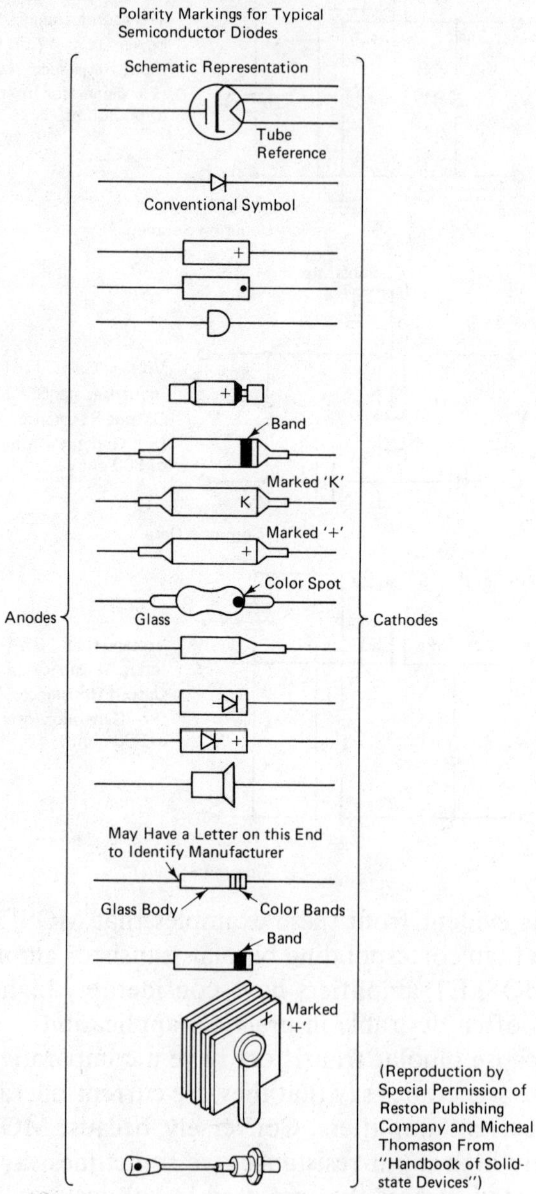

(Reproduction by Special Permission of Reston Publishing Company and Micheal Thomason From "Handbook of Solid-state Devices")

Note that light-emitting diodes (LEDs) will exhibit a useful front-to-back ratio when checked with a high-sensitivity VOM such as the Radio Shack 50,000 ohms-per-volt multitester. An LED will not exhibit a recognizable front-to-back ratio on a DVM. Observe also that although its front-to-back ratio can be checked with a high-sensitivity VOM, the LED does not glow during the test inasmuch as insufficient test current flows through the diode. (Higher forward voltage must be applied to the LED to make it glow.)

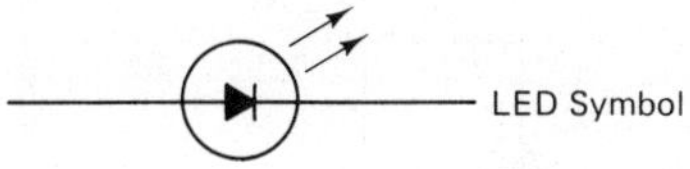

LED Symbol

It was previously noted that about 0.25 V must be applied to a germanium diode before it will conduct, and that about 0.6 V must be applied to a silicon diode before it will conduct. Observe that about 1.7 V must be applied to an LED before it will conduct, and that even somewhat higher voltage is necessary to make the LED glow. Note also that there is a certain variation in forward operating voltages for different kinds of LEDs.

Practical situations can frequently be facilitated by the possible series arrangements of zener diodes to obtain a desired operating voltage. Observe also that these arrangements may include germanium diodes or silicon diodes as 0.25-V and 0.6-V dc level shifters. In other words, the forward-voltage drop across a germanium diode will be about 0.25 V, and the forward-voltage drop across a silicon diode will be about 0.6 V.

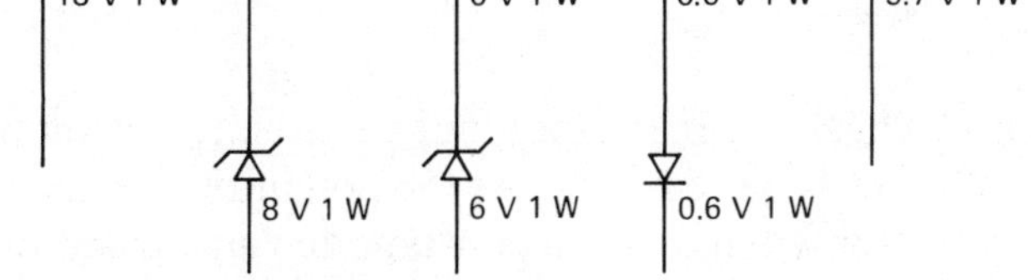

18 V 1 W = 10 V 1 W, 8 V 1 W = 6 V 1 W, 6 V 1 W, 6 V 1 W

4.5 V 1 W, 0.6 V 1 W, 0.6 V 1 W = 5.7 V 1 W

Practical Methods to Change Zener Voltages

(Reproduced by Permission of Prentice-Hall, Inc, From "Electronic Workshop Manual and Guide," by Carl G. Grolle)

As previously noted, a zener diode is reverse-biased in normal operation, whereas a germanium or silicon diode is forward-biased in normal operation. Construction of a simple shunt-regulated zener power supply was previously noted. The effective power rating of a zener diode can be increased by connecting the diode in series with a transistor, as shown below. Thus, a ½-W zener diode operates as a 5-W zener diode when it is connected in series with a 5-W power transistor.

+
+
12 V $\frac{1}{2}$ W
=
12 V 5 W (Approx.)
5 W

A zener-transistor circuit is sometimes noisy. To suppress the noise, connect a 100 μF electrolytic capacitor across the zener diode.

The previous basic arrangement is connected to an unregulated power supply as shown below in order to provide a regulated power supply. Choose a suitable value for *R* to avoid passing excessive current through the zener diode. (The diode will draw maximum current when the power supply is unloaded.)

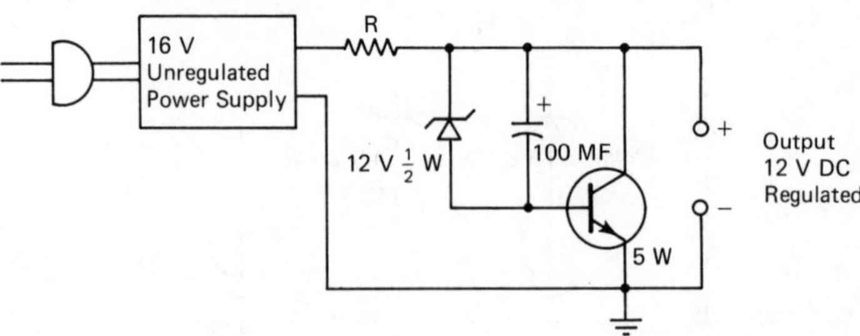

If you wish to check a zener diode out-of-circuit, the simple arrangement shown below is quite handy. As the potentiometer resistance is reduced, the voltmeter will indicate a constant dc value past the zener point.

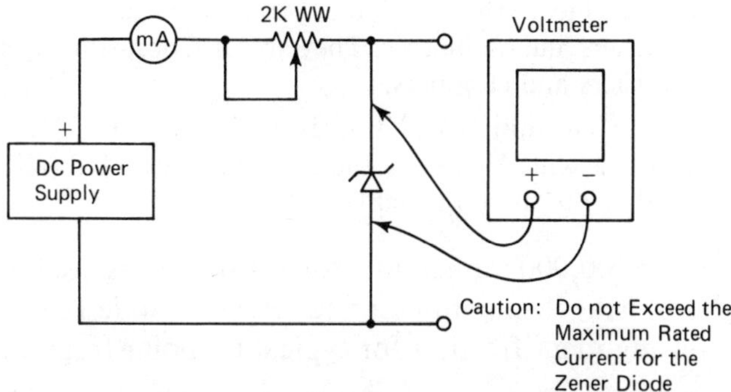

Basing identification for typical bipolar transistors is as follows:

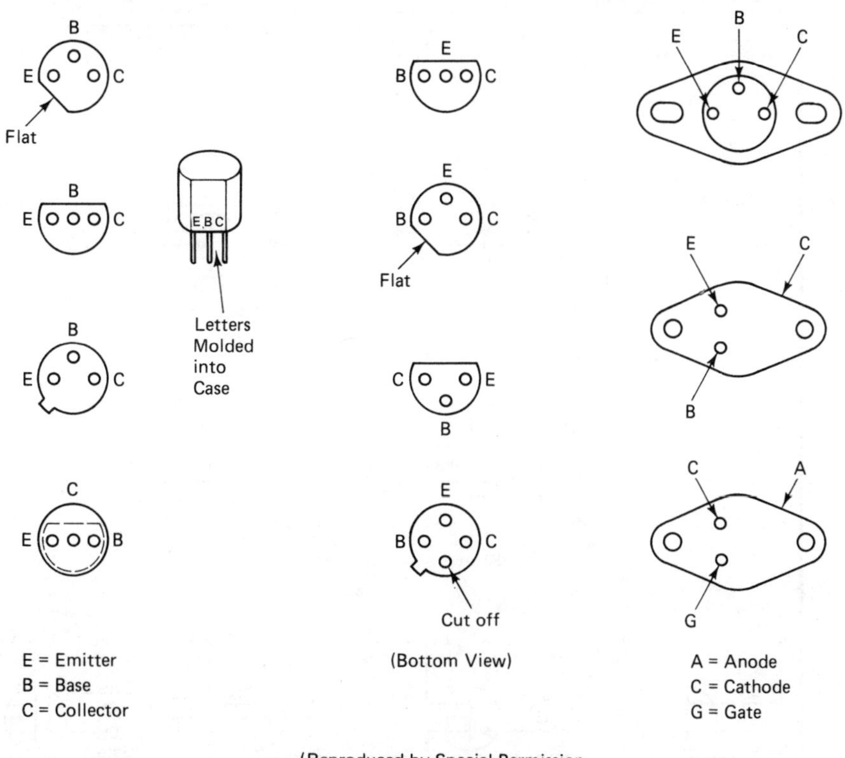

(Reproduced by Special Permission of Reston Publishing Company and Walter Folger From "Radio, TV, and Sound System Diagnosis and Repair")

Out-of-circuit tests and in-circuit quick checks for bipolar transistors have been previously described and illustrated.

Note 1: Field-effect (unipolar) transistors are employed chiefly as RF amplifiers and oscillators. They are used occasionally as general-purpose amplifiers and choppers.

Note 2: The unijunction transistor (UJT or JFET) is now used to a very limited extent. However, the insulated-gate FET (IGFET or MOSFET) is utilized to an appreciable extent.

Over 100,000 semiconductor substitutions are listed in the Archer (Radio Shack) Semiconductor Reference Guide No. 276-4006.

Basing identification for typical unipolar transistors is as follows:

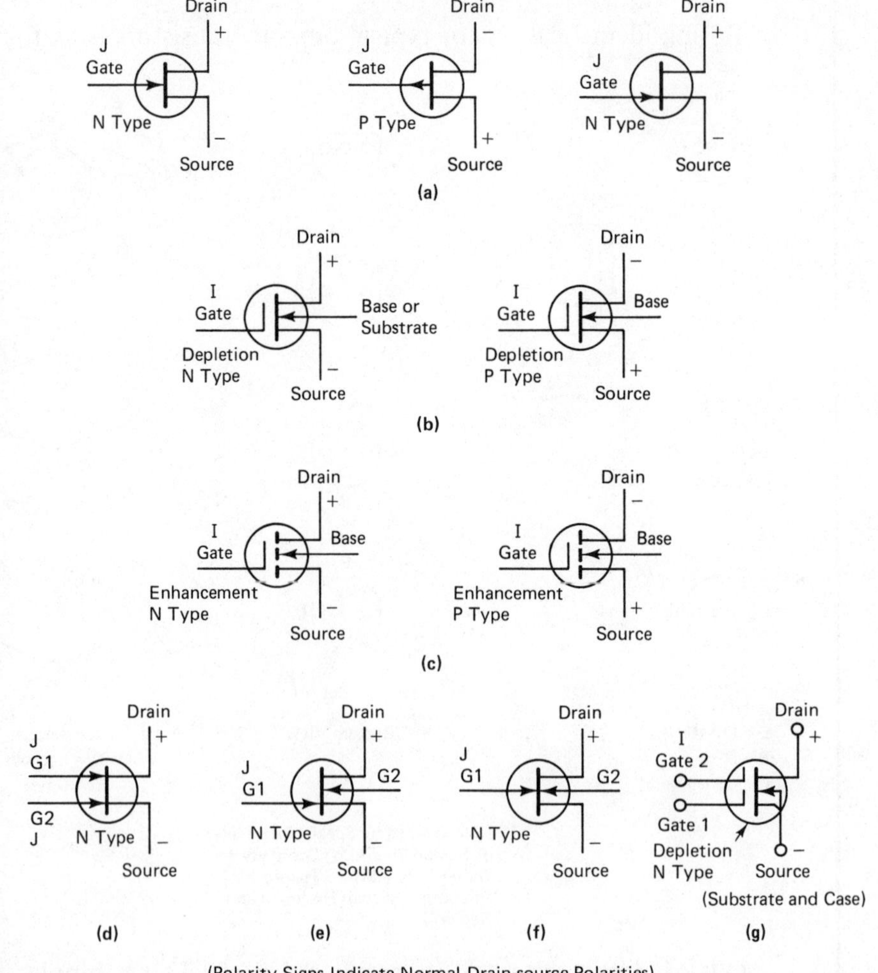

(Polarity Signs Indicate Normal Drain-source Polarities)

(Reproduced by Special Permission of Reston Publishing Company and Michael Thomason From "Handbook of Solid State Devices")

Field-effect transistors. (a) N and P types of JFETS—arrow points to N-type substance, away from P-type substance. (b) Depletion type MOSFETS—arrow points to N-type substrate, away from P-type substrate. (c) Enhancement-type MOSFETS—arrow points to N-type substrate, away from P-type substrate. (d) Dual-gate N-channel FET. (e) Nonsymmetrical N-channel FET. (f) Alternate symbol for (e). (g) Dual-gate depletion N-type MOSFET.

Caution: *JFETS are comparatively rugged and can be handled and tested in much the same manner as bipolar transistors. On the other hand, MOSFETs are quickly damaged by static electricity whenever they are not in their circuits. Ground your wrist before you handle a MOSFET; ground the tip of your soldering iron before you solder a MOSFET into a circuit. Use heat sinks on both bipolar and unipolar transistor leads to avoid damage from overheating.*

AUDIO SIGNAL-TRACER/LEVEL INDICATOR

When weak-output or no-output trouble symptoms are being tackled in an audio system, we are not concerned with distortion, but instead with progressive gain, or signal levels. For this purpose, the amp-speaker/dB-meter arrangement shown in Fig. 5-2 is very handy. It is compact, has comparatively high gain for quick checks in low-level circuitry, plus a VOM with dB scales. It is usually advantageous to have dB readout available, inasmuch as the ear is not a particularly critical judge of sound levels.

Observe also that audio sections are sometimes rated for normal dB gain, and this type of tracer/level-indicator provides a quantitative check of relative dB level. The audible output is quite informative—if the signal is noisy or contaminated by hum interference, these facts are immediately evident. Note that the mini-amp functions also as a preamp for the dB meter, thereby making relative dB measurements in low-level circuits practical. A coupling capacitor is included in series with the probe-input lead to the signal tracer in Fig. 5-2. It blocks flow of dc current into the tracer from the circuit under test. A nonpolarized capacitor is required inasmuch as either positive and negative dc voltages may be present.

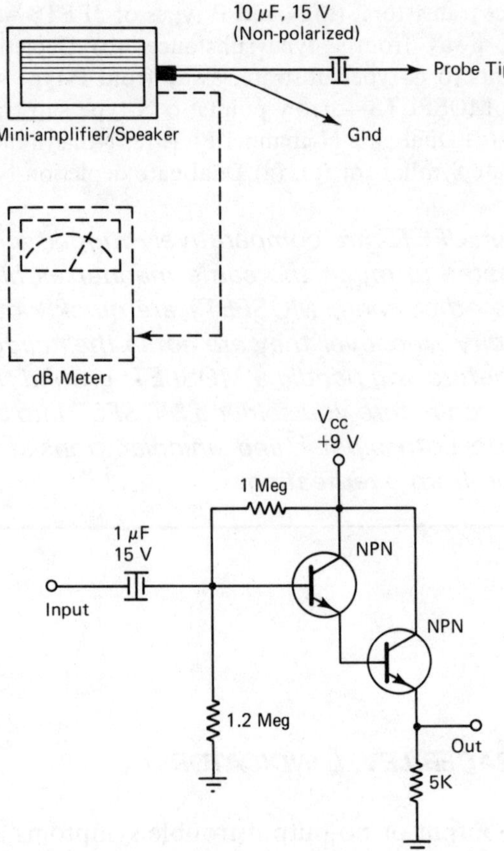

Note: The mini-amp/speaker may be an Archer (Radio Shack) No. 277-1008. A gain of 1,700 times is provided, with an input impedance of 5 kΩ. A 1-mV input signal produces a 200 mW output.

(Most VOMs have a dB scale for use on the ac voltage ranges). The dB meter is plugged into the earphone jack on the miniamp housing. Observe that an input impedance of 5 kΩ will not objectionably load the majority of audio circuitry that we encounter. However, if excessive loading happens to occur in MOSFET circuitry, for example, the emitter follower shown in the diagram can be placed ahead of the signal tracer, and the input impedance thereby stepped up to 0.5 MΩ.

Figure 5-2 A time-saving audio signal-tracer/level indicator.

SIGNAL TRACING AT HIGHER FREQUENCIES

Signal tracing at frequencies above the audio range requires test circuitry that is appropriate for the chosen frequency range. An ac DVM serves as an elementary signal tracer at frequencies somewhat above the audio band, although it is not applicable in the AM radio frequency band, for example. A basic peak-rectifier diode probe was noted previously; this is the simplest practical signal-tracing probe arrangement for RF circuitry. The output from the probe is applied to a dc DVM, and, with a suitable value of isolating resistor between the probe circuit and its output cable, the DVM will read the rms voltage of the radio-frequency signal.

Effectively, a peak-rectifier diode probe extends the ac voltmeter capability of a DVM into the megahertz region. This capability greatly facilitates preliminary troubleshooting procedures in radio-frequency (RF) circuitry. However, a simple peak-rectifier diode probe has some (often) serious limitations. For example:

1. A simple probe has comparatively low input impedance and tends to objectionably load typical RF circuitry.
2. The simple probe has appreciable input capacitance and tends to detune resonant circuits in RF equipment.
3. Only moderate to high RF signal levels can be checked with a simple probe; it has negligible response to weak signals. (See also Chart 5–2.)

Chart 5–2

WORKING WITH RADIO FREQUENCIES

We have seen that signal-tracing at radio frequencies often requires somewhat elaborated testers, compared with signal-tracing at audio frequencies. First, we will find that typical ac DVMs indicate incorrectly and are virtually unusable at radio frequencies. For example if a 300-mV 1-MHz signal is applied to a service-type autoranging DVM, a reading of 8 mV is obtained. In other words, the DVM indicates less than 3 percent of the true voltage value.

Various models of DVMs will indicate other subnormal values (or zero) when a 1-MHz signal voltage is applied. As an illustration, another model of service-type auto-ranging DVM indicated zero when a 300-mV 1-MHz signal was applied.

Observe also that top-of-the-line service-type DVMs are similarly unsuited for signal tracing at radio frequencies. For example, a top-rated DVM indicated inconsistent and subnormal voltage values from range-to-range in response to a 300-mV 1-MHz input signal as follows:

On the 200-mV range, 0.7 mV was indicated
On the 2-V range, 20 mV was indicated
On the 20-V range, 40 mV was indicated

Accordingly, a service-type DVM cannot be used directly in RF signal-tracing procedures. However, when used with a peak-rectifier diode probe, the DVM will indicate the rms value of the RF signal voltage (provided that the value of the isolating resistor in the probe has been chosen for proper scale calibration).

We have also seen that a peak-rectifier probe requires at least a moderate input signal level, and is not suitable for signal-tracing in low-level RF circuitry. Note that a semiconductor diode has a barrier potential, such as about 0.2 V for a germaniun diode. This means that the peak-rectifier probe will produce zero output at input signal-voltage levels less than 0.2 V (200 mV). Observe also that at input signal-voltage levels a bit higher than 200 mV, the semiconductor diode has such high internal resistance that the full value of the input voltage is not indicated. This is just another way of saying that a peak-rectifier probe requires the use of an RF pre-amp in order to function as an RF signal-tracer in low-level circuitry.

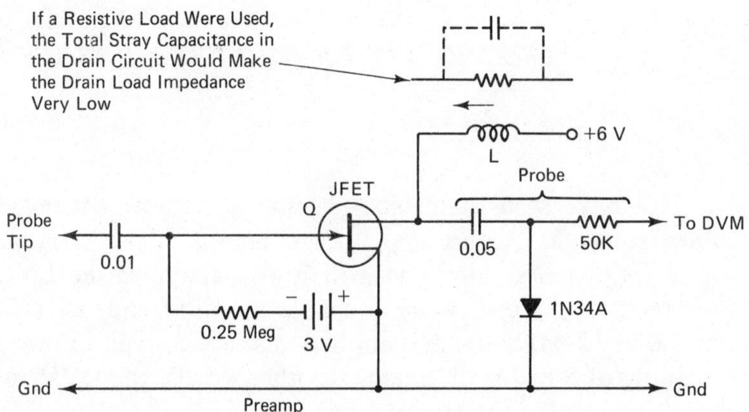

The most important fact to note in regard to the pre-amp configuration exemplified above is that the JFET has drain junction capacitance plus stray wiring capacitance in the drain circuit. If we utilized a resistor as the drain load, its stray resistance would also need to be taken into account (shown by the dotted lines in the diagram). The total effective shunt capacitance in the drain circuit would function as a bypass to ground, and very little signal voltage would be applied across the probe.

However, if we employ a tuned coil *L* as the drain load, the total effective shunt capacitance then functions as tuning capacitance across *L,* and contributes to its resonant frequency. Observe that this is a parallel-resonant load that develops a high impedance at its resonant frequency. In turn, practically all of the output signal voltage is now applied to the probe, so that efficient pre-amplification of the RF signal voltage is realized.

Next, the dyed-in-the-wool experimenter will ask why a tuned drain load is required in the RF probe pre-amp, whereas the emitter-follower stage depicted in Fig. 5-3 does not require a tuned emitter load. (If we experiment with a tuned emitter load, we will find that it provides an advantage of only 1 or 2 mV in the signal output level.)

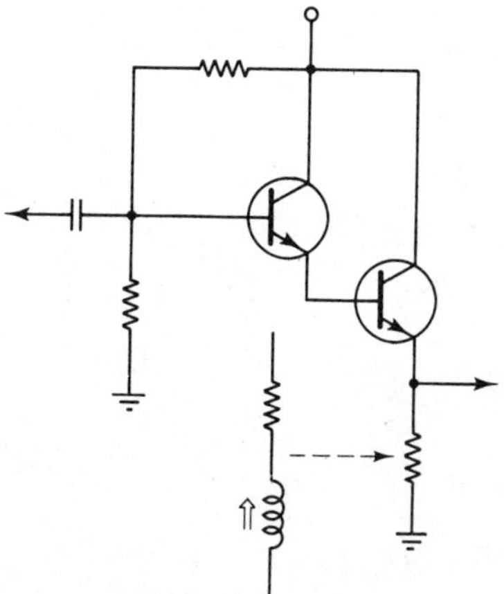

The emitter junction of the transister has junction capacitance, the emitter circuit has stray wiring capacitance, and the emitter load resistor has a small amount of stray capacitance. This total shunt capacitance from emitter to ground tends to bypass the output signal voltage to ground. However, in this kind of circuitry, the total shunt capacitance from emitter to ground can be practically neglected, for the following reasons.

Observe that the emitter-follower arrangement has a very low output impedance. This low output impedance is the result of utilizing a pair of transistors in a Darlington-connected emitter-follower circuit. An emitter follower has high current gain, and the Darlington configuration has very much higher current gain (the current gain of the circuit is equal to the product of the current gains of the individual transistors).

Since the output impedance of the emitter-follower stage is very low, the total shunt capacitance from emitter to ground is only a minor factor in signal attenuation. If we include a tuned emitter load arrangement as indicated by the dotted arrow, we will obtain only a very minor improvement in output signal level. This is just another way of saying that the emitter follower has almost unity voltage gain at 1 MHz.

Again, the dyed-in-the-wool experimenter will be interested in constructing the Darlington RF pre-amp shown next, which provides a maximum available gain of approximately 500 times.

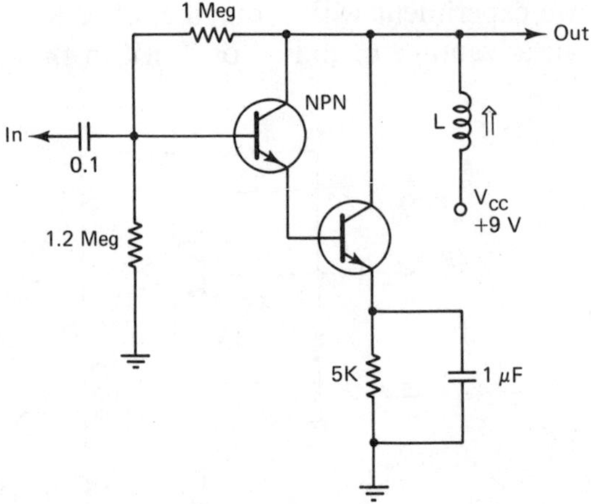

Note: *This Darlington-connection RF pre-amp is configured with a tuned collector load. The emitter resistor is bypassed. A maximum available voltage gain of approximately 500 times is provided. The 5 K emitter resistor contributes to bias stability. Observe that the 1-MΩ and 1.2-MΩ resistors bias the transistors for class-A operation and also provide some negative feedback that assists in linearizing the amplifier. Note that if the emitter resistor is not bypassed, the amplifier gain becomes extremely small.*

The maximum available gain may not necessarily be usable in some signal-tracing situations. As previously noted, when this type of pre-amp is applied in a high-Q inductive circuit, instability can develop and result in a false output when the circuit under test is "dead." In such a case, the technician should connect sufficient resistance in series with the pre-amp input lead to stabilize operation.

Fortunately, there are readily available means for overcoming the foregoing disadvantages, whereby much time can be saved in preliminary RF troubleshooting procedures. The low input impedance of a simple peak-rectifier diode probe can be avoided by the use of an emitter-follower input circuit, as shown in Fig. 5-3. Although the emitter follower helps to step up the input impedance and to reduce the input capacitance, it contributes nothing to the ability of the probe to trace weak signals. Weak-signal tracing requires a voltage pre-amp for the probe in order to overcome the poor efficiency of the semiconductor diode at low signal levels. An effective voltage pre-amp is shown in Fig. 5-4. A maximum available gain of 45 times is provided, which can be a prime time-saver in preliminary troubleshooting of low-level RF circuitry.

Observe that the probe depicted in Fig. 5-4 has a disadvantage in that it imposes approximately the same loading on the circuit under test as does a simple passive probe. In turn, the technician may prefer to employ an active probe that has somewhat less gain, but which provides higher input impedance.

For example, the arrangement shown in Fig. 5-5 utilizes a JFET instead of a bipolar transistor. This type of active probe offers substantial voltage gain and higher input impedance than a bipolar-transistor probe. Note that its lower potential gain may be more than offset by its reduced circuit loading in high-impedance RF circuitry.

UNMODULATED RF SIGNALS MADE AUDIBLE

A disadvantage of the audible signal tracer shown in Fig. 5-4(b) is that it cannot indicate the presence of an unmodulated radio-frequency (CW) signal. However, this function can be easily provided by means of an auxiliary signal generator, as depicted in Fig. 5-4(c). Observe that the output from the auxiliary signal generator is injected into the circuit under test through an isolating capacitor, and the generator is tuned to the vicinity of the signal being traced. In turn, a beat note is heard from the mini-amp/speaker. On the other hand, if the signal being traced is absent, there is no sound output.

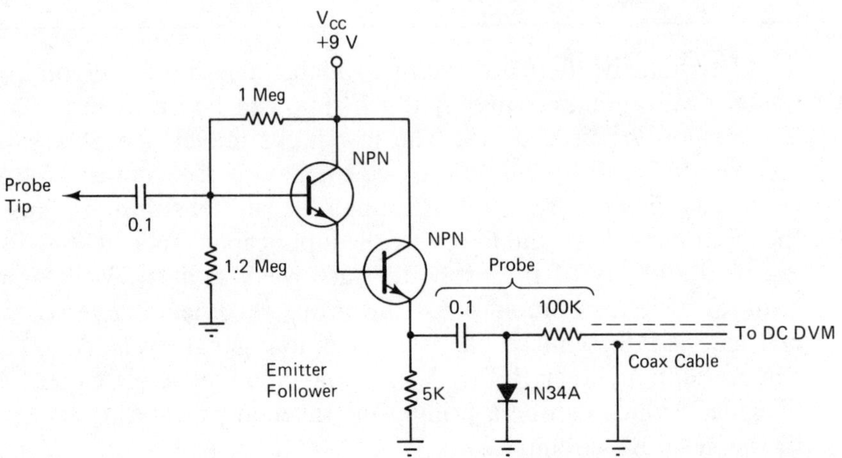

Note 1: This is a peak-rectifier probe circuit with a preceding emitter-follower input section. The emitter follower helps to step up the input impedance and to reduce the input capacitance. RF-type transistors must be used in the emitter-follower circuit. Although there is some disadvantage in employing an elaborated tester with a V_{CC} battery, this complication is often well justified from the standpoint of reduction in circuit loading, particularly at high radio frequencies.

Note 2: Observe that a coax cable is needed from the probe to the DVM inasmuch as open test leads will often pick up stray 60-Hz hum fields and upset the reading on the DVM. It is also good practice to enclose the probe in a metal case such as an old-style IF can.

Figure 5–3 Emitter follower input stage for a peak-rectifier diode probe.

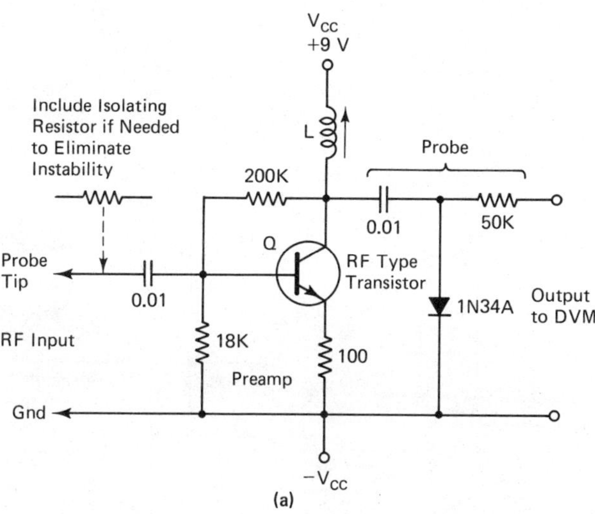

Note: This active RF or IF signal-tracing probe consists of a high-frequency amplifier that drives a peak detector. It is much more effective in low-level circuits than a peak detector probe followed by an audio amplifier because it operates the detector diode more efficiently. In other words, a diode has poor rectification efficiency at low signal levels. This limitation in diode operation is avoided by amplifying the high-frequency signal before detection in the active probe.

Transistor *Q* may be a Radio Shack 276-1603, or equivalent. *L* is a broadcast RF or IF coil. For RF signal tracing, *L* may be adjusted to 1 MHz. For IF signal tracing, *L* is adjusted to 455 kHz. Note that the active probe will develop maximum gain only when the signal frequency is the same as the resonant frequency of *L.* (The 18 K bias resistor may need to have a slightly lower or slightly higher value for optimum operation, depending upon the characteristics of the particular transistor that is utilized.)

Note that when the active probe is directly connected across a high-*Q* coil which is tuned to the same frequency as *L,* instability may occur (the DVM might indicate a signal voltage when no signal is present). In such a case, connect an isolating resistor of suitable value in series with the probe tip. (The isolating resistor should have a sufficiently high value that the DVM indicates zero signal voltage when no signal is present.)

Example: When *L* is a 455-kHz IF coil, the active probe typically provides a gain of 45 times. Next, when the probe is directly connected across a high-*Q* IF coil, instability may occur, but the probe will be stabilized by an

Figure 5–4 (a) Active RF or IF signal-tracing probe. (b) Active RF or IF signal-tracing probe with audible output. (c) Adapting the classical RF signal tracer for tracing unmodulated RF (CW) signals.

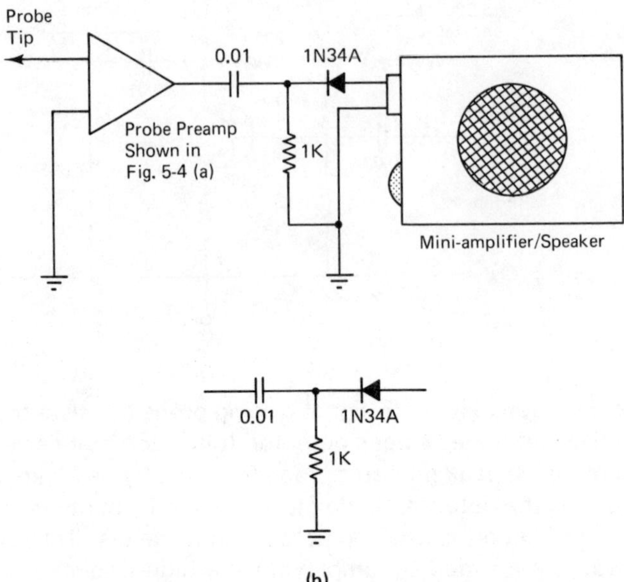

(b)

isolating resistor of approximately 1,700 Ω. This isolating resistor reduces the gain of the probe to 15 times—a gain that still provides a great advantage in signal-tracing low-level or weak circuits.

Note 1: This arrangement provides audible output, as in a classical radio signal tracer, instead of indication on a DVM. The probe preamp provides an output signal level sufficient to drive the 1N34A amplitude detector diode and the mini-amp/speaker. Observe that this type of signal tracer necessarily operates with respect to a modulated-RF signal (it has no response to a CW signal).

Note 2: Observe that the probe preamp includes a 0.01 μF series capacitor in its output circuit, so that the amplitude-detector circuit has the following configuration:

Technical Point: An amplitude detector differs from a peak-reading RF probe in that the amplitude detector has a comparative short time-constant. In turn, an amplitude detector reproduces the modulated component of the RF input signal, whereas a peak-reading RF probe virtually wipes out the modulated component.

Figure 5-4 (*cont.*)

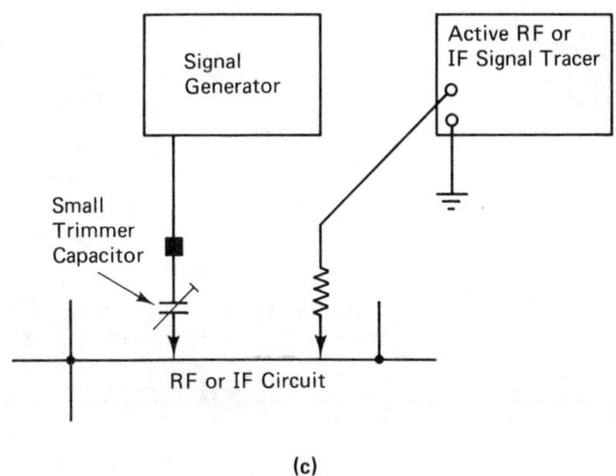

(c)

Note: An auxiliary signal generator can be utilized to make CW (or FM) signals audible with a conventional rf signal tracer. When the signal generator is tuned within 500 or 1,000 Hz of the signal being traced, a beat tone is heard from the signal tracer. This is a trick of the trade that can occasionally speed up preliminary troubleshooting procedures.

Figure 5–4 (*cont.*)

Experiment: With reference to Fig. 5–6, connect a 1,000-Ω ac load across the emitter-follower output, as indicated. Measure the unloaded output at 1 MHz, and then measure the loaded output voltage at the same frequency. Then calculate the output resistance of the emitter follower. For example:

Unloaded output voltage = 317 mV rms
Loaded output voltage = 278 mV rms
Load resistance = 1,000 Ω
Unloaded voltage minus loaded voltage = 39 mV rms
Load current through 1 K resistor = 0.278 mV rms
Output resistance of emitter follower = 140 Ω

Although a superficial look at the emitter-follower circuit might lead to the conclusion that its output resistance is about 5,000 Ω, this is not so. The reason that its output resistance is only 140 Ω is that there is a very large amount of current feedback in the emitter circuit. This negative feedback makes the output resistance of the emitter follower much less than the value of the emitter resistor by itself.

Experiment: Next, it is interesting to measure the input impedance (resistance) of the emitter-follower configuration, as shown in Fig. 5–7. The first measurement is made at a comparatively low frequency of 5 kHz. The emitter-follower output voltage is measured with an ac DVM, and then remeasured with a

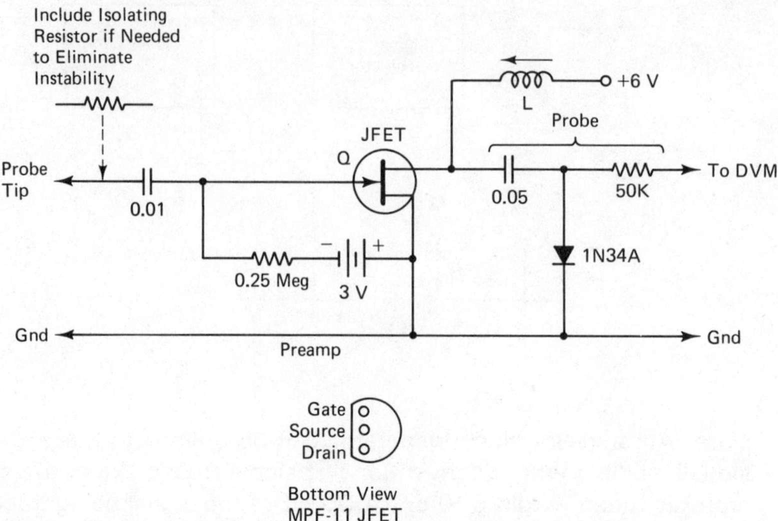

Note 1: Transistor *Q* may be a type MPF-11 N-channel JFET, or equivalent. *L* is an AM broadcast-band inductor. The active signal-tracing probe consists of an RF-amplifier stage driving a peak detector.

When the probe is directly applied across a high-*Q* tuned coil with the same resonant frequency as *L,* instability could occur (the DVM could indicate a signal voltage when no signal is present). In such a case, connect an isolating resistor of suitable value in series with the probe tip (the isolating resistor should have a sufficiently high value that the DVM indicates zero signal when no signal is present).

Note 2: Although the JFET pre-amp has less maximum available gain than a bipolar transistor pre-amp, this disadvantage may be more than offset in testing high-impedance circuitry due to the comparatively high input impedance of the JFET arrangement.

Figure 5–5 Active AM broadcast high-impedance RF signal-tracing probe.

170,000-Ω resistor connected in series with the input. The input resistance to the emitter follower can then be calculated as exmplified in the following:

Initial output voltage = 305 mV rms
Output voltage with 170 K resistor = 214 mV rms

$$0.305 \times R_{in}/R_{in} + 170{,}000) = 0.214$$

Input resistance = 400,000 Ω, approx.

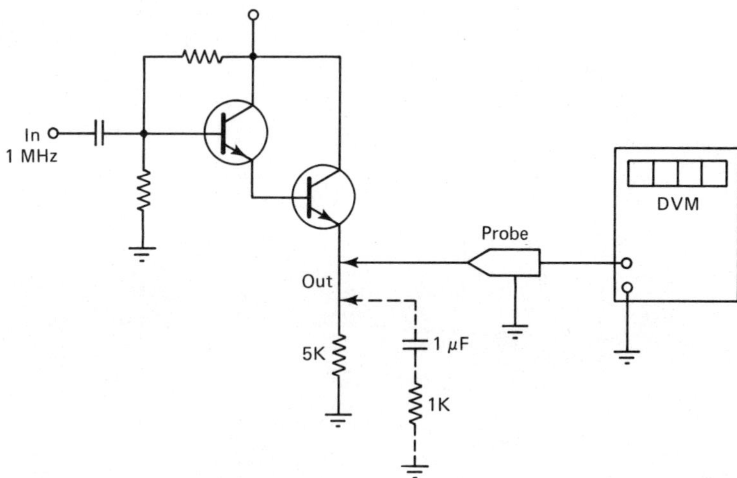

Note: When a 300-mV 1-MHz sine-wave voltage is inputted to the emitter follower, it appears at almost the same value across the 5 K emitter resistor. The voltage levels are checked with the half-wave rectifier probe and DVM operated on its dc-voltage function. Then, when the 1,000-Ω load (consisting of the 1 μF blocking capacitor and 1 K resistor) is shunted across the 5 K emitter resistor, the RF output voltage level will drop to about 88 percent of its unloaded value. In turn, the output resistance of the emitter-follower circuit is approximately 140 Ω, in accordance with Ohm's Law.

Figure 5–6 Measurement of the RF output impedance (resistance) of the emitter-follower circuit.

The second measurement is made at a frequency of 1 MHz. Now, the DVM can no longer be directly connected to the emitter-follower output (a service-type DVM cannot measure ac voltage at 1 MHz). Consequently, we must employ a peak-reading RF probe, as shown in Fig. 5–8. The input resistance to the emitter follower at 1 MHz can then be calculated as follows:

Initial output voltage (measured) = 300 mV peak
Initial output voltage (corrected) = 500 mV peak
Output voltage with 170 K resistor (measured) = 16 mV peak
Output voltage with 170 K resistor (corrected) = 216 mV peak
$0.5 \times R_{in}/(R_{in} + 170{,}000) = 0.216$
Input resistance = 130,000 Ω, approx.

Observe that the input resistance at 5 kHz was found to be 400,000 Ω, whereas the input resistance at 1 MHz was found to be 130,000 Ω.

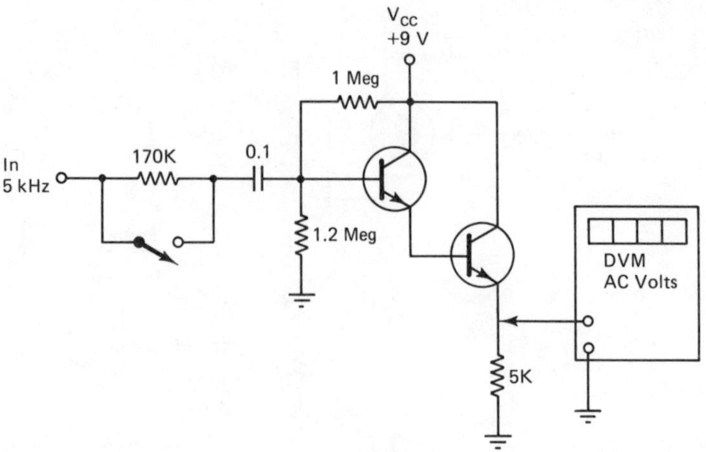

Note: This method of measuring the input resistance of the emitter follower consists of measuring the reduction in output voltage resulting from insertion of a certain value of resistance in series with the input to the emitter follower. In turn, the value of the input resistance is calculated in accordance with Ohm's Law. When this measurement is made at 5 kHz, the input resistance of the emitter follower is found to be approximately 400,000 Ω.

Observe that the input resistance to the emitter follower is effectively in shunt to its input terminals. In turn, we form a voltage divider when the 170-kΩ resistor is connected in series with the input circuit. We calculate the value of the input resistance in terms of this voltage-divider action.

Figure 5–7 Measurement of input impedance (resistance) to emitter follower at 5 kHz.

Bottom Line: The input resistance to the emitter follower decreases as the operating frequency increases. This decrease in input resistance is primarily due to the bypassing effects of the stray capacitances and the transistor junction capacitances in the emitter-follower circuit.

It is also important to note that our measurement at 5 kHz was comparatively precise, inasmuch as we connected the DVM directly to the output of the emitter follower. On the other hand, we had to use a diode rectifier

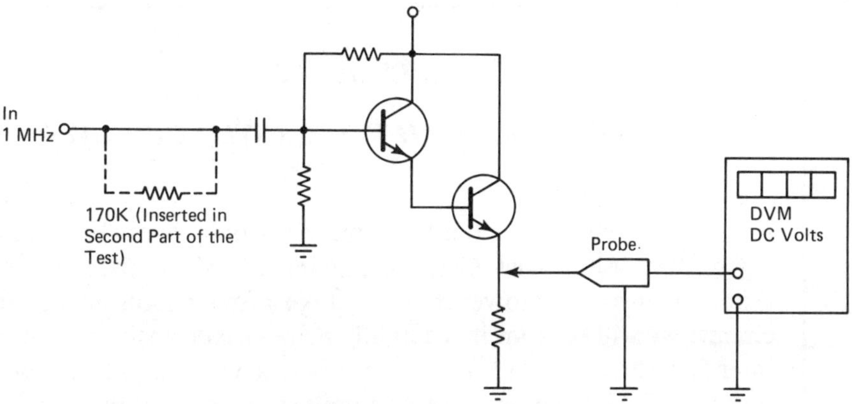

Note 1: This method provides a practical measurement of the input impedance to the emitter follower at 1 MHz. However, it is only an approximate measurement, inasmuch as an assumed value of barrier potential must be assigned to the diode in the peak-reading half-wave rectifier probe.

Note 2: If you have a signal generator with a high output such as 1 V rms, you can improve the precision of the foregoing measurement by using a 1 V signal instead of 300 mV signal. That is, a higher signal level reduces the effect of the barrier-potential factor. Only a few types of AM signal generators, however, provide a 1 V rms output. Some service-type AM signal generators have a maximum output of only 0.1 V rms, and these cannot be used in this type of RF measurement.

Figure 5–8 Measurement of input impedance (resistance) to the emitter follower at 1 MHz.

probe in our measurement at 1 MHz. Since the diode in the probe has a barrier potential of approximately 0.2 V, we used this value to calculate the corrected values of output voltage. Although this is a sufficiently accurate assumption for practical work, a more precise measurement could have been made if we had used an elaborated probe to avoid the barrier-potential uncertainty. (See also Chart 5–3.)

Chart 5–3

WORKING WITH INTEGRATED CIRCUITS

Integrated circuits tend to appear formidable to the beginning experimenter or hobbyist. Certainly, an integrated circuit is more complex than a transistor. However, if we take a close look at an integrated circuit, we will see that it is actually a prewired transistor configuration intended for use as an audio amplifier, a video amplifier, an IF amplifier, an FM detector, or other familiar electronic unit. The integrated circuit contains transistors, diodes, and resistors. Other components such as capacitors and potentiometers are externally connected to the pins on the IC package.

The first question asked by the experimenter, hobbyist, technician, troubleshooter, or student is: "How do I find out what the IC is intended to do, and how do I connect it to external components?"

I would recommend starting with the *National Semiconductor Linear Databook*. It is a comprehensive reference to all types of analog integrated circuits, and it also answers your "how to" questions in detail. For example, suppose that we are interested in the 387 low-noise dual preamplifier. We will find:

A pin-connection diagram for the IC.
A general description explaining what it does.
Specifications (operating features).
Typical applications.

The diagrams provided in this example are as follows:

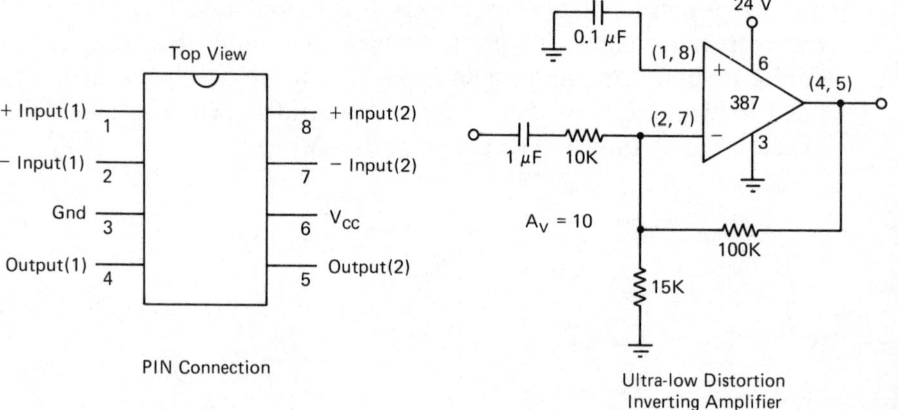

PIN Connection

Ultra-low Distortion Inverting Amplifier

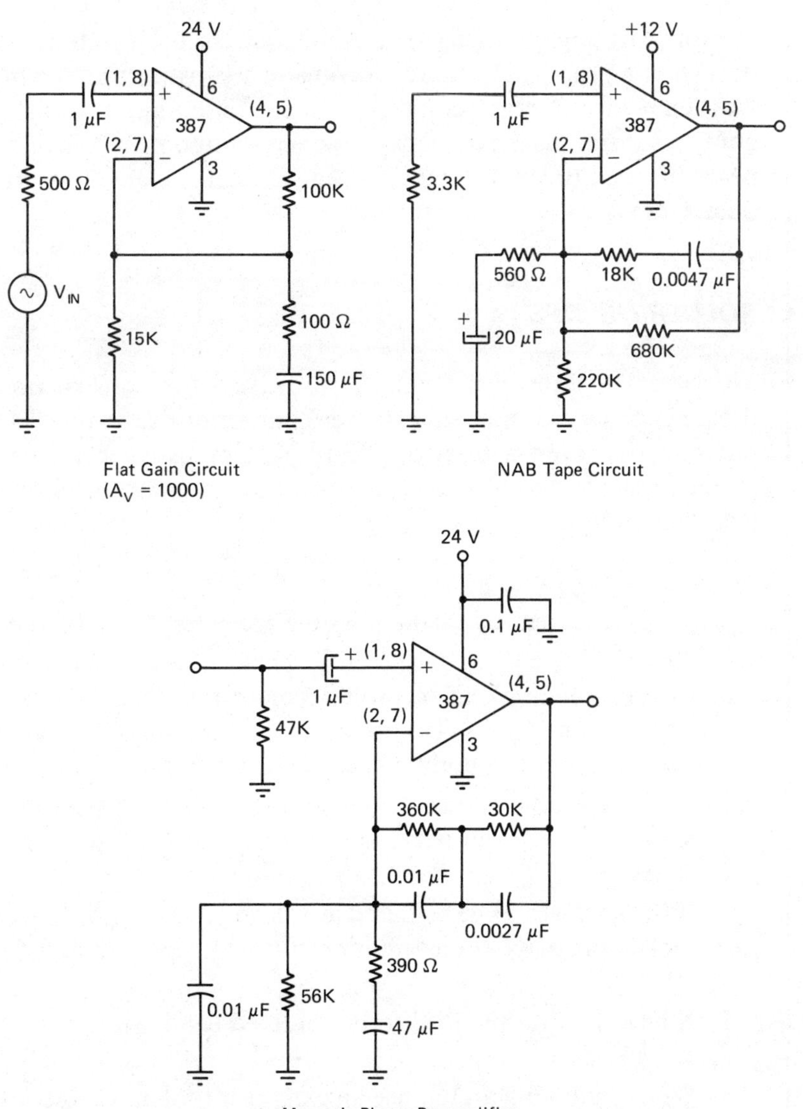

Flat Gain Circuit
(A_V = 1000)

NAB Tape Circuit

Magnetic Phono Preamplifier

When you construct an IC project, consider the possible use of an IC socket, instead of soldering the connections directly to the IC pins. A socket will sometimes provide a desirable flexibility in experimentation.

When installing an integrated circuit to replace an original equipment type in an FM tuner or other circuitry operating at frequencies in the VHF or UHF regions, it is very important not to change any of the lead lengths or positioning in the original arrangement. Otherwise, improper tuning characteristics AND/OR circuit instability is likely to occur. As a practical note, IC or transistor substitution in a tuned high-frequency circuit may require alignment touch-up, although the foregoing precautions have been observed. Alignment touch-up can be necessary due to small tolerances on junction capacitances in semiconductor devices.

SOLDERING TIPS

Much wasted time (and expense) can be saved in construction projects if care is observed in making solder connections to semiconductors. In other words, even a brief application of excessive heat, or extended application of normally tolerable heat, can damage or destroy the device. For example:

1. Apply the heat as far as possible from the body of the semiconductor.
2. Avoid applying heat or molten solder to a lead or a terminal for longer than 10 secs. Do not apply solder to a point closer than 1/16 inch from the body of the IC or transistor.
3. It is preferable to use a low wattage soldering iron (30 watts or less) manufactured particularly for use with IC or transistor circuitry.
4. If the surfaces to be soldered are clean, and the tip of the soldering tool is properly tinned, connections can be soldered faster than otherwise.
5. A heat sink on the lead to be soldered is a basic device "life preserver."
6. When removing an IC, use an extractor tool and a specialized slotted-bar desoldering tip for your soldering iron.

MOSFET ICs require some additional consideration to avoid damage. Like MOSFET transistors, they are susceptible to damage from static electricity. You are likely to have enough static electricity built up on your hands and body to instantly ruin a MOSFET IC. Here are the basic rules to follow:

1. The leads (pins) of a MOSFET IC should always be in contact with a conductive substance to avoid exposure to static electricity, except when they are being tested or when they are being installed in a circuit.
2. You should ground the tip of your soldering iron and your tools as well as metal parts of fixtures, and loop a ground wire around your wrist before you touch a MOSFET IC terminal.
3. Like MOSFET transistors, ICs can be damaged if they are inserted or removed with power applied to the circuit.
4. It is possible to damage a MOSFET IC by applying signal voltage to its input terminal(s) with the power turned off.

INTEGRATED-CIRCUIT QUICK CHECKS

Troubleshooting integrated circuitry often involves signal tracing followed by dc-voltage measurements. However, there is a time-saving quick-check procedure that you can use, provided that you have a similar unit that is in normal working condition available. This quick check has a great advantage in preliminary troubleshooting procedure in that you do not need to have the circuit diagram for the unit, nor any service data. It is a basic comparison test.

The quick check is made with a pair of DIP IC test clips and a DVM. One test clip is attached to a suspected IC in the unit under test, and the other test clip is attached to the corresponding IC in the reference unit. Now, all IC terminals are conveniently accessible at the tops of the clips. Set your DVM to its dc voltage function, and connect one test lead to the first terminal on one test clip, and connect the other test lead to the first terminal on the other test clip. Observe whether any significant reading is displayed on the DVM. If a zero or near-zero reading occurs, no trouble is indicated. Proceed to the second terminals on the test clips, and again observe whether any significant reading is displayed on the DVM. In case you come to a pair of terminals that produce a substantial reading on the DVM, a trouble condition is indicated in the associated circuit.

At this point, we would not know whether the IC is defective, or whether an external component in the associated circuit is defective. Therefore, we would follow up our quick test by checking out the particular circuit in detail.

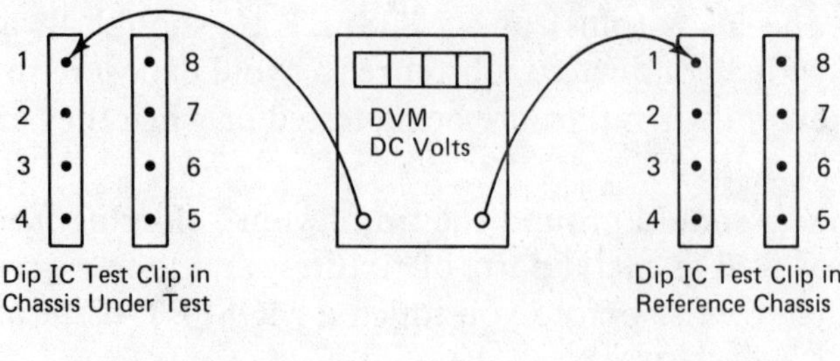

This example illustrates the use of a pair of 8-pin DIP IC test clips. However, you can use clips with as many pins as necessary to test larger ICs.

6

Signal Injectors and Analyzers

GENERAL DISCUSSION

Signal injectors and analyzers (signal substituters) occupy an extensive area of electronic test equipment utilized in preliminary troubleshooting procedures. They rank with the most important types of quick-checkers for saving time at the bench. A very simple signal injector is essentially a go/no-go tester; it merely serves to show whether a signal can pass through the circuit under test. On the other hand, more elaborate types of signal injectors are also circuit analyzers and fault indicators. Some signal injectors are designed for checking ''dead'' circuitry, whereas other signal injectors are designed for checking ''live'' circuitry.

A signal injector may operate at any frequency from the audio range through the ultra-high-frequency range; it may provide a single-frequency output, a two-tone frequency output, or it may provide a combination of many frequencies. For example, an AM signal generator provides a CW output or a modulated CW output; a ''noise'' generator supplies an extremely large number of frequencies. A square-wave generator is sometimes used as a signal injector; it provides a large array of harmonically-related frequencies.

TWO-TONE SIGNAL INJECTOR

A two-tone signal injector is not only a go/no-go type of tester—it is also a frequency-response quick checker. As shown in Fig. 6–1, a two-tone signal injector may be arranged as a pair of audio oscillators with their outputs connected in parallel. One of the audio oscillators is set to a low frequency such as 100 Hz, and the other is set to a high frequency such as 10 kHz. The two-tone signal injected at any chosen test point in an audio system, and the resulting sound output (if any) is analyzed for its relative treble and bass proportions.

This frequency-response quick checker speeds up frequency-distortion localization in an audio system. In other words, we go back from the speaker(s) step-by-step toward the input end to determine just where the frequency distortion is occurring. As an illustration, if the trouble symptom

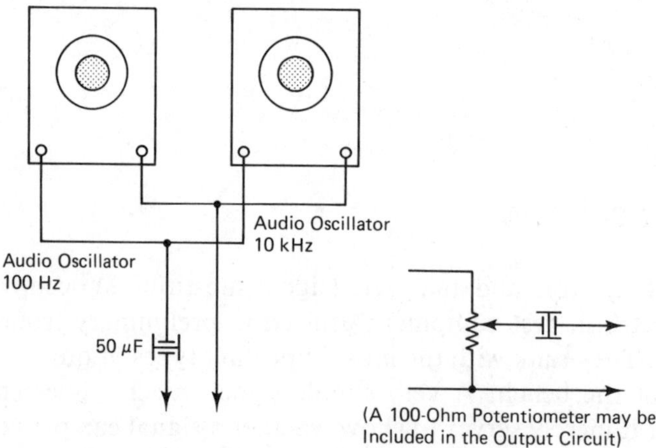

Note: This two-tone signal-injector arrangement provides a quick check of bass/treble balance for any section in an audio system. Observe that the output levels of the audio oscillators should be set for a normal bass/treble balance. After the operator has acquired some experience with the two-tone signal injector it will serve as a substantial time-saver in preliminary troubleshooting of frequency-distortion problems.

Observe that a potentiometer level control may be included in the output leads. This feature permits the level of the injected signal to be varied by means of a single control.

Figure 6–1 A two-tone signal injector serves as a frequency-distortion quick-checker.

is weak bass output, a quick check with a two-tone signal injector might show that although the bass/treble balance is good in the output section, that the bass becomes attenuated in the driver section. In turn, follow-up tests would be made to localize the fault in the driver circuitry.

The two-tone signal injector can also be used to quick-check for frequency distortion in RF circuitry as shown in Fig. 6–2. In this application,

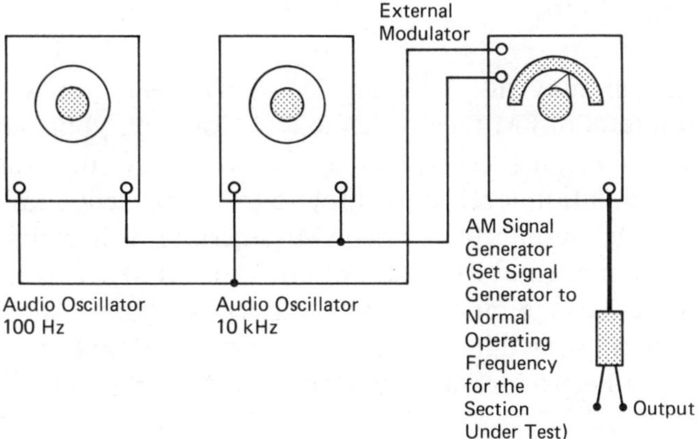

Note 1: This is a two-tone amplitude-modulated RF signal source for frequency-distortion quick-checks in RF or IF circuitry. It tests for bass-treble balance in the RF-amplifier section or in the IF-amplifier section of a radio receiver, for example. Observe that the audio level input to the external-modulation terminals of the signal generator should be set to provide about 30 percent modulation. (The instruction manual for the signal generator will specify the proper voltage for application to the external-modulation terminals).

Observe that if test results show that the bass-treble balance is satisfactory in the audio section of the radio receiver, for example, but that the bass-treble balance is poor when a signal is injected in the IF section, it is indicated that an IF circuit fault is probably present. However, if there is no IF circuit fault, then the IF amplifier is in need of realignment.

Note 2: Some AM signal generators have a built-in blocking capacitor in series with the output cable, and others do not. If your generator does not contain a blocking capacitor, connect a 500-pF capacitor in series with the output lead to avoid bias-voltage drain-off in the circuit under test.

Figure 6–2 Frequency distortion in the RF or IF circuits of a radio receiver can be quick-checked with this signal-injection arrangement.

the output from the audio oscillators is used for external modulation of the RF oscillator in the AM generator. This arrangement permits the troubleshooter to quick-check the IF and RF sections of a radio receiver for localization of frequency distortion.

IFM SIGNAL INJECTOR

You may or may not have an FM signal generator available for signal-injection tests in FM receiver circuitry. However, your service-type AM signal generator can pinch-hit as an FM signal injector. Just advance the internal-modulation level control for 75 to 100 percent amplitude modulation. In turn, the output from the AM generator will contain an appreciable amount of incidental frequency modulation (IFM). This IFM is a by-product of high percentage modulation in the RF oscillator. When this "artificial" FM signal is injected into the IF or IF circuitry of an FM receiver, the speaker will normally produce a tone output with a pitch equal to the modulating frequency.

Observe that although service-type AM signal generators do not provide a frequency range extending into the FM band, harmonics are present that can be utilized for signal injection in 88–108 MHz circuitry. For example, the third harmonic of 30 MHz is 90 MHz, which is within the FM band. The FM IF section operates at 10.7 MHz, and this frequency is provided as a fundamental output by standard AM signal generators.

SPEEDY STAGE IDENTIFIER

When an RF/IF network is being "buzzed out" in a "dead" receiver, preliminary troubleshooting procedures can be speeded up by employing a resonance probe such as shown in Fig. 6–3. This type of resonance probe includes a DVM with a built-in analog indicator. The provision of an analog indicator simplifies the test procedure in this kind of application. However, a conventional DVM can be used, if desired. From a functional viewpoint, this resonance probe can be regarded as a "reverse dip meter" which identifies a tuned circuit as an RF stage, IF stage, intercarrier stage, or chroma stage, for example.

Observe that the resonance probe operates at test voltages less than 0.5 V peak, so that the probe does not "turn on" transistors or diodes that may

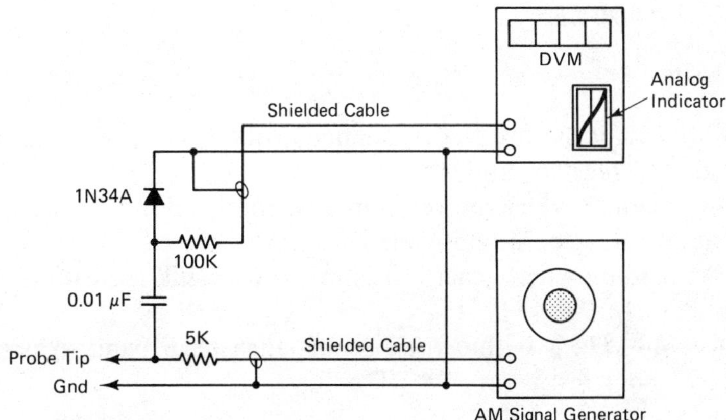

Note: Various types of analog indicators are provided on different models of elaborate DVMs. In the example above, the analog indicator consists of a rotating cylinder with a line that rises progressively from the bottom to the top of the cylinder. Thus, as the input voltage increases, the hairline on the "window" indicates a higher point of intersection with the line on the rotating cylinder.

Operating Note: The resonance probe is a form of signal injector combined with a point-of-test voltage indicator. Since the impedance of the coil under test varies with frequency, an indication of its resonant frequency is obtained that serves to identify whether the stage under test is an RF, IF, AM, FM, intercarrier, or chroma stage.

Observe also that the resonance probe has a certain amount of input capacitance that tends to detune the circuit under test. In other words, the normal operating frequency of the coil under test will be a bit higher than the value indicated by the signal generator.

Equipment Note: As previously mentioned, not all signal generators provide a sufficiently high-level output to operate this type of test equipment. Therefore, it is necessary to select a generator that will produce an indication of at least 50 or 100 mV on the DVM.

Figure 6–3 DVM with analog indicator is a time-saver in preliminary troubleshooting procedures.

be connected to the tuned circuit under test. Inasmuch as a 1N34A germanium diode is utilized in the probe rectifier circuit, the probe can respond to test voltages as low as 0.2 V. This is just another way of saying that a silicon diode should not be used in construction of the probe.

PRACTICAL EXAMPLES

Example 1

DVM reading before the resonance probe is connected across a coil in the "dead" receiver under test: 48 mV.

DVM reading when the resonance probe is connected across the coil, and before the generator is tuned: 13 mV.

DVM reading when generator is tuned for peak indication: 72 mV at a frequency of 748 kHz.

Since 748 kHz is within the AM broadcast RF band, it is evident that the coil under test is an AM RF coil.

Note: the analog indicator provides considerable convenience in making the test, inasmuch as we are only concerned with the point on the generator dial at which peak voltage occurs across the coil under test. (It is confusing to watch the rapidly changing numerals on the DVM readout as the generator dial is adjusted.) On the other hand, it is easy to observe the analog indicator rise up to a peak height and then start to fall.

Example 2

DVM reading before the resonance probe is connected across a coil in the "dead" receiver under test: 56 mV.

DVM reading when the resonance probe is connected across the coil, and before the generator is tuned: 1.5 mV.

DVM reading when generator is tuned for peak indication: 47 mV at a frequency of 450 kHz.

Since 450 kHz is within the AM IF region, it is evident that the coil under test is an AM IF coil.

As before, the analog indicator provides convenience in making the test. We will find that the generator tuning is more critical in the case of an IF coil than an RF coil, and it is helpful to watch the analog indicator instead of the DVM numerical readout as the generator dial is turned.

Technical note: In the first test, the DVM reading before the resonance probe was connected across the coil was 48 mV. However, when the generator was tuned for peak indication with the resonance probe connected across the coil, the DVM reading was then 74 mV. This is an increase (voltage magnification) of 24 mV, and represents an effective Q value of 1.5 under the test conditions.

Example 3

DVM reading before the resonance probe is connected across a coil in the "dead" receiver under test: 33 mV.

DVM reading when the resonance probe is connected across the coil, and before the generator is tuned: 3 mV.

DVM reading when generator is tuned for peak indication: 23 mV at a frequency of 1.95 MHz.

Since 1.95 MHz is within the range of local-oscillator frequencies for an AM broadcast converter stage, it is evident that the coil under test is an AM local-oscillator coil.

TRANSISTOR CLICK TEST

We will observe that simple and effective signal-injection tests may be made by causing the tested circuit to generate its own signal. This procedure can be a significant time-saver in preliminary troubleshooting procedures. For example, if we are looking for a dead section in a radio receiver that is stopping the signal, we can quickly localize the dead section by means of transistor click tests. This test is made by starting with the output transistor(s) and working back step-by-step, as follows:

A short-circuit is applied between emitter and base of the last transistor in the audio section. Normally, a click will be heard from the speaker, due to the step-voltage signal that is generated when the short-circuit is applied.

If a click is not heard from the speaker, we proceed to make a detailed checkout of the audio-output stage. (The output transistor might be defective, the V_{CC} supply might be subnormal or zero, or the speaker might be defective.)

In case that the output stage passes the click test, we proceed to the driver stage, and so on. The click test produces a useful signal in the IF or RF sections, as well as in the audio section.

The only exception occurs in a class-B output section, wherein the output transistors are operated with zero bias. However, almost all push-pull output sections are operated in class AB, and can be click-tested.

STAGE-GAIN MEASUREMENT BY SIGNAL INJECTION

Stage-gain or section-gain measurements in a radio receiver, for example, are made as shown in Fig. 6–4. The voltage gain of a section is measured by noting the change in DVM reading as the signal-injection point is transferred from the input to the output of a section. Thus, the voltage gain is equal to the first reading divided by the second reading. (The injected signal voltage is the same in both measurements.) As indicated in the diagram, each section has an appropriate test frequency.

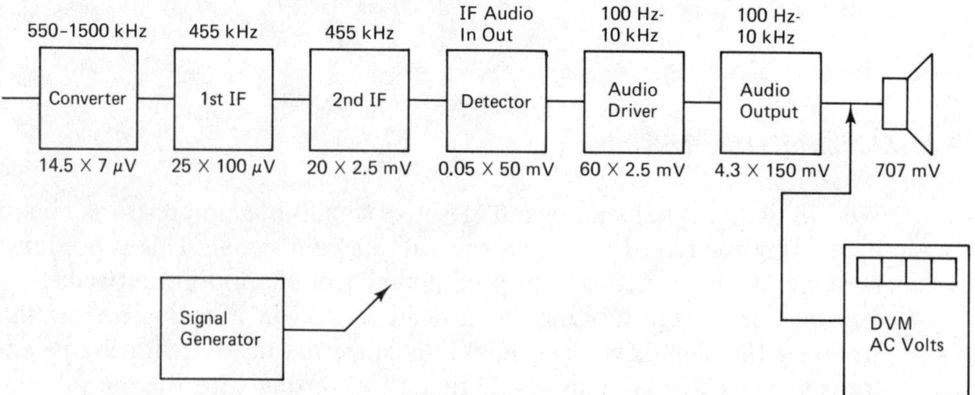

Note 1: These are typical section gains for a widely used AM radio receiver. A 7 μV input signal normally produces a 707 mV output signal. In this example, the speaker has an impedance of 10 Ω, and the 707-mV output signal supplies 50 mW of audio power to the speaker.

Observe that a diode detector is utilized in this example, and it imposes an insertion loss; a 50-mV input to the detector corresponds to 2.5 mV output to the audio driver. Some receivers provide a transistor detector; in turn an insertion gain of approximately 2× is obtained.

Section gains can be measured without knowing the output voltage from the signal generator by noting the change in DVM reading when the signal-injection is transferred from the input to the output of a section. (The AGC circuit should be disabled during gain measurements.)

Note 2: The voltage-gain values exemplified above are maximum-available-gain ratings. Maximum available gain is obtained when the IF transistors are forward-biased 0.65 V and the volume control is advanced to maximum.

Figure 6–4 Stage-gain measurement by signal injection.

Observe that amplitude-modulated output must be used from the signal generator to make voltage-gain measurements in the IF and RF (converter) sections. It is standard practice to employ 30 percent amplitude modulation. The modulating frequency is usually 1 kHz, although it may have any value in the range from 100 Hz to 10 kHz. It is important not to overload the stage into which the signal is being injected. Otherwise, the gain measurement will be in error. However, any signal amplitude below the overloading level may be utilized.

A quick check to determine whether overloading may be occurring is to back off slightly on the output attenuator of the signal generator. If the DVM indicates the same output voltage, overloading is present. The attenuator on the generator should be turned down sufficiently that any small change in setting produces a change in the DVM reading.

Time can be saved and measurements made more easily if the attenuator on the generator is adjusted to produce an indication of 100 mV, for example, on the DVM when the test signal is injected at the input of a section. Then, when the same test signal is injected at the output of the section, the gain can be calculated more easily; as an illustration, if the second reading is 4 mV, it is evident that the section gain is 25 times, inasmuch as 4 "goes into" 100 twenty-five times.

OUT-OF-PLACE SIGNAL VOLTAGES

When a fault occurs in an electronic system, signal voltages may appear at unexpected points. In turn, identification of these abnormal "hot" points can speed up localization of the fault. As illustrated in Fig. 6–5, when we inject a signal into the IF section of a radio or TV receiver, we would not expect to find signal voltage on the AGC line. However, in case that the AGC delay capacitor is open, we will then find IF signal voltage on the AGC line.

This out-of-place signal voltage can cause various trouble symptoms such as "birdies," distorted signal output, or self-oscillation with signal blocking.

A short-cut method of checking for out-of-place signal voltages is shown in Fig. 6–5. Thus, if an IF signal injected into the IF section does not normally gain entry into the AGC line, then an IF signal injected into the AGC line will not normally gain entry into the IF section and produce an output from the receiver. In case that the receiver does produce an output, the troubleshooter proceeds to check out the AGC section in detail—particularly for open capacitors.

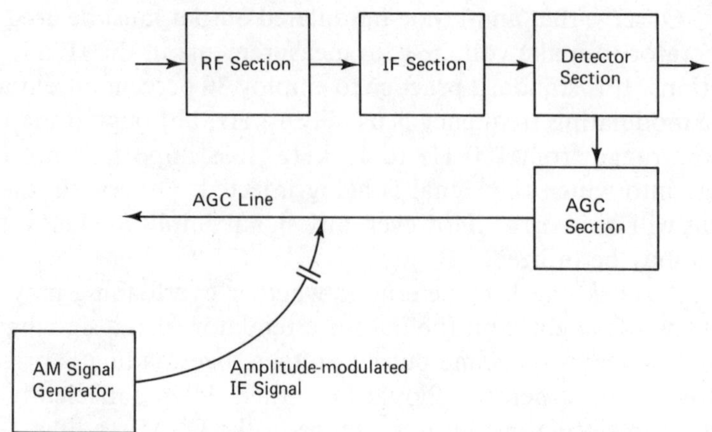

Note: Signals in the IF section of a receiver do not normally gain entry into the AGC line. By the same token, an IF signal injected into the AGC line will not produce any output from the receiver in normal operation. Accordingly, a quick check for the possibility of a ''hot'' AGC line is to inject an IF signal into the line in order to determine whether there is any resulting output from the receiver.

Figure 6–5 An injected signal voltage may gain entry into unexpected circuits under fault conditions.

Another useful quick-check that can sometimes be a substantial time-saver in ''tough-dog'' situations is to inject RF, IF, and AF signal voltages into the V_{CC} circuit to determine whether there is any resulting output from the receiver, as shown in Fig. 6–6. In the event that the receiver does develop an output, the troubleshooter then proceeds to look for open capacitors or capacitors with a poor power factor in the V_{CC} network. A tricky point about a ''hot'' V_{CC} network is that it can simulate trouble symptoms that seem to point elsewhere.

QUICK CHECK FOR AM REJECTION BY FM DETECTOR

An FM detector normally rejects any amplitude modulation that may be present in an FM signal. Although it might be supposed that we could check for normal AM rejection by merely injecting an AM signal into the IF section of an FM receiver, this is not a feasible method for use with service-type generators. In other words, most service-type generators contain amplitude-modulated oscillators, and in turn the oscillator generates inci-

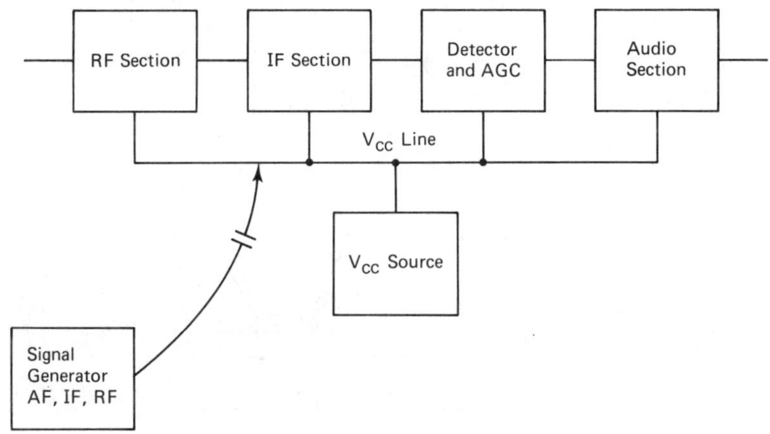

Note: The V_{CC} line is common to all of the receiver sections. In normal operation, the V_{CC} line is "cold," inasmuch as RC decoupling circuits are provided in each branch. In other words, we do not normally find RF, IF, or AF signal voltages on the V_{CC} line. Signal voltages cannot gain entry into the V_{CC} circuitry unless there is a bypassing or decoupling fault in one of the receiver sections. However, if signal voltage from the IF section, for example, can gain entry into the V_{CC} circuitry, then an IF signal injected into the V_{CC} circuitry will result in an output from the receiver.

Figure 6–6 Out-of-place signal voltages may gain entry into the V_{cc} network.

dental frequency modulation (IFM) in combination with the amplitude-modulated output.

However, a quick check for AM rejection by an FM detector can be made by using a simple external amplitude modulator, as illustrated in Fig. 6–7. The semiconductor diode is a nonlinear device that produces an amplitude-modulated signal from a mixture of IF and AF voltages. In other words, we feed a CW signal at IF frequency into the diode with a 1-kHz signal from the audio oscillator. In turn, the 10.7 MHz CW signal, for example, will be amplitude modulated at 1 kHz. The peak voltage of the AF signal should be about 1/3 the peak voltage of the IF signal.

TV ANALYZER

Although TV analyzers are not new, they are substantial time-savers and merit attention here. With reference to Fig. 6–8, a television analyzer is a

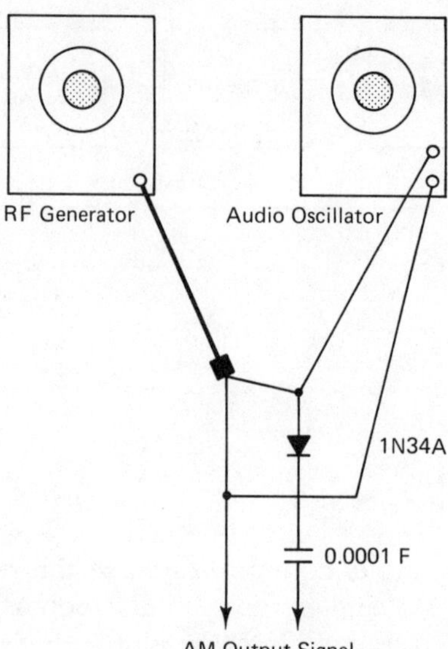

Note: This external-modulator arrangement provides isolation between the oscillator in the RF generator and the modulating signal. In turn, the AM output signal is virtually free from incidental frequency modulation. When quick-checking an FM receiver for AM rejection, for example, the RF generator may be set to 10.7 MHz and the audio oscillator may be set to 1 kHz. In turn, the AM output signal may be injected into the IF circuitry to determine whether the 1 kHz amplitude-modulating tone is rejected or passed by the FM detector.

Figure 6–7 External modulator avoids incidental frequency modulation of the test signal.

specialized type of signal generator, and is particularly adapted for signal-substitution tests in black-and-white or color television receivers. The exemplified analyzer employs 10 oscillators comprising UHF, VHF, IF, vertical, horizontal, sound (1 kHz), intercarrier (4.5 mHz), color (3.563 MHz), bar (189 kHz), and shorted-turn (inductor checking) oscillators. In addition, a test-pattern generator is provided. This unit consists of a flying-spot scanner that forms a video signal from a slide transparency. The essential components in the flying-spot scanner arrangement are a cathode-ray tube and a photomultiplier tube.

A TV analyzer is basically a quick-checker that is utilized to inject appropriate signal voltages at various points in the signal channels of a TV

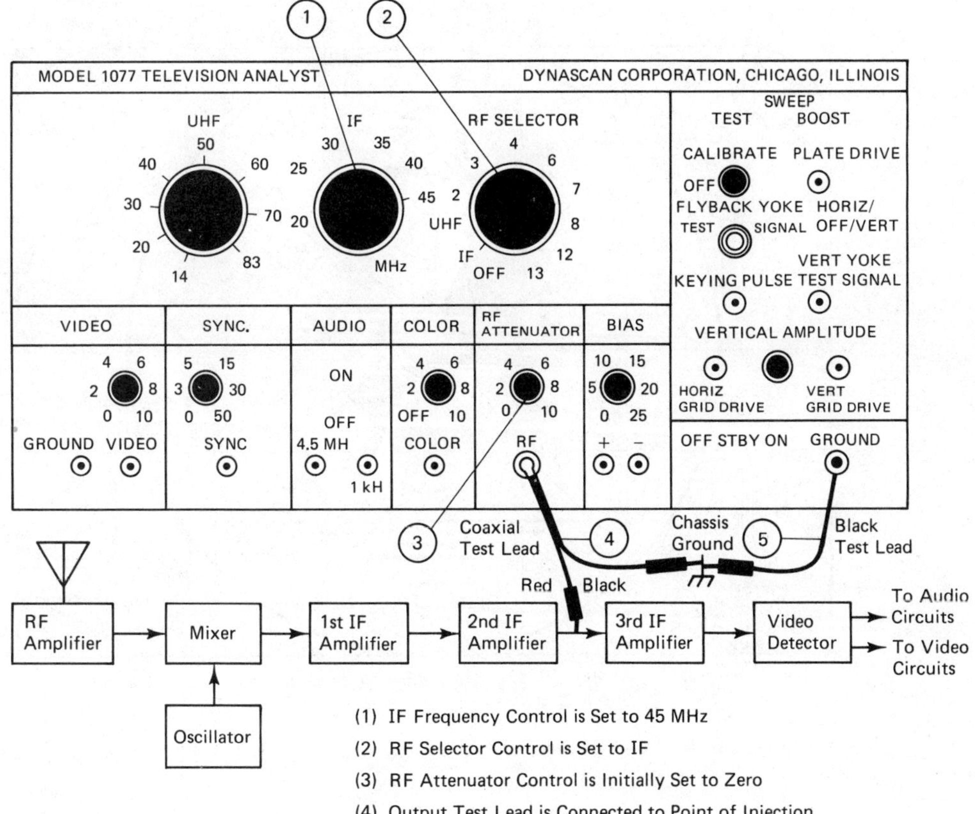

(Reproduced by Special Permission of Reston Publishing Company, Inc. From "Color Television")

Figure 6–8 A standard TV analyzer, showing control settings and cable connections for an IF signal-substitution test.

receiver. It is also applied to inject standard internally-generated waveforms associated with sync and sweep circuitry. The picture tube and/or the speaker in the receiver under test are employed as go/no-go indicators. When the UHF, VHF, or IF oscillators are used, a test-pattern video signal is amplitude-modulated on the CW output from the oscillator. In turn, a test pattern such as illustrated in Fig. 6–9 will normally be displayed on the picture-tube screen. Considerable data concerning receiver circuit action is provided by a test pattern, as follows:

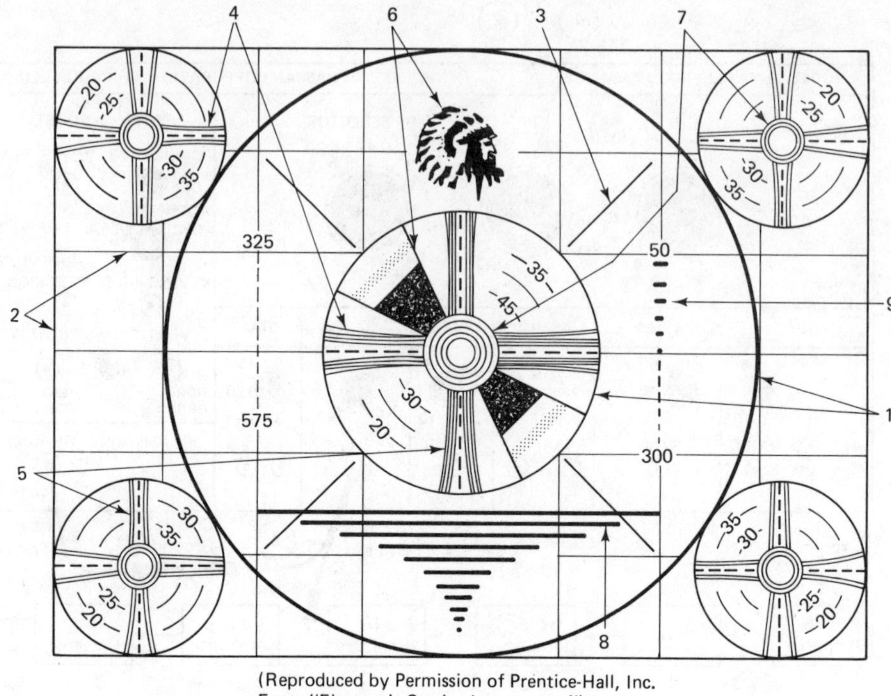

(Reproduced by Permission of Prentice-Hall, Inc.
From "Electronic Service Instruments")

Figure 6–9 A standard test pattern. (See text for explanation of numbered features.)

1. Height, width, linearity, and centering adjustments are indicated by the shapes of the circles. All circles are normally round, within practical tolerances, and not more than ¾ inch is cut off from the circles by the edges of the screen.
2. Aspect ratio, pincushion distortion, or barrel distortion are shown by the vertical and horizontal lines, which normally form squares. True squares correspond to the standard 4:3 aspect ratio. These lines are normally straight; if curved, the presence of pincushion or barrel distortion is indicated. Again, if the lines are not parallel, keystoning is present. Or, if the lines are parallel but unevenly spaced, nonlinear deflection is occurring.
3. Poor interlacing is indicated by a jagged or sawtooth display of the normally straight diagonal lines. All four lines are affected by interlacing faults.

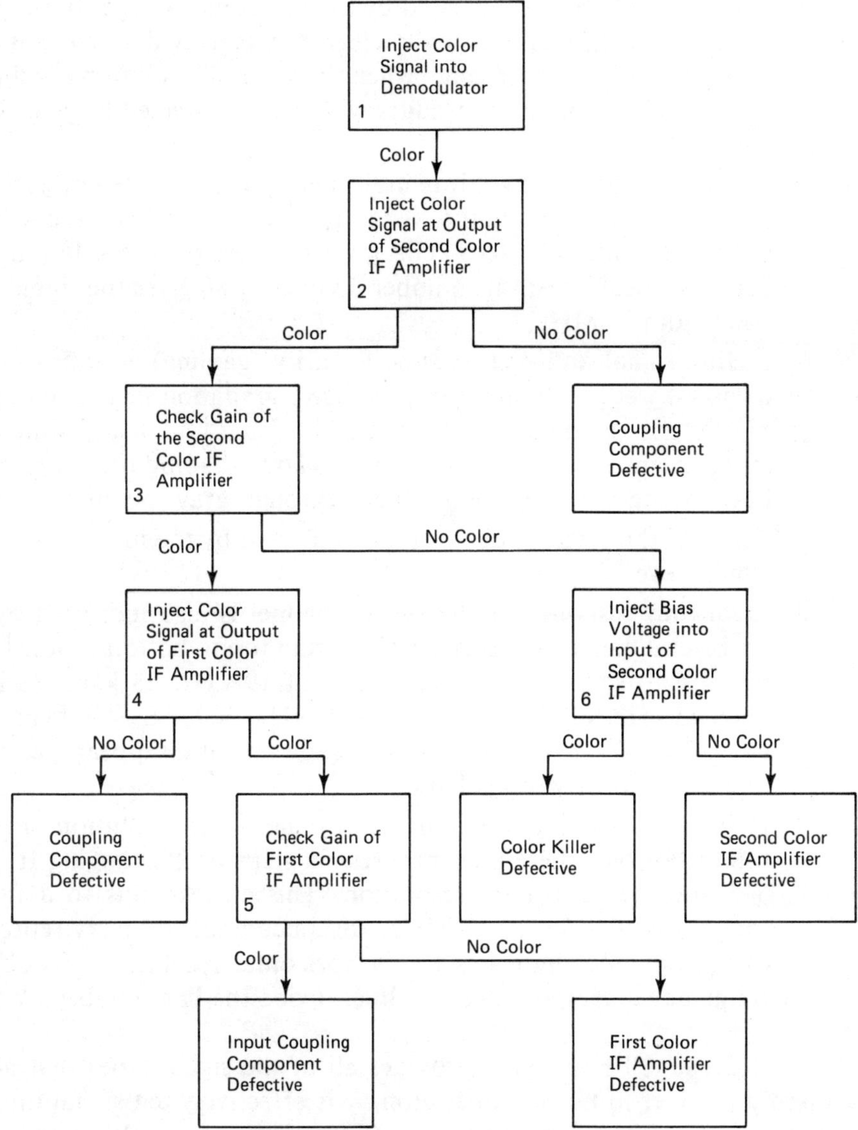

(Reproduced by Permission of Prentice-Hall, Inc., From "Electronic Service Instruments")

Figure 6–10 TV analyzer signal-injection procedures for localizing a no-color trouble symptom to a circuit section.

4. Vertical resolution is indicated by the horizontal wedges in the test pattern. The wedges are normally clear and sharply defined, but become indistinct and "washed out" when the vertical resolution is subnormal. Note that the horizontal wedges will also be affected by poor interlacing.
5. Signal-channel bandwidth is indicated by the vertical wedges. We observe the number along the wedge at which the vertical wedge becomes indistinct, and add a zero. This gives the horizontal resolution in number of vertical lines; this number divided by 80 gives the signal channel bandwidth in MHz.
6. Picture-signal and picture-tube linearity (gamma) is indicated by the diagonal wedges. Nonuniform contrast gradation in wedge display indicates poor gamma. These wedges are also useful in adjusting the contrast, brightness, and AGC controls. Four shading tones are provided in the wedges, in steps from black through gray to white.
7. Focus of the picture-tube beam is indicated by the innermost circles or "bull's-eye."
8. Frequency response of the signal channel is also indicated by the 11 horizontal bars of various lengths. From top to bottom, these bars correspond to square-wave frequencies of 19 kHz, 28 kHz, 38 kHz, 56 kHz, 75 kHz, 113 kHz, 150 kHz, 225 kHz, 300 kHz, 450 kHz, and 600 kHz. Poor low-frequency response shows up as disappearance or blurry reproduction of various bars.
9. Horizontal resolution is indicated by the single resolution lines. From top to bottom, these lines progress in steps of 25, ranging from 50 to 575 lines of horizontal resolution. This corresponds to a frequency range from 600 kHz to 7 MHz. Disappearance or blurry reproduction of various lines indicates faulty horizontal resolution. To obtain the frequency corresponding to a line, divide the line number by 80.

Because a TV analyzer provides all of the signal types that are processed a receiver in normal operation, it is effectively ten signal injectors in one instrument. In turn, it is a speedy preliminary-troubleshooting signal injector in both black-and-white and color-TV servicing. As an example of application, the procedure for localizing a no-color trouble symptom to a circuit section is shown in Fig. 6–10.

DIGITAL SECTION

1

Looking Ahead

If you are unfamiliar with the logic circuit in a unit of digital equipment, and do not know what integrated circuits are used in the circuit . . .

- *You can usually "spot" a faulty circuit area, or faulty device easily and quickly,* using the techniques and test equipment described in this section.
- *If an integrated circuit is unmarked, you can usually identify the IC package,* using the test methods explained in this book.
- New digital test equipment that you can build, and new ways to troubleshoot digital equipment without service data are detailed in the following pages.

GATE RECOGNITION

Digital troubleshooters are concerned with many kinds of devices. Nevertheless, even the most complex devices are arrangements of gates—the digital "atoms." There are 14 forms of gates, as shown in Fig. 1-1. However, the AND gate has the same function as the negated-NOR gate; the OR gate has the same function as the negated-NAND gate; the NAND gate has the

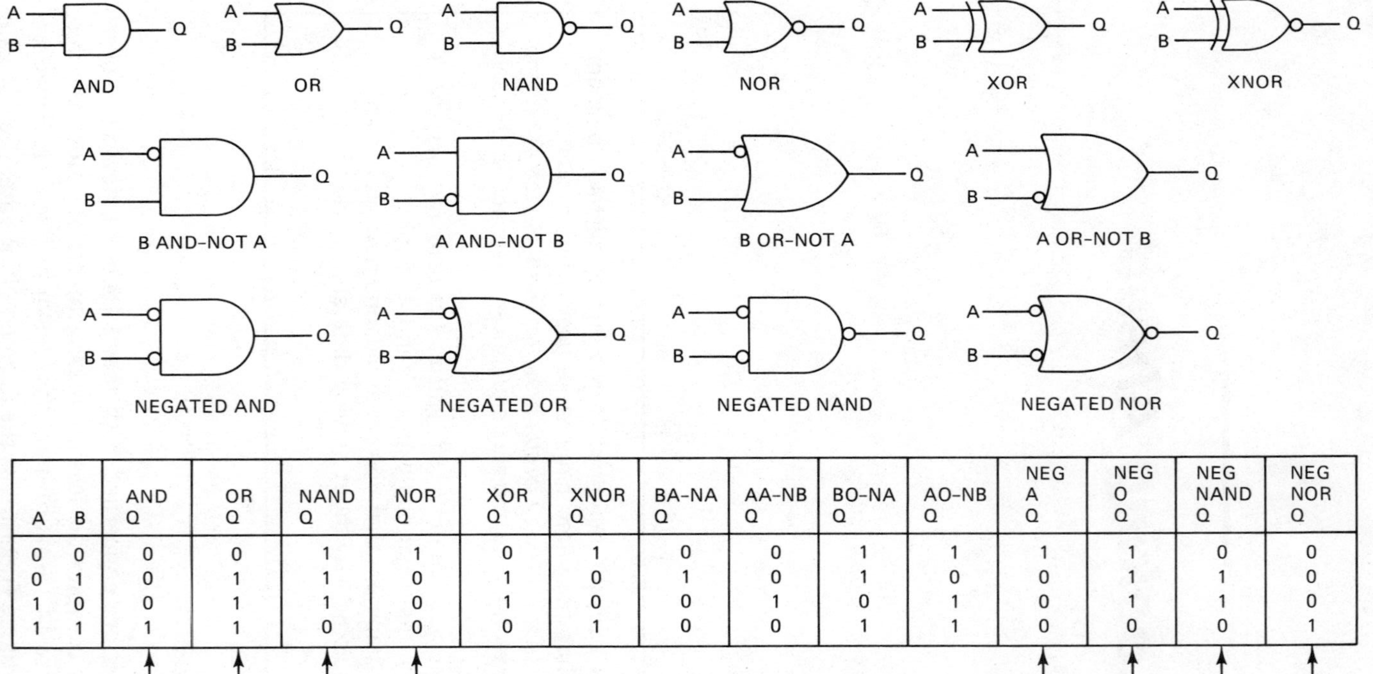

A	B	AND Q	OR Q	NAND Q	NOR Q	XOR Q	XNOR Q	BA-NA Q	AA-NB Q	BO-NA Q	AO-NB Q	NEG A Q	NEG O Q	NEG NAND Q	NEG NOR Q
0	0	0	0	1	1	0	1	0	0	1	1	1	1	0	0
0	1	0	1	1	0	1	0	1	0	1	0	0	1	1	0
1	0	0	1	1	0	1	0	0	1	0	1	0	1	1	0
1	1	1	1	0	0	0	1	0	0	1	1	0	0	0	1

Note: Basic test procedures for quick-checking any gate are shown in Fig. 1–3.

Note: An AND gate has the same truth table as a negated-NOR gate; an OR gate has the same truth table as a negated-NAND gate; a NAND gate has the same truth table as a negated-OR gate; a NOR gate has the same truth table as a negated-AND gate.

Figure 1–1 Gate symbols and truth tables.

same function as the negated-OR gate; the NOR gate has the same function as the negated-AND gate.

Important note: To save time and take shortcuts in digital troubleshooting procedures, we need to keep the following points in mind:

1. Nearly all digital circuits are merely combinations of gates.
2. A gate is simply an electronic switch; the gate is either "on" or it is "off."
3. Any gate waits for a certain combination of "high" and/or "low" input signals before it switches.
4. Digital time-savers and shortcuts are based chiefly on signal injection at inputs and signal indication at outputs.
5. Other time-savers and shortcuts are based on resistance quick checks with an ordinary ohmmeter.

Gates with one or more negated inputs are often encountered in digital circuitry. Each basic gate has a standard symbol, and the normal input/output relations of each gate is listed in a truth table, as seen in Fig. 1–1. The troubleshooter usually knows the AND, OR, NAND, and NOR truth tables "by heart." Gates with negated inputs are ordinarily viewed in terms of cross-relations, such as negated-AND is equal to NOR.

HIGH, LOW, AND BAD LOGIC LEVELS

A gate terminal is normally either logic-high or logic-low. However, under a fault condition, a gate terminal might well be checked out in the "bad region" (in between the logic-high level and the logic-low level). This is a common trouble symptom. Here, it is important to make a distinction between "white-and-black" trouble symptoms, and "gray area" trouble symptoms. For example, if the logic-low threshold is +0.4 V, and the logic-high threshold is +2.4 V, a potential of +1.5 V is clearly in the bad region.

However, if the troubleshooter measures +0.41 V, this is not certainly a bad-level value—it depends upon whether the circuit "sees" +0.41 V as a low level or as a bad level. Therefore, this is an example of a "gray area" value that requires further investigation. Marginal voltage troubleshooting is detailed subsequently.

A representative circuit board with integrated circuits for a scanner-monitor radio is depicted in Fig. 1–2. Each of the IC packages contains four NAND gates (in this example). Gate action is often tested in-circuit by means

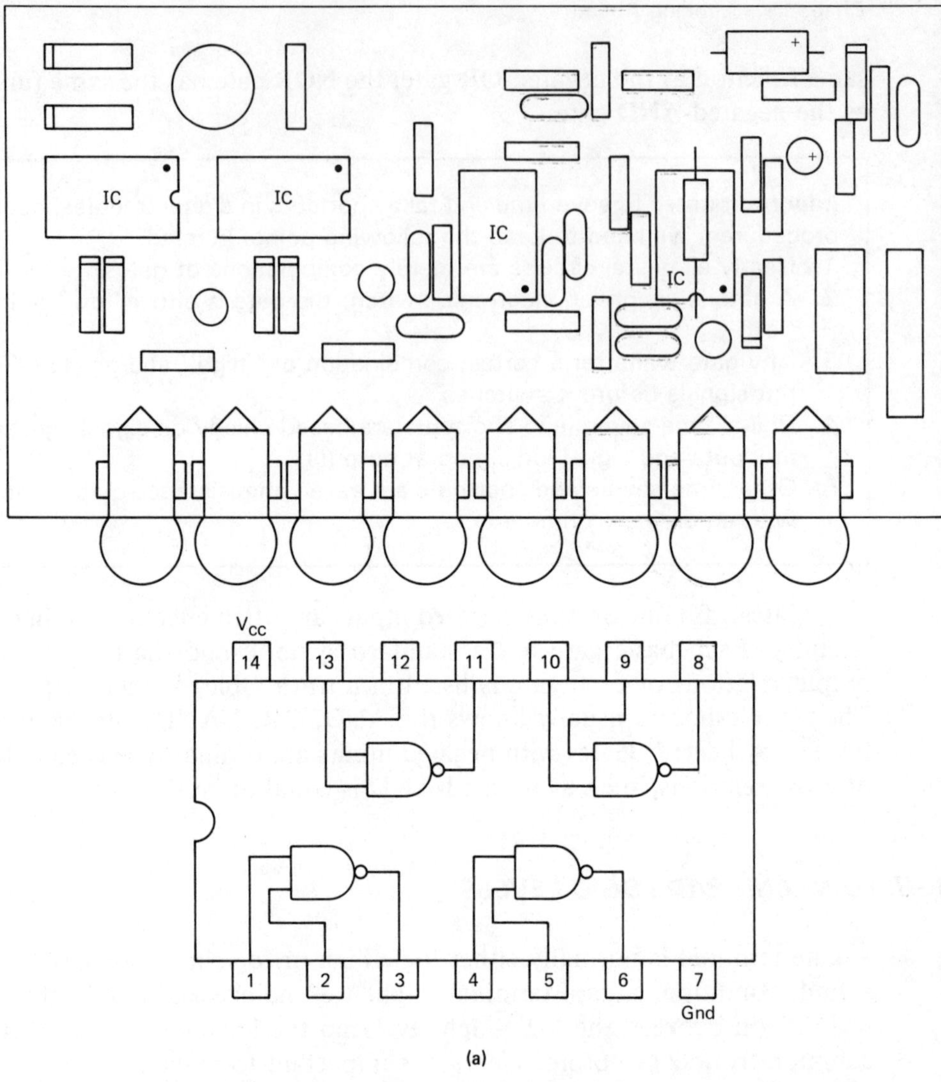

(a)

"You can't know too much about digital circuit action when troubleshooting without service data."

Digital circuit action is based on truth tables.

Note 1: If a NAND gate output becomes short-circuited to ground, the output current flow increases greatly—typically to 25 mA. In turn, the temperature of the IC package rises 7°C, approximately.

Note 2: Each integrated circuit contains four NAND gates:

Figure 1–2 (a) Typical circuit board with integrated circuits for a scanner-monitor radio. (b) Basic plan of a scanner-monitor radio.

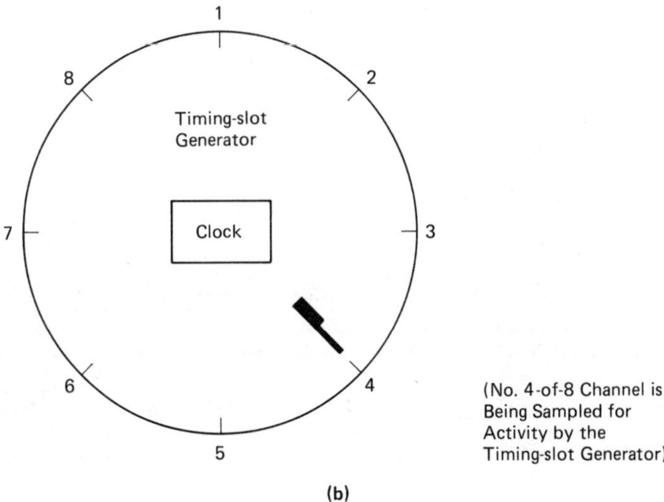

(b)

Note 1: The No. 4 output of the timing-slot generator is logic-high, and the No. 4-of-8 receiving channel is being sampled for activity. If it is active, the clock is automatically stopped until transmission activity ceases. Then, the clock will start, and the No. 5-of-8 channel will be briefly sampled for activity, and so on. The output logic-high level from the timing-slot generator turns on a diode switch associated with the particular channel; this switching action also turns on an LED associated with the particular channel. (Timing-slot generators and related logic circuits are explained in the following chapters).

Note 2: The clock is an astable multivibrator that drives the timing-slot generator. A timing-slot generator is also called a ring counter or a recirculating shift register.

Figure 1–2 *(cont.)*

of go/no-go logic pulsers and probes, as illustrated in Fig. 1–3. This is a highly informative preliminary troubleshooting procedure; its chief disadvantage is that the troubleshooter must know (or somehow determine) the gate input and output terminals.

GENERALIZED TROUBLESHOOTING PROCEDURES AND DATA STORAGE

We will find that there are various time-saving troubleshooting procedures which have an outstanding advantage when checking out digital equipment

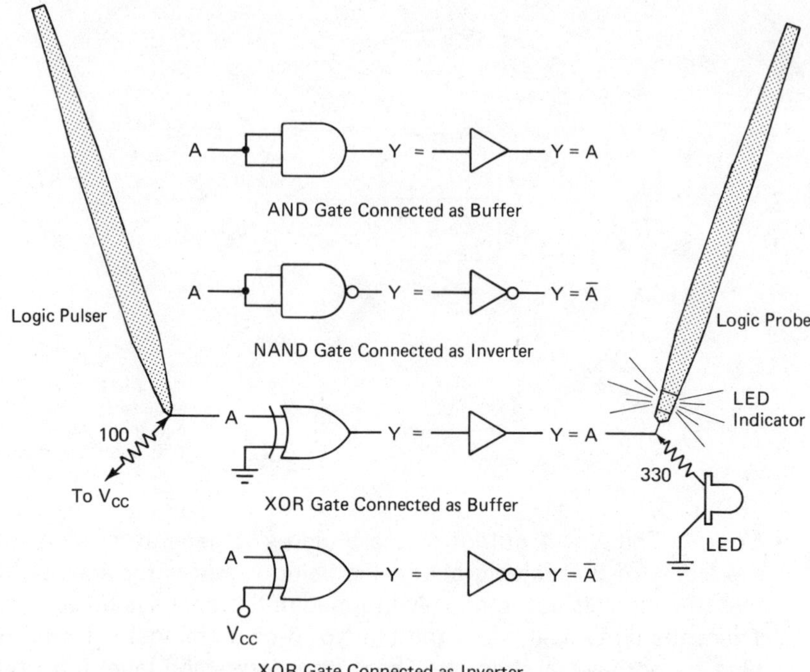

Note 1: Two basic test procedures for checking gate action are shown below. You can use a logic pulser and a logic probe, or you can use a pair of resistors and an LED. The latter provides a static test, which is usually adequate in preliminary troubleshooting procedures.

Note 2: Digital troubleshooters commonly check gates and related devices with a logic pulser and a logic probe. As detailed subsequently, a logic pulser is a miniature pulse generator, and a logic probe provides visual indication of logic-high and logic-low states.

Figure 1–3 Common gate arrangements for buffer or inverter action.

without service data. This is just another way of saying that a completely generalized troubleshooting procedure can be successfully used without any knowledge of the type of IC which is under test, nor of the kind of circuit in which it operates. It is good practice to apply generalized troubleshooting procedures first, and to then follow up with appropriate specialized troubleshooting procedures, if necessary.

Digital troubleshooters can also employ various semi-generalized troubleshooting procedures in related fault environments. As an illustration,

DECIMAL AND BINARY NUMBERS

Decimal	*Binary*
0	0000
1	0001
2	0010
3	0011
4	0100
5	0101
6	0110
7	0111
8	1000
9	1001
10	1010
11	1011
12	1100
13	1101
14	1110
15	1111
16	0001 0000
17	0001 0001
|	| |
30	0001 1110
31	0001 1111

Figure 1–3 (*cont.*)

basic thermal tests are completely generalized. On the other hand, related "thermal signature" tests are semi-generalized and require that the troubleshooter know what type of IC is under test. Note, however, that thermal-signature tests are completely generalized if made on a comparison basis. The advantage of this approach is obvious in the case of complex equipment, such as the personal-computer logic board depicted in Fig. 1–4.

Experiment: Standard TTL NAND-gate and inverter circuitry is exemplified in Fig. 1–5. Beginning digital troubleshooting will find it helpful to construct the inverter arrangement shown in Fig. 1–5(b) using NPN transistors (the clamp diode may be omitted in this experiment). The arrangement may be assembled on an experimenter's socket, such as the Archer (Radio Shack) 276–174. De-

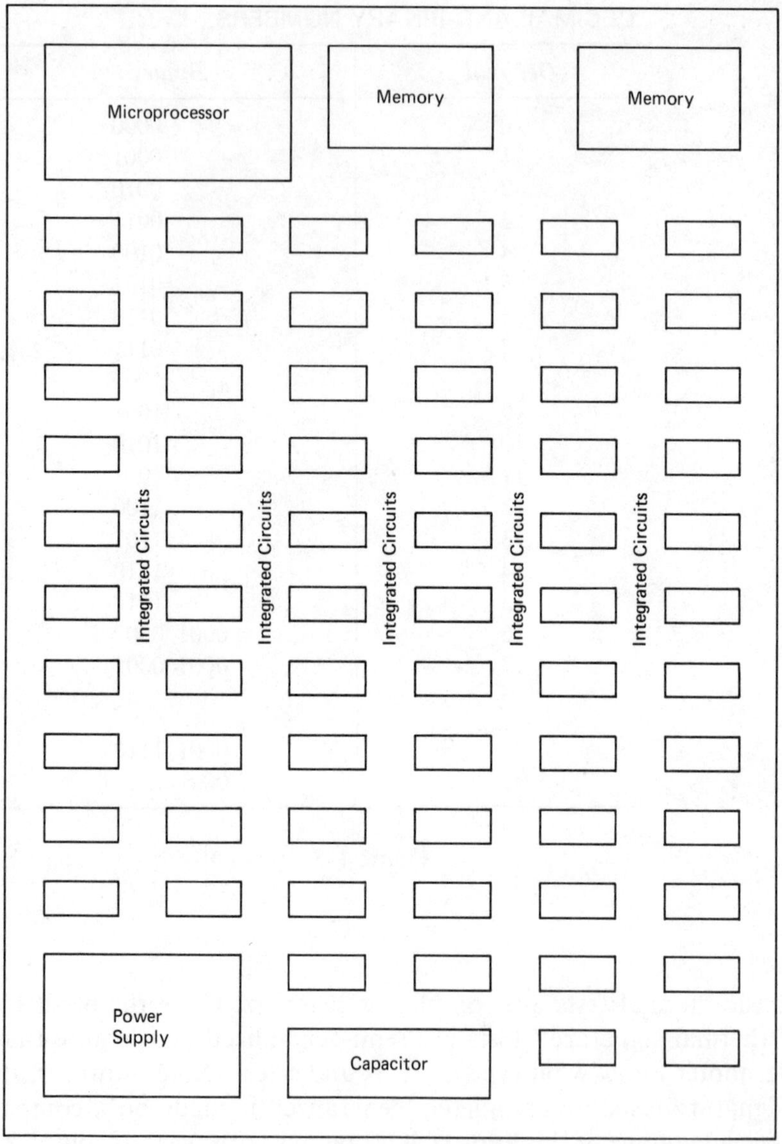

(a)

Note: If you are unfamiliar with the ICs used in digital equipment, refer to pages 113 through 344 in *Encyclopedia of Integrated Circuits* by Walter H. Buchsbaum, Sc.D. (Prentice-Hall, 1980).

Figure 1–4 A logic board layout for a personal computer that employs 80 integrated circuits. (a) Plan of logic board; (b) A data-stream/clock-line frequency quick checker.

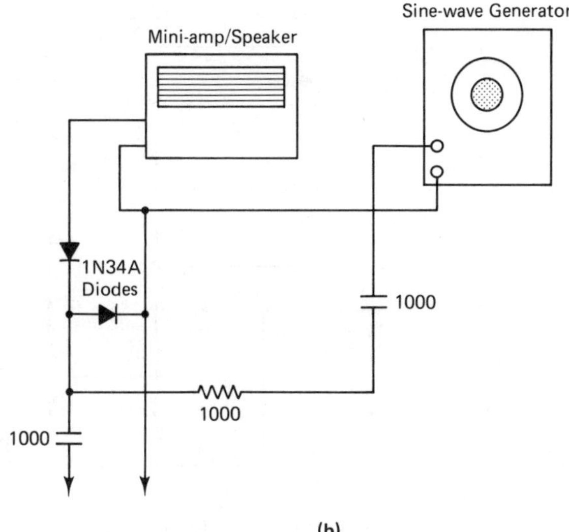

(b)

Note: This is a very handy speed-up data-stream/clock-line frequency quick checker for use in preliminary digital troubleshooting procedures. It operates on the basis of beat-frequency ''birdies.'' In other words, when the test leads are applied between a node and ground on a logic board, the digital data sequence is heterodyned with the output from the sine-wave generator. In turn, when you tune the generator through the fundamental data-stream frequency, you will hear a ''birdie'' outputted by the mini-amp/speaker.

Activity indicator: Observe that this quick checker is also a digital activity indicator. In other words, it provides a clear distinction between a data stream and a static logic-high state.

Figure 1–4 (*cont.*)

vices and components may be easily plugged into the rows and columns of holes provided on the socket, as shown in Fig. 1–6. The related V_{CC} voltage for gates in an IC package is typically a small amount over +5.1 V; as noted in Fig. 1–7, application of excessive V_{CC} voltage will quickly ruin gates in an IC package, with ''blowout'' and a noxious odor.

After the inverter dipicted in Fig. 1–5 has been assembled and checked out, the beginning troubleshooter will find it helpful to obtain a commercial NAND gate IC package, and to observe its responses to logic-low and logic-high outputs applied to one of the gates. (Don't forget that a disconnected or ''floating'' input ''looks like'' a logic-high input.) These experiments provide a ''feel''

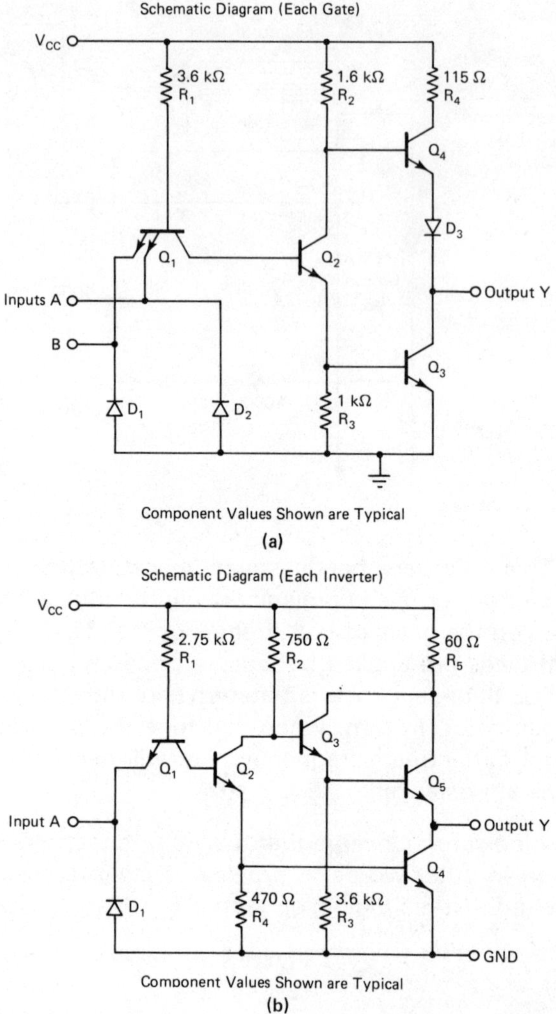

Note 1: Clamp diodes are provided to ensure that a negative input voltage (if it should occur) will be bypassed to ground. If negative pulses (unwanted negative inputs) were permitted to gain entry into the logic circuitry, erroneous output responses could occur.

Note 2: A typical quad 2-input gate IC package draws 17 mA from V_{CC} in normal operation.

Figure 1–5 Examples of standard NAND-gate and inverter logic circuitry. (a) Schematic diagram, each gate. (b) Schematic diagram, each inverter.

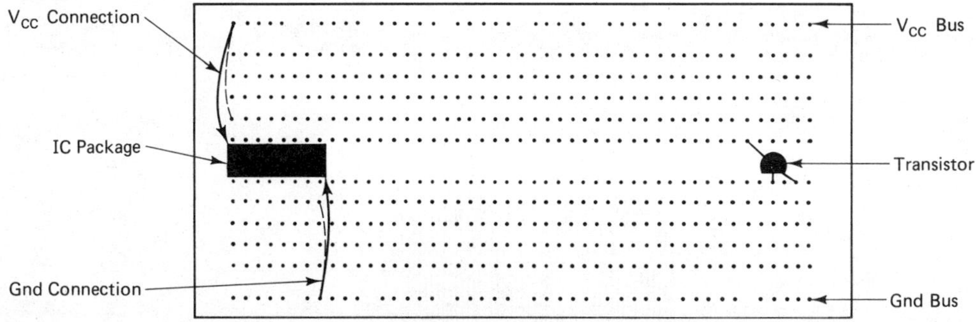

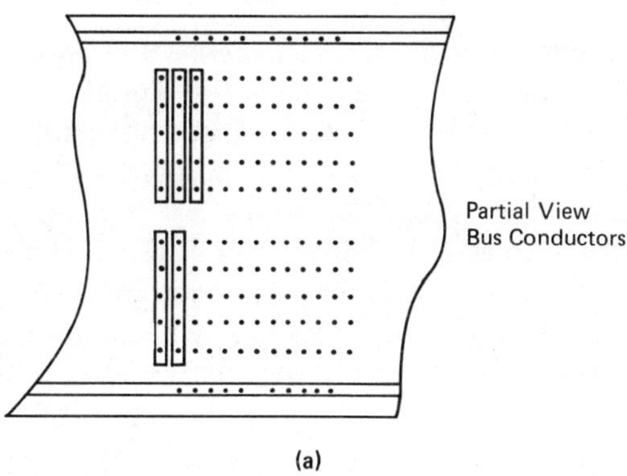

(a)

Note: This experimenter's socket board provides 560 holes into which device and component pigtails may be plugged for interconnection. Underneath the socket board, conductors are provided as shown in the partial view. These conductors provide interconnection of each column of five holes, and also provide interconnection of each row of holes along the top and bottom of the socket board.

Figure 1–6 Experimenter's socket board speeds up test procedures. (a) Arrangement of socket. (b) Experimental data-stream memory.

for logic circuitry and help to give the beginner some much-needed self-confidence. A DVM should be used for measuring voltages.

The experimenter should review what he has observed, and consider how he could use a dc voltmeter to determine whether an IC package contains AND gates, OR gates, NAND gates, or NOR gates. *This is the first step in learning*

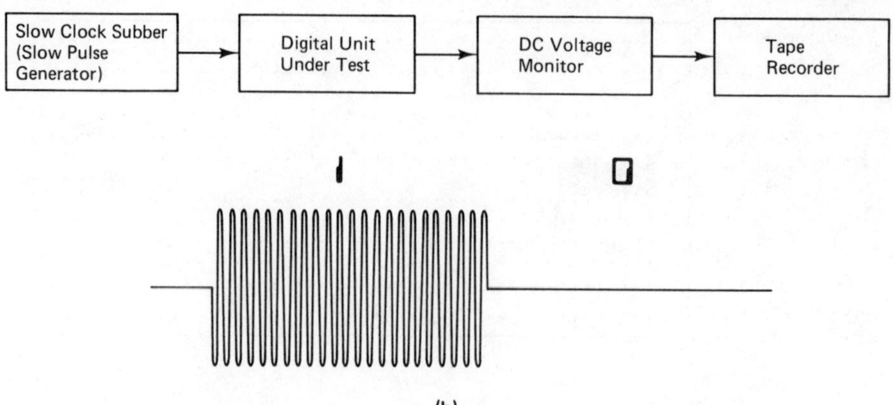

(b)

Note: Preliminary digital troubleshooting procedures are often facilitated by means for slow clocking of the digital unit under test, and recording various data streams for subsequent playback, comparison, and analysis. The slow clock subber may be operated at 2 Hz/s, for example. Then, the recorded data stream can easily be "played by ear." Observe that it is impractical to directly record the output from the digital unit under test, inasmuch as conventional tape recorders have practically no response at very low frequencies. Therefore, you must employ a dc voltage monitor between the digital unit under test and the tape recorder.

Figure 1–6 *(cont.)*

how to troubleshoot digital equipment without service data, and saving time with shortcuts.

Experiment: This experiment illustrates the troubleshooting advantage provided by a termperature probe. With reference to Fig. 1–8, connect an OR-gate IC package to a +5.1 V source, and measure the package temperature with a temperature probe and DVM. Then locate one of the gate output terminals, and short-circuit the terminal to ground. Observe the rise in package temperature. Locate another gate output terminal and short-circuit the terminal to ground; again, observe the rise in package temperature that results.

Review the experiment in your mind and consider how you could locate a short-circuited output to ground in a unit of digital equipment, without any consideration of the system circuitry. *This is the second step in learning how*

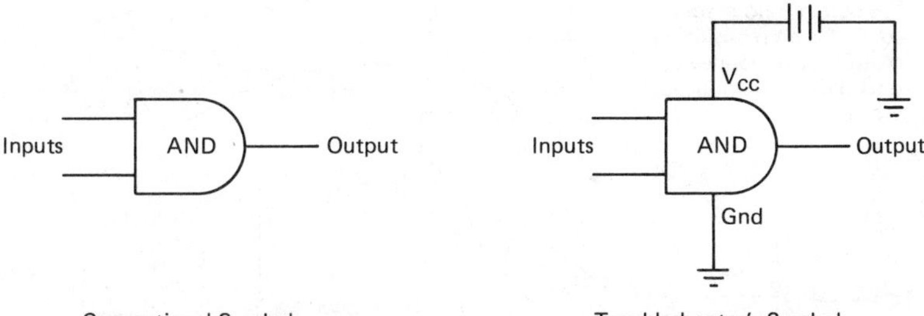

Note 1: It is helpful to observe that a typical NAND gate draws about 3.6 mA from V_{CC} when its inputs are logic-high, and draws about 1.07 mA from V_{CC} when its inputs are logic-low. However, different varieties of gates may have considerably different I_{CC} ratings. (Consult the manufacturer's data book.)

CAUTION

Note 2: If excessive V_{CC} voltage is accidentally applied to a gate IC package, burn-out can occur, with rupture of the package on the underside. Smoke from the opening has the odor of rotten eggs.

Figure 1-7 Conventional gate symbol, and "troubleshooter's symbol."

to troubleshoot digital equipment without service data, using time-saver short-cuts.

Note that an inverter is not a gate in the strict sense of the term, because an inverter has only one input. Technically, an inverter is an operator. It is also called a NOT circuit.

Troubleshooters should also note that quad 2-input gate packages are quite common, although some 14-pin packages may contain fewer gates. For example, a 3-input NOR gate IC package contains three gates. Again, a dual 4-input AND gate IC package contains two gates. As another example, an 8-input NAND gate IC package contains one gate. Observe that when a 14-pin IC package contains more than one gate, all gates will be of the same type.

To continue this temperature experiment, observe the tests depicted in Fig. 1-9. An IC operating in a given circuit will have some pins running at a bit higher temperature than other pins. This is a "thermal signature," and it can be quite informative on occasion. Thermal signatures are most helpful when checked on a comparison basis with respect to a similar unit of digital equipment which is in good working condition. Even if a comparison test is not feasible, there are "ballpark" values which will not be exceeded in normal operation.

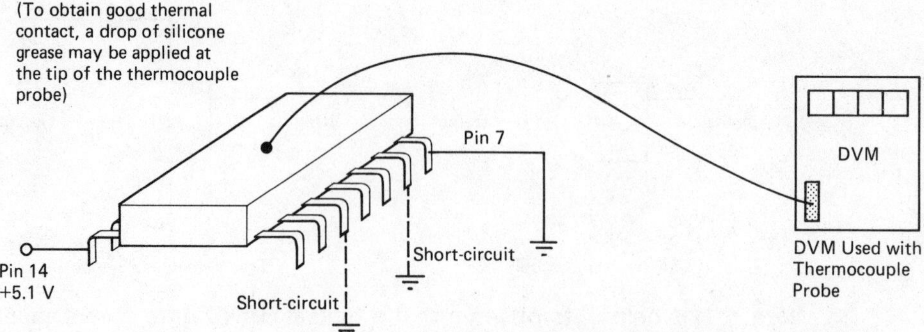

Note 1: Temperature measurements with a thermocouple probe are most useful when a catastrophic failure is being tracked down. However, when a marginal failure has occurred, other techniques should be used, as explained in Chapter 11.

Note 2: Pin 14 is commonly the V_{CC} terminal, and pin 7 is commonly the ground terminal in TTL gate IC packages. Pin 14 connects to all of the gate V_{CC} terminals inside of the package, and pin 7 connects to all of the gate ground terminals inside the package. With an ambient temperature of 21°C, the typical "idling" temperature for the IC package is 25°. If one output is shorted to ground, the package temperature typically rises to 28°. If two outputs are shorted to ground, the package temperature typically rises to 31°.

Note 3: If there is an open circuit in the power line, either inside or outside of the IC, the temperature of the IC package remains at ambient temperature (21° in this example).

Note 4: A thermocouple probe test can often be usefully supplemented by a digital stethoscope test, as explained in Chapter 6.

Figure 1-8 Temperature checks are of great importance in troubleshooting without service data.

Although not generally recognized, a gate normally radiates a pulse of RF energy while changing state, as described in Fig. 1-10. The frequency of this RFI depends on the circuit in which the gate is operating. An open circuit branch, or a short-circuit to an adjacent PC conductor, for example, will change the frequency of the RFI. Of course, a catastrophic fault will "kill" the RFI from the associated gate. In turn, the RFI field from an IC package has potential value as a troubleshooting parameter.

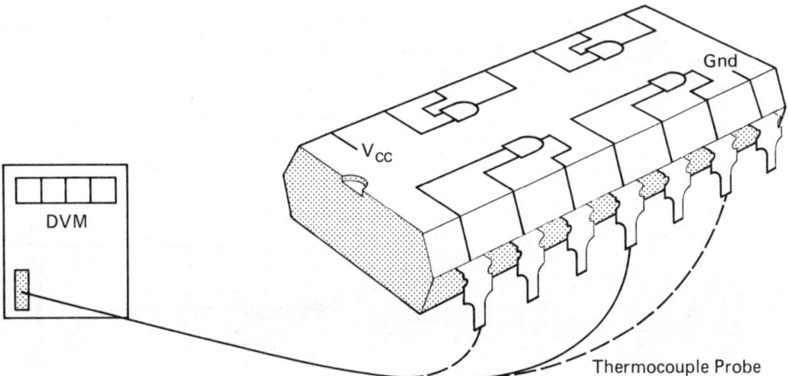

Note 1: Different pins on a digital IC package may normally operate at different temperatures. For example, one pin may measure 19° C, whereas an adjacent pin measures 18° C. This is the temperature "signature" of the IC in its associated circuit. If a short-circuit or an open-circuit occurs inside the IC package, its temperature "signature" changes, and this change is immediately evident in a comparison test on a similar unit of digital equipment which is in normal working condition.

This is a helpful preliminary check when troubleshooting digital equipment without service data because the technician does not need to know the type or function of the IC under test, nor the kind of circuit in which it is operating.

Note 2: Temperature checks are frequently more informative when recorded over an appropriate period of time. Thermocouple probes can be used in combination with an easily constructed recording voltmeter, as explained subsequently.

Figure 1–9 Digital IC temperature "signatures" provide helpful clues in preliminary troubleshooting procedures.

FAN-IN AND FAN-OUT

Digital troubleshooters frequently speak of the fan-out in a circuit, and may also speak of the fan-in. With reference to Fig. 1–11, note that fan-in denotes the number of inputs that are tied together. It is a common observation to find two inputs connected in parallel, and three inputs may occasionally be tied together. Some gates are rated for a fan-in of 1; other gates do not

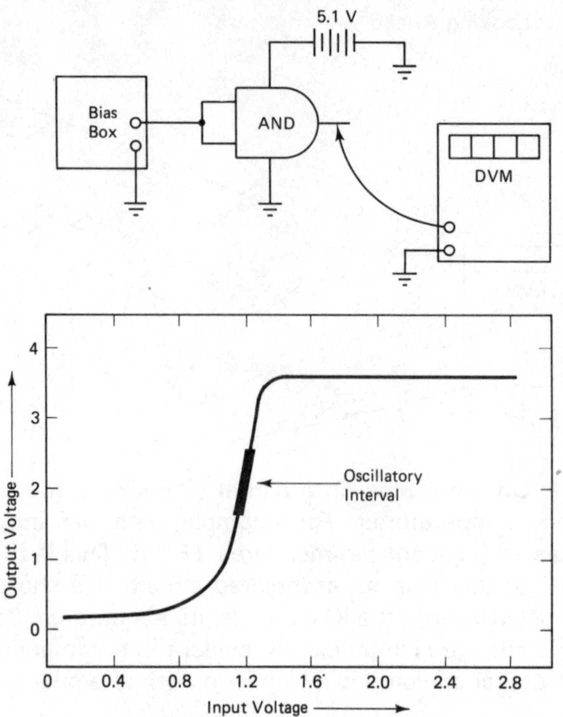

Experiment: This experiment shows the voltage transfer characteristic for an AND gate. In other words, it shows the relation of output voltage to input voltage.

A typical voltage transfer characteristic is shown in the plot. Observe that the output normally goes logic-high rapidly, although not instantaneously. Stated otherwise, the switching characteristics of the AND gate traverses the ''bad region'' rapidly as the input voltage increases.

Both the logic-high and the logic-low levels of the gate are flat, because these levels correspond to saturated transistors in the gate circuitry.

Note: A typical TTL gate has a gain of 25 dB over a portion of the ''bad region.'' Any fluctuation within this region will be amplified about 10 times at the gate output. Consequently, the troubleshooter will find that self-oscillation (parasitic oscillation) generally occurs for a brief instant as a gate goes from logic-low to logic-high, or vice versa. The frequency of this parasitic oscillation depends upon the stray inductance and capacitance associated with the gate circuit. If a short-wave radio receiver is provided with an antenna probe lead brought near the IC package, it will be observed that this parasitic oscillation in a typical gate circuit during data processing has a fundamental frequency of 10 MHz, with second-harmonic output at 20 MHz.

Figure 1–10 Transfer characteristic and oscillatory interval of an AND gate.

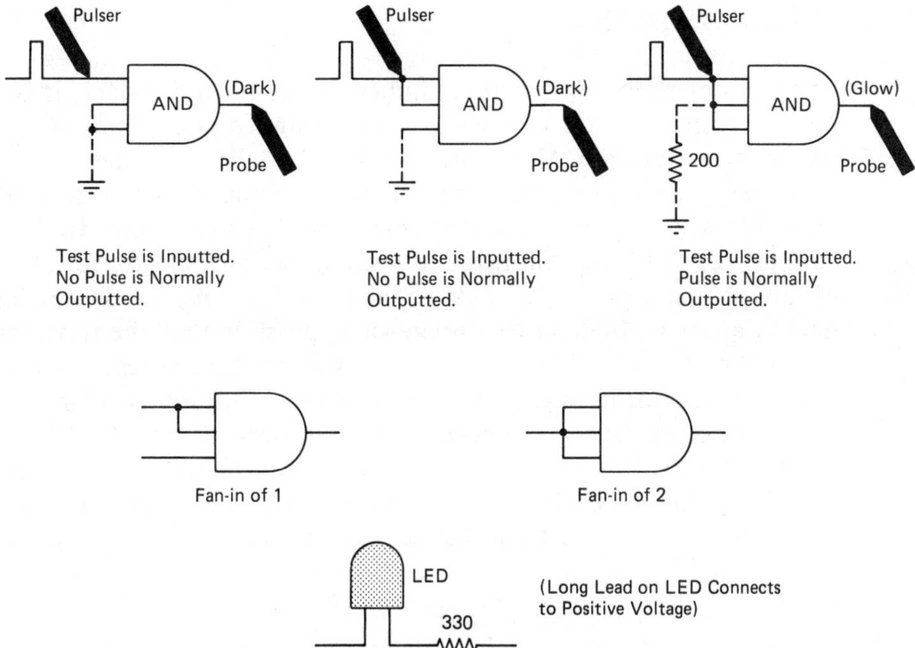

Note 1: Digital troubleshooters may refer to paralleled inputs as the fan-in of a gate.

Note 2: Instead of using a commercial logic probe in simple experimental applications, an LED may be connected in series with a 330-Ω resistor. When the gate output is logic-high, the LED will glow. An Archer (Radio Shack) 276-026 miniature red LED is suitable.

Note 3: To avoid floating inputs when checking a gate, the inputs may be returned to ground through 200-Ω resistors.

Figure 1–11 Three basic circuit situations for a 3-input AND gate.

have a fan-in rating, and any number of inputs may be paralleled without resulting operating trouble.

Fan-out denotes the number of gates which are driven. For example, if a NAND gate output terminal drives two gates, it is said to have a fan-out of 2, or, the NAND gate has two unit loads (ULs). Typical TTL gates are rated for a fan-out of 10. Note that if the rated fan-out of a gate is exceeded, the associated node is likely to fall in the bad region when it should be in the logic-low state.

THE 7/11 GROUND RULE

When troubleshooting digital equipment without service data, it is helpful to know which pins on IC packages are grounded pins. Many IC packages have 14 pins, and the "7/11 Ground Rule" applies. In other words, a 14-pin IC package employs either the Pinout A or the Pinout B standard. Pinout A has a ground pin 7, and Pinout B has a ground pin 11. This is just another way of saying that if the troubleshooter happens to find that pin 11 is grounded on a 14-pin IC package, he does not immediately conclude that there is a circuit fault at this node—it is possible that the (less common) Pinout B may used on the particular IC. In turn, further tests must be made to determine whether a fault is indeed present. (See Fig. 1–12.)

Observe that the 7/11 Ground Rule has a correlated "4/14 V_{CC} Rule." In other words, if a 14-pin IC package has a ground pin 7, it will have a V_{CC} pin 14. Or, if the 14-pin IC package has a ground pin 11, it will have a V_{CC} pin 4. This is just another way of saying that if the troubleshooter

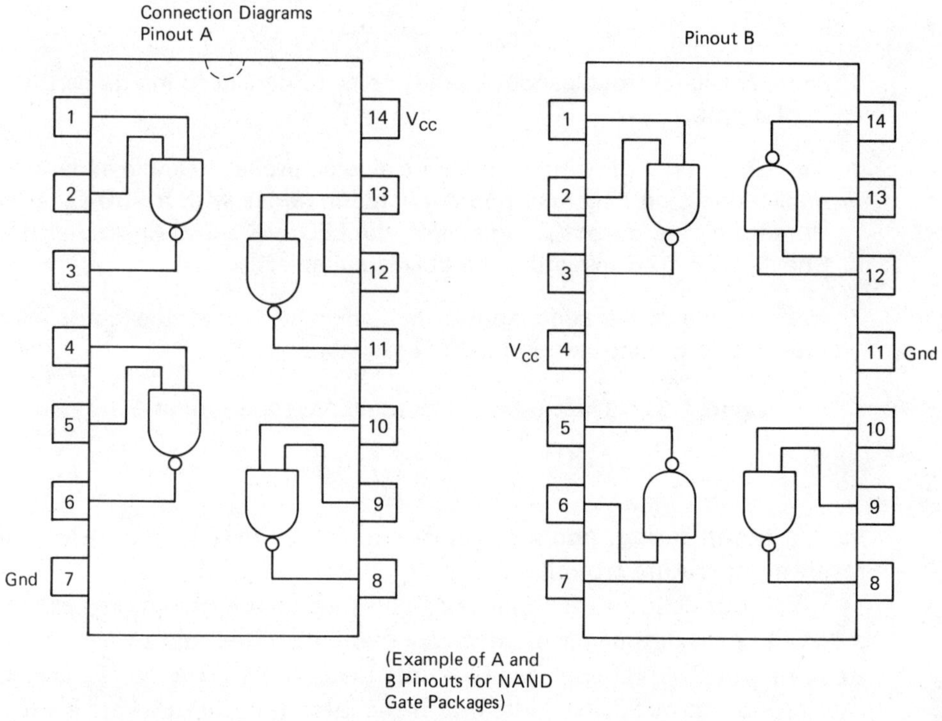

Figure 1–12 A familiar example of the 7/11 ground rule in NAND-gate A and B pinouts.

happens to find that pin 4 is at V_{CC} potential on a 14-pin IC package, he does not immediately conclude that there is a short to V_{CC} at this node—it is possible that the (less common) Pinout B may be used on the particular IC. Accordingly, further tests are required in this situation to determine whether a fault is actually present. (See also Chart 1-1.)

Chart 1-1

CURRENT SPIKE QUICK CHECKER

When troubleshooting digital equipment without service data, obscure (''tough dog'') trouble symptoms can arise due to decoupling faults on a circuit board. TTL totem-pole circuitry is prone to current ''spikes'' (switching transients) in the V_{CC} and ground conductors. Voltage spiking is comparative low-profile, but current spiking is high-profile and can cause malfunctioning in circuitry that employs common V_{CC} and ground conductors. Consequently, decoupling capacitors are employed to minimize current spiking. For example, a circuit board with 25 IC packages typically includes ten 0.02-μF ceramic disc capacitors connected between the V_{CC} and ground conductors at various points. A 10-μF tantalum capacitor is generally included.

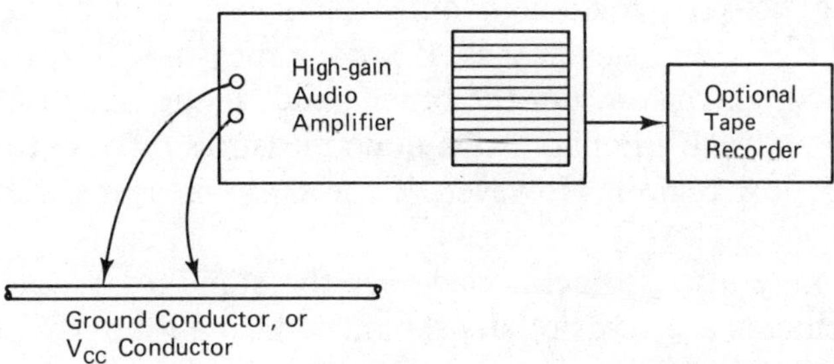

Open test leads with pointed test prods may be used in this arrangement, inasmuch as the source impedance is very low. Note that you will usually set the volume control to maximum during the test for current spiking. However, before the test leads are removed from the conductor, the volume control should be turned to minimum.

This test should be made with a data input to the digital equipment from a digital word generator or equivalent source that has a low audio-frequency repetition rate. The digital equipment under test may also be clocked by a clock subber set to a low audio-frequency rate.

When the troubleshooter suspects that malfunction is being caused by current spikes, a high-gain audio amplifier and speaker can be used to make an informative quick-check, as depicted above. This test is definitive when made on a comparison basis, and is also quite useful on a "ballpark" basis. The input leads to the audio amplifier are applied across various sections of the ground and V_{CC} conductors. If the equipment is clocked and operated at an audio frequency from a clock subber, the current spike pattern is heard from the speaker.

The previous current spiking test is based on the fact that each current spike produces an IR drop along a ground (or V_{CC}) conductor. Although this IR value is quite small, it corresponds to a large current spike, inasmuch as the ground (or V_{CC}) conductor has a very low resistance. A high-gain audio amplifier responds to the IR drops produced by the current spikes, and reproduces them as transient clicks. The digital troubleshooter can soon establish "ballpark" levels as he becomes familiar with the test technique. For example, if the test leads are applied along a 2-inch interval of ground conductor, a reference "ballpark" sound output will be established if the equipment under test is operating normally.

Note in passing that a DVM is the most basic instrument for distinguishing between logic-high and V_{CC} levels, and between logic-low and ground levels. In other words, if a node measures 3 V, for example, it is in the logic-high region. On the other hand, if the node measures 5.1 V, it is at the V_{CC} level. Similarly, if a node measures 0.06 V, for example, it is in the logic-low region. However, if the node measures zero, it is at ground level.

It may also be mentioned that the standard Pinout-A and Pinout-B arrangements are occasionally expanded into a Pinout-A, Pinout-B, Pinout-C, and Pinout-D arrangement. Observe that when a 14-pin IC package is available with 4-version standard pinouts, that the 7/11 Ground Rule still holds—the distinction in this case is that ground-pin 11 packages are then referred to as the Pinout-C or Pinout-D versions. The bottom line is that the 7/11 Ground Rule is valid, regardless of the pinout letter designation that may apply to the particular package.

THE 8/12 GROUND RULE

Digital troubleshooters are frequently concerned with 16-pin IC packages. Here, the "8/12 Ground Rule" usually applies. That is, a 16-pin IC package employs either pin 8 or pin 12 as a ground pin (with rare exceptions). (See Fig. 1–13.) Accordingly, if the troubleshooter happens to find that pin 12 is grounded on a 16-pin IC package, he does not immediately conclude that there is a circuit fault at this node—it is possible that the (less common) pin-12 ground arrangement is present. In this situation, further tests are required to determine whether the grounded pin 12 is a normal or abnormal condition.

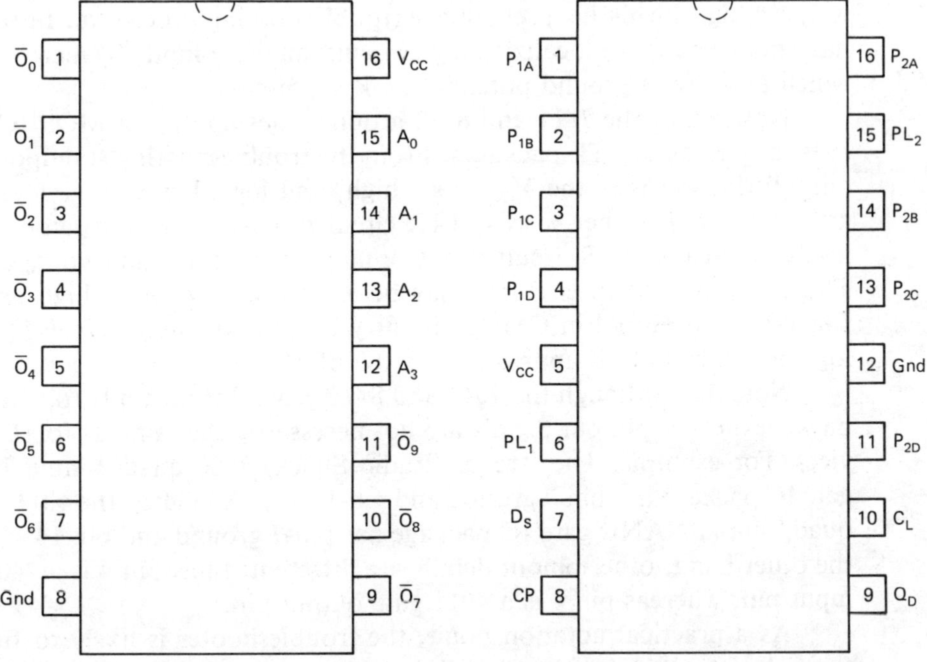

Practical Note: Some digital integrated circuits and PC boards tend to develop trouble symptoms if they are stored for a long period of time in a damp location. If you are servicing a digital unit that has been exposed to high humidity over a substantial length of time, start by operating the malfunctioning unit for several days in a dry and warm location. Case histories show that the trouble symptoms may clear up before the drying-out procedure is completed.

Figure 1–13 Typical examples of the 8/12 ground rule.

A rare exception to the 8/12 Ground Rule employs pin 13 as the ground pin in a 16-pin IC package. (Sort of a February 29th rule, except that it is a rarer exception.) The 8/12 Ground Rule has an accompanying 5/16 V_{CC} rule. Stated otherwise, a 16-pin IC package employs either pin 5 or pin 16 as the V_{CC} pin. The 5/16 rule holds true, even in the event that pin 13 might happen to be the ground pin, instead of pin 12.

Thus, if the troubleshooter happens to find that pin 5 is at V_{CC} potential on a 16-pin IC package, he does not immediately conclude that there is a short to V_{CC} at this node—it is possible that the (less common) pin-5 V_{CC} arrangement is present. In such a case, further tests are required to determine whether the V_{CC} level at pin 5 is a normal or abnormal condition.

The advantage of the ground rules is that the troubleshooter does not need to know what type of IC is under test, nor in what kind of circuit it is operating, during his preliminary troubleshooting procedure. In turn, he has a good chance of localizing logic circuit faults promptly by merely noting which pins are at ground potential or V_{CC} potential.

Notice that the 7/11 and 8/12 ground rules apply to CMOS IC packages, as well as to TTL packages. From the troubleshooting standpoint, the only distinction is in the V_{CC}, logic-high, and logic-low voltage levels that are involved. In other words, TTL circuitry always has the same voltage levels, whereas CMOS circuitry may or may not have the same voltage levels. Thus, V_{CC} in CMOS circuitry may be as low as 3 V, or as high as 15 V. Logic-high threshold in CMOS circuitry is approximately 0.7 of V_{CC}, and logic-low threshold is approximately 0.3 of V_{CC}.

Note that although the 7/11 and 8/12 ground rules apply to CMOS IC packages, other pin-out details are not necessarily the same as for TTL devices. For example, The Archer (Radio Shack) 7400 quad 2-input NAND gate IC package has pin-7 ground and pin-14 V_{CC}. Similarly, the 4011 CMOS quad 2-input NAND gate IC package has pin-7 ground and pin 14 V_{CC}. On the other hand, other pinout details are different; thus, pin 4 is a 7400 gate input pin, whereas pin 4 is a 4011 gate output pin.

As a practical notation point, the troubleshooter is likely to find the ground pin called V_{CC} in CMOS data sheets. Again, the V_{CC} pin may be called V_{DD}.

NC PINS

"All pins are connected to something" in most digital IC packages. Accordingly, it is a general rule that if a quick test shows that a pin is open-circuited, the IC is defective. This is a very useful rule-of-thumb, although the troubleshooter should keep in mind that there are a few exceptions. For example, the 7420 dual 4-input NAND gate is provided in a 14-pin package. (It conforms to the 7/11 ground rule, and pin 7 is grounded.) However, pin 11 is an NC (no-connection) pin in this particular IC. Pin 3 is also an NC pin.

Since pins 3 and 11 "are connected to nothing," they appear to be open circuits insofar as quick checks are concerned. Unless further checks are made in this situation, the IC would be needlessly replaced. An easy cross-check that does not require identification of the IC package is to inspect the pins on the solder side of the PC board. In other words, if the IC package is a 7420, for example, it will be seen that there are no PC conductors running to pins 3 and 11. Accordingly, the suspicion of an open-circuit fault would be cleared.

UNEXPECTED GROUNDS

It is a general rule that any pins other than 7/11:8/12 will not be connected to ground in a digital network. Therefore, if the troubleshooter finds an unexpected ground when quick-checking IC pins, he immediately suspects a short-to-ground fault. There is an occasional exception to this general rule; for example, it is possible that an XOR-gate input terminal could be returned to ground (gate operated as a buffer). Accordingly, an unexpected ground is very strong, but not absolutely conclusive, evidence of a short-to-ground fault.

"NORMAL" BAD LEVEL

In nearly all cases, when a bad-level voltage is found in a quick check, it indicates that a fault is present. However, this is not absolutely conclusive evidence of a device or circuit fault. That is, there is a small possibility that a preset or reset pin may have been left floating, instead of connecting it with a PC conductor to V_{CC}. Therefore, the troubleshooter should cross-check on the solder side of the PC board, to determine whether or not a PC conductor is connected to the bad-level pin.

Note in passing that although a floating input pin "looks" logic-high to the following circuitry, that it is not considered good design practice to float a terminal instead of connecting it to V_{CC}. A floating input tends to be noisier than an input tied to V_{CC}, and tends to raise the residual noise level in the system. In turn, operating reliability is reduced to some extent.

THE RULE OF 2

The Rule of 2 is helpful in preliminary quick-check procedures. This rule states that only one pin on an IC package will be at V_{CC} potential, and only one pin on the package will be at ground potential. In turn, if quick checks show that there are two (or more) pins at V_{CC} potential on the same IC package, the troubleshooter recognizes that there is a strong suspicion of trouble either in the IC package, or in its associated circuitry.

Similarly if quick checks show that there are two (or more) pins at ground potential on the same IC package, it is indicated that a fault is probably present, either in the IC package or in its associated circuitry. It follows from previous discussion that there is an occasional network in which a gate input may be tied to V_{CC}, or tied to ground. In such a case, a rare exception to the Rule of 2 occurs. If a comparison test is feasible, doubtful situations can be promptly resolved.

Bottom Line: We can save a lot of time in preliminary digital troubleshooting procedures by using shortcuts that are based on generalized digital principles. (See also Chart 1–2.)

Chart 1–2

OSCILLOSCOPE BASICS

In various tough-dog situations, considerable time can be saved by applying an oscilloscope to supplement the test data that can be provided by logic pulsers and probes. An oscilloscope is a form of voltmeter that displays a varying voltage as a function of time. (This is called a time-domain display.) An elaborated type of oscilloscope displays binary number sequences as a function of the clock timing in the digital equipment under test. (This is called a data-domain display.)

An oscilloscope is more informative in many situations than is a basic voltmeter. As an illustration, an oscilloscope provides measurement of rise time, waveform period, pulse width, duty cycle, repetition rate, peak voltage, peak-to-peak voltage, phase, percentage modulation, damping time, and various other electrical parameters. Some high-performance oscilloscopes contain microprocessor circuitry whereby digital readout is provided for various waveform parameters.

Oscilloscopes are basically classified as free-running or triggered-sweep types. An oscilloscope with a free-running time base is adequate for considerable troubleshooting requirements in analog circuitry. On the other hand, oscilloscopes with triggered-sweep time bases and calibrated sweeps are a basic necessity in digital circuit troubleshooting procedures. The beginner will find it helpful to note the following descriptive overview of a typical wide-band triggered-sweep oscilloscope.

A modern solid-state triggered-sweep oscilloscope. (*Courtesy,* B & K Precision, Div. of Dynascan Corp.)

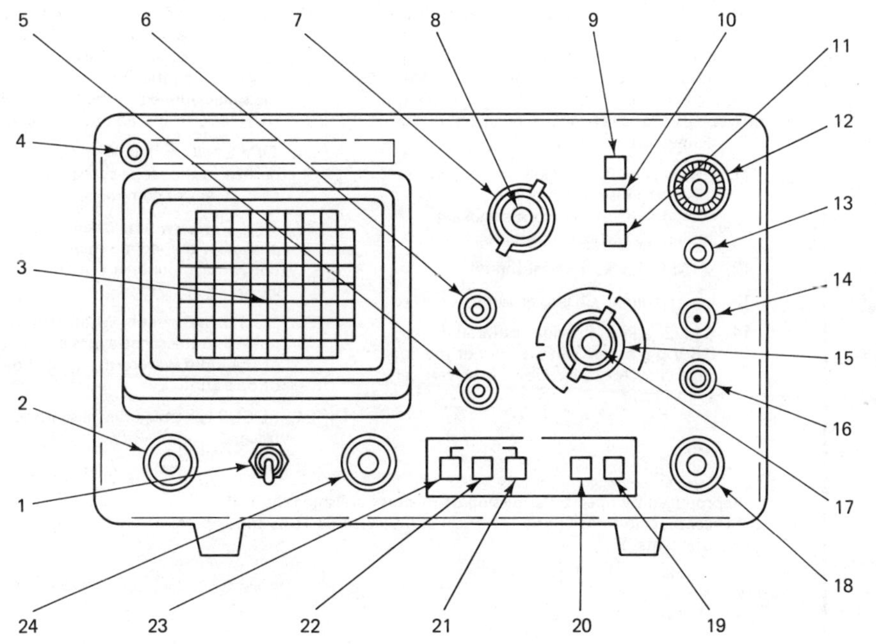

1. POWER ON toggle switch. Applies power to oscilloscope.
2. INTENSITY control. Adjusts brightness of trace.
3. Scale. Provides calibration marks for voltage and time measurements.
4. Pilot lamp. Lights when power is applied to oscilloscope.
5. ◀▶ POSITION control. Rotation adjusts horizontal position of trace. Push-pull switch selects 5X magnification when pulled out; normal when pushed in.
6. ▲▼ POSITION control. Rotation adjusts vertical position of trace.
7. VOLTS/DIV switch. Vertical attenuator. Coarse adjustment of vertical sensitivity. Vertical sensitivity is calibrated in 11 steps from 0.01 to 20 volts per division when VARIABLE 8 is set to the CAL position.
8. VARIABLE control. Vertical attenuator adjustment. Fine control of vertical sensitivity. In the extreme clockwise (CAL) position, the vertical attenuator is calibrated.
9. AC vertical input selector switch. When this button is pushed in the dc component of the input signal is eliminated.
10. GND vertical input selector switch. When this button is pushed in the input signal path is opened and the vertical amplifier input is grounded. This provides a zero-signal base line, the position of which can be used as a reference when performing dc measurements.
11. DC vertical input selector switch. When this button is pushed in the ac and dc components of the input signal are applied to vertical amplifier.
12. V INPUT jack. Vertical Input.
13. ⏚ terminal. Chassis ground.
14. CAL ⎍ jack. Provides calibrated 0.8 V p-p square wave output at the line frequency for calibration of the vertical amplifier.
15. SWEEP TIME/DIV switch. Horizontal coarse sweep time selector. Selects calibrated sweep times of 0.5 μ SEC/DIV to 0.5 SEC/DIV in 19 steps when VAR/HOR GAIN control 17 is set to CAL. Selects proper sweep time for television composite video waveforms in TVH (television horizontal) and TVV (television vertical) positions. Disables internal sweep generator and displays external horizontal input in EXT position.
16. EXT SYNC/HOR jack. Input terminal for external sync or external horizontal input.
17. VAR/HOR GAIN control. Fine sweep time adjustment (horizontal gain adjustment when SWEEP TIME/DIV switch 15 is in EXT position). In the extreme clockwise position (CAL) the sweep time is calibrated.
18. TRIG LEVEL control. Sync level adjustment determines point on waveform slope where sweep starts. In fully counterclockwise (AUTO) position, sweep is automatically synchronized to the average level of the waveform.
19. TRIGGERING SLOPE switch. Selects sync polarity (+), button pushed in, or (−), button out.
20. TRIGGERING SOURCE switch. When the button is pushed in, INT, the waveform being observed is used as the sync trigger. When the button is out, EXT, the signal applied to the EXT SYNC/HOR jack 16 is used as the sync trigger.
21. TVV SYNC switch. When button is pushed in the scope syncs on the vertical component of composite video.
22. TVH SYNC switch. When button is pushed in the scope syncs on the horizontal component of composite video.
23. NOR SYNC switch. When button is pushed in the scope syncs on a portion of the input waveform. Normal mode of operation.
24. FOCUS control. Adjusts sharpness of trace.

A basic summary of oscilloscope applications in both analog and digital troubleshooting fields is shown in the following:

- Waveform Analysis
  - Voltage
  - Current
  - Waveshape Identification and Classification
  - Time Measurements
  - Rise-time Measurements
  - Pulse-width Measurements
  - Duty-cycle Measurements
  - Repetition Rate Measurements
  - Tilt Measurements
  - Bandwidth Measurements
  - Transient Characteristics
  - Steady State Characteristics
- Amplitude Measurements
  - Peak-to-peak Voltage Measurements
  - Peak Voltage Measurements
- Frequency Measurements
  - Lissajous Figures
  - Linear Time Base
  - Beat Marker
  - Absorption Marker
  - Pulse Marker
  - Visual Alignment
- Phase Measurements
  - Lissajous Figures
  - Dual Trace
- Waveform Storage
  - Time-frequency Domain
  - Data Domain
- Distortion Measurements
- Jitter Measurement
- Damping-time Measurement Q Values, etc.
- Bandwidth Measurements
- Modulation Monitoring
- Spectrum Analysis
- Coincidence Tests
- Sum Waveforms
- Difference Waveforms
- Product Waveforms
- Quotient Waveforms
- Involute Waveforms
- Evolute Waveforms
- Power-factor Measurement
- True Power Measurement
- Reactive Power Measurement
- Peak Power Measurement

Observe the following example of pulse display on high-speed sweep. This pulse is displayed with a triggered-sweep oscilloscope that has a calibrated time base. The exemplified pulse has a rise time of approximately 20 nanoseconds (ns) and has a width of 20 ms (μs). Note that when the pulse waveform is displayed at a sweep speed of 0.02 ms/cm, the leading and trailing edges of the pulse appear to rise and fall instantaneously. In turn, the finer detail of the waveform remains invisible.

As the sweep speed is increased to 0.2 μs/cm, it becomes apparent that the leading edge does not rise instantaneously, and that the corners of the pulse are somewhat rounded. Finally, at a sweep speed of 0.04 μs/cm, it is seen that the rise time of the pulse is about 0.02 μs. This type of progressive waveform expansion is sometimes puzzling to the beginner, because the human eye is unable to resolve the waveform detail at slow sweep speed. This is just another way of saying that the oscilloscope is a technological extension of the troubleshooter's sense of sight, whereby high-speed sweep action reveals waveform detail that would otherwise remain invisible.

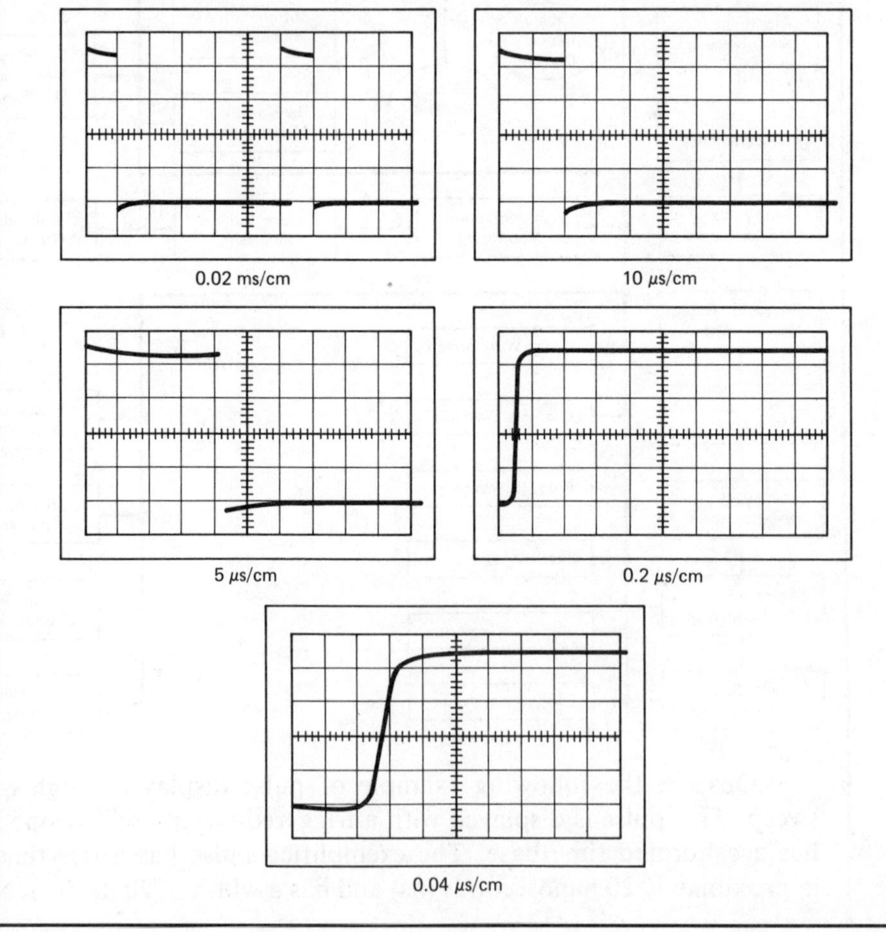

2

Progressive Digital Troubleshooting Procedures

TROUBLESHOOTING LATCHES

Digital troubleshooters are routinely concerned with various types of latches and flip-flops.* A latch is essentially a device (gate) arrangement with positive feedback which retains a logic-high or a logic-low state. Most latches employ bistable multivibrator circuitry which "locks up" in either a logic-high or a logic-low state. The most basic form of latch is called a reset-set (RS) latch; it is usually configured from NOR gates or NAND gates, as shown in Fig. 2–1. Preliminary troubleshooting procedures are outlined in Chart 2–1.

Latch characteristics differ from gate characteristics in that the output from a gate depends only upon the prevailing input states, whereas the output from a latch depends not only upon the prevailing input states, but also upon the previous input state. This is just another way of saying that a latch is an elementary storage device, and functions as a memory. Observe in Fig. 2–1 that the *RS* inputs are simultaneously applied in complementary form. The *RS* latch stores the resulting response until another input pulse drives the latch into opposite complementary states. Thereupon, the former stored datum is erased, and the new datum is stored in complementary form.

*A latch is not clocked; a flip-flop is clocked.

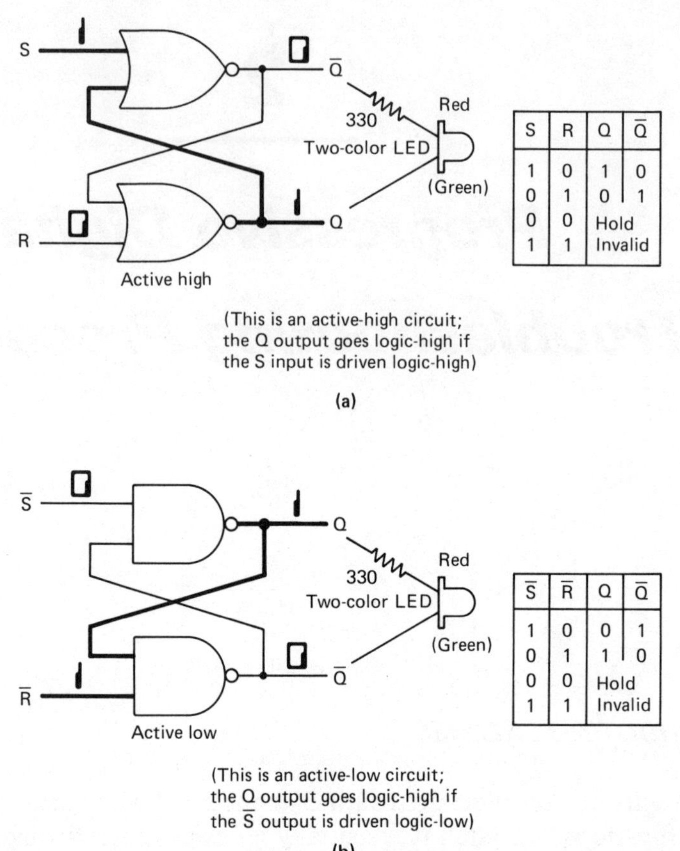

Note 1: The current demand of an IC package can be checked on a comparative basis without opening the circuit as explained in Chapter 8.

Note 2: The two-color LED glows red if the Q terminal goes logic-high; it glows green if the Q terminal goes logic-low. The LED glows yellow if a rapid pulse train is applied to the S and R inputs. Troubleshooters state that an RS latch is ''set'' when its Q output is logic-high. In the NOR-gate version of the RS latch, the ''set'' condition corresponds to $S = 1$ and $R = 0$. Or, $Q = 1$ when $S = 1$. However, in the NAND-gate version of the RS latch, the ''set'' condition corresponds to $\bar{S} = 0$ and $\bar{R} = 1$. Or, $Q = 1$ when $\bar{S} = 0$. This is just another way of saying that the NOR-gate RS latch is an active-high latch, and that the NAND-gate RS latch is an active-low latch. Active-low response is indicated by $\bar{R}$ and $\bar{S}$ inputs.

Figure 2–1 Basic *RS* latch arrangements. (a) Active-high circuit using NOR gates. (b) Active-low circuit using NAND gates. (c) 54LS commercial *RS* latch IC package.

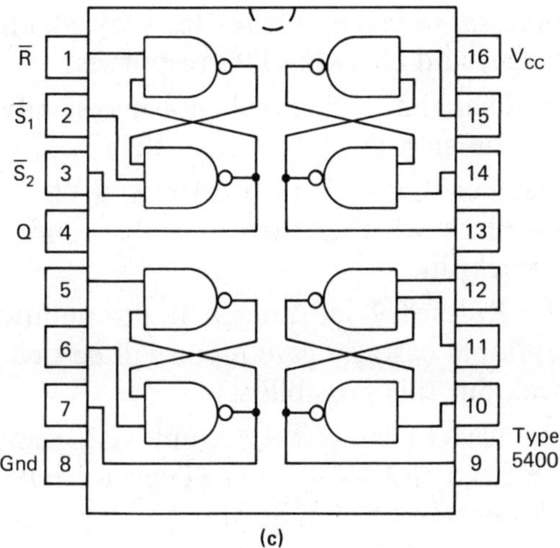

(c)

Figure 2–1 (*cont.*)

Chart 2–1

PRELIMINARY TROUBLESHOOTING PROCEDURE (WITHOUT SERVICE DATA)

1. Measure V_{CC}.
2. Measure IC temperatures with thermocouple probe.
3. Make "wipe" tests of IC pins with ohmmeter (or equivalent tester).
4. Observe the 7/11, 8/12, and 10/12/14 ground rules.
5. Observe the 4/14 and 5/16 V_{CC} rules.
6. Observe the Rule of 2.
7. Be alert for NC pins and "false-alarm" nodes.
8. If possible, make comparative voltage and temperature measurements on a şimilar unit of equipment.
9. In marginal situations, compare temperature signatures.

10. If necessary, make further tests by identifying the suspect IC package, and checking I/O responses.
11. Remember that the fault is not necessarily in the IC package—it may be an associated circuit defect.
12. Remember that on rare occasions, a "bad-level" voltage is normal—check whether there is a PC conductor connected to the bad-level pin.
13. If the Rule of 2 indicates a fault condition, remember that in exceptional cases, a gate input will be tied to V_{CC} or to ground. (Check out this possibility.)
14. When CMOS circuitry is employed, remember that V_{CC} may range from +3 to +15 V. (Typical MOS-to-TTL and TTL-to-MOS interfacing circuits are shown below).

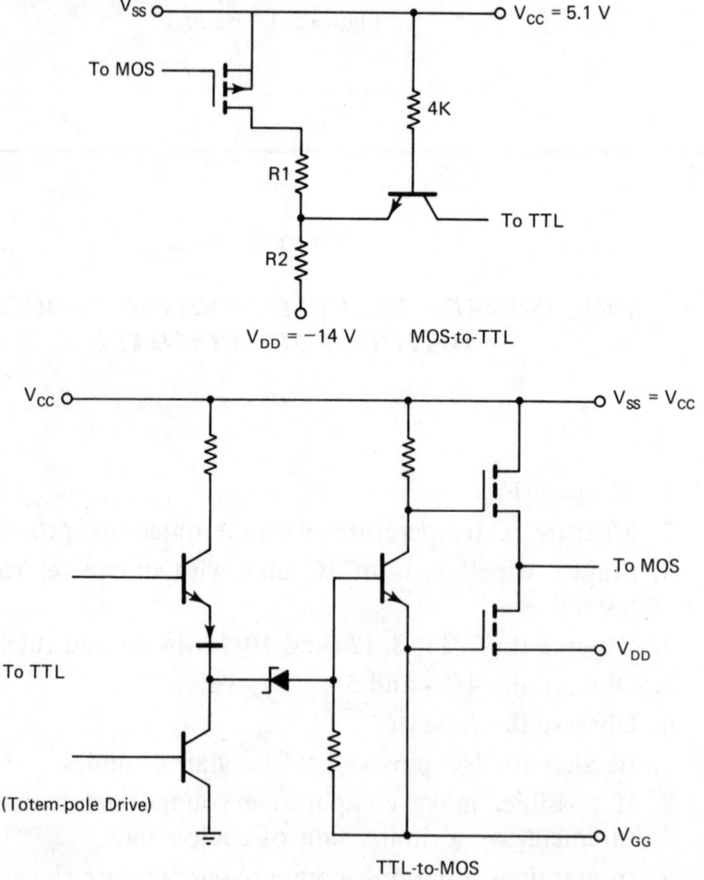

Two-Color "Probe"

Observe that a red-green LED indicator is used in Fig. 2–1 to indicate whether the Q output is logic-high or logic-low. This indicator may be an Archer (Radio Shack) 276-035 tri-color LED, or equivalent. Thus, if Q is high, the LED glows red; on the other hand, if Q is low, the LED glows green. This two-color probe is particularly useful for checking latches, because Q and $\bar{Q}$ terminals are available. (The probe provides red-green indication only if it is applied to complementary outputs, or inputs.)*

As seen in the truth tables for the *RS* latches in Fig. 2–1, a 0,0 input signal may be applied, with the result that the latch will remain in its prevailing state. However, a 1,1 input signal is invalid, or "forbidden." The reason for this prohibition is that a 1,1 input triggers conflicting circuit actions in an *RS* latch, with the result that circuit response is unpredictable. Stated otherwise, the prevailing state might or might not change in response to a 1,1 input, depending upon circuit tolerances.

Experiment: Obtain a NOR-gate IC package such as the Archer (Radio Shack) 7402 quad 2-input NOR gate, and a tri-color LED such as the Archer (Radio Shack) 276-035. Assemble a NOR-gate latch on an Experimenter Socket such as the Archer (Radio Shack) 276-174. Operate the latch from a 5.1-V power supply. Connect the tri-color LED in series with a 330-Ω resistor across the latch output terminals.

Proceed as follows:

1. Connect the R input to Ground, and connect the S input to V_{CC}. Observe the LED indication; measure the Q and $\bar{Q}$ voltages.
2. Connect the S input to Ground, and connect the R input to V_{CC}. Observe the LED indication; measure the Q and $\bar{Q}$ voltages.
3. Connect the R input to Ground, and let the S input "float." Observe the LED indication; measure the Q and $\bar{Q}$ voltages; measure the "float" voltage at the S input.
4. Let both the R input and the S input "float." Observe the LED indication; measure the Q and $\bar{Q}$ voltages. Explain the reason for your experimental findings.

Consider next the active-high and active-low *RS* latch implementations with OR-NOT and NEGATED-OR gates depicted in Fig. 2–2. Circuit analysis will show that when a logic-high input is applied to the S line of the active-high latch, (and a logic-low input applied to the R line), the Q output will go

*A digital paraphase indicator is described subsequently.

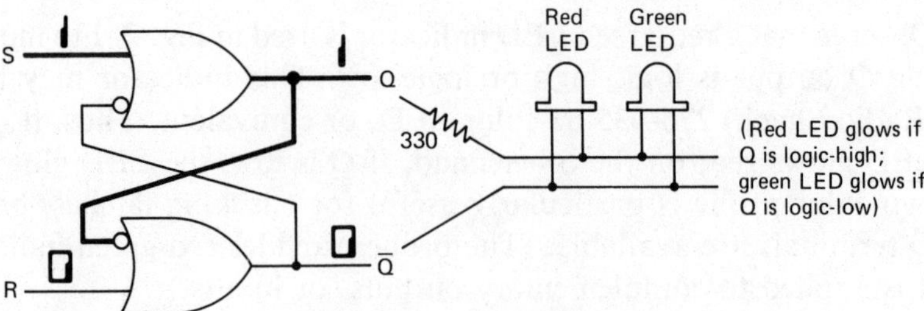

Active High (When S is Driven Logic-high, and R is Driven Logic-low, Q Goes Logic-high)

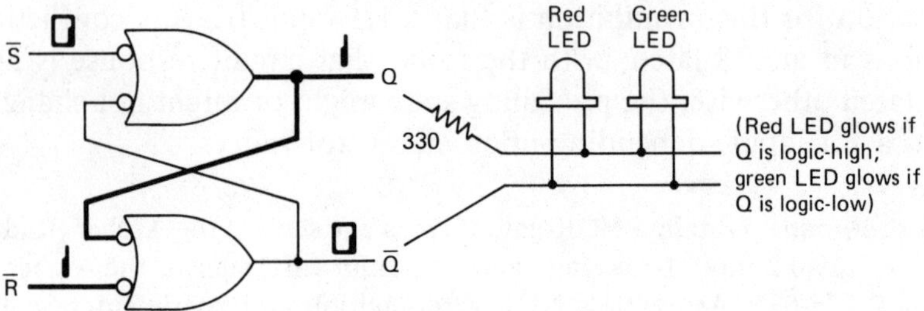

Active Low (When $\overline{S}$ is Driven Logic-low, and $\overline{R}$ is Driven Logic-high, Q Goes Logic-high)

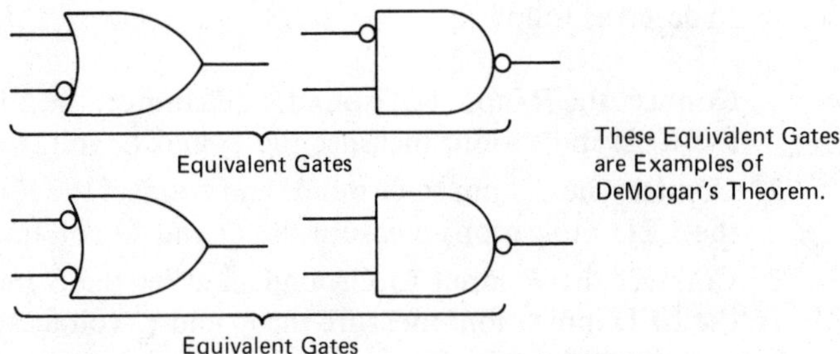

Note: Active-high and active-low *RS* latches may be configured as shown above. Observe that the active-high latch could also be configured from a pair of *A* NAND-NOT *B* GATES. An active-low *RS* latch may also be configured from a pair of *A* AND-NOT *B* gates.

Figure 2–2 Active-high and active-low latch implementations with OR-NOT and NEGATED-OR gates.

logic-high. On the other hand, when a logic-low input is applied to the *S* line of the active-low latch (and a logic-high input applied to the *R* line), the *Q* output will go logic high.

Observe also that an active-high latch will function as a logic-low latch if its *Q* output is regarded as a $\bar{Q}$ output, and its $\bar{Q}$ output is regarded as a *Q* output. With this reversal of viewpoint, we call the *S* input $\bar{S}$, and call the *R* input $\bar{R}$, as a reminder that the latch now has active-low function. Stated otherwise, there is no "hardware" or interconnection distinction involved in this reversal of viewpoint—it is merely a change in designation of the operating mode. An example of latch application is seen in Fig. 2–3.

TROUBLESHOOTING THE BASIC D LATCH

Digital troubleshooters encounter *D* latches in many logic circuits. The basic *D*-latch arrangement is shown in Fig. 2–4. Like the *RS* latch, the *D* latch is a simple form of logic-storage device. Observe that if a logic-high input is applied, this datum will be stored in complementary form: the *Q* output will go logic-high, and the $\bar{Q}$ output will go logic-low. Later, if a logic-low input is applied, the stored datum is erased: the *Q* output goes logic-low, and the $\bar{Q}$ output goes logic-high.

CHECKOUT OF THE GATED (TRANSPARENT) LATCH

An elaborated type of *D* latch, called a gated, strobed, or transparent latch, is also encountered by the digital troubleshooter. The basic arrangement is shown in Fig. 2–5. Observe that the latch has a strobed input AND gate which provides a "window" for timed data entry. As shown in the diagram, the strobed input functions to lock out any data from the *D* line, except for a brief duration of the strobe pulse. (It is called a transparent latch because the *Q* output follows the *D* input for the duration of the strobe pulse.)

Note the back-to-back piezo buzzers in Fig. 2–5. These are high-pitched and low-pitched buzzers which are used to indicate whether the *Q* output terminal is logic-high or logic-low. They may be a pair of Archer (Radio Shack) piezo buzzers. The high-pitched piezo buzzer is connected with its red lead to the *Q* terminal, and with its black lead connected to the $\bar{Q}$ terminal. The low-pitched piezo buzzer is connected with its black lead to the *Q* terminal, and with its red lead connected to the $\bar{Q}$ terminal.

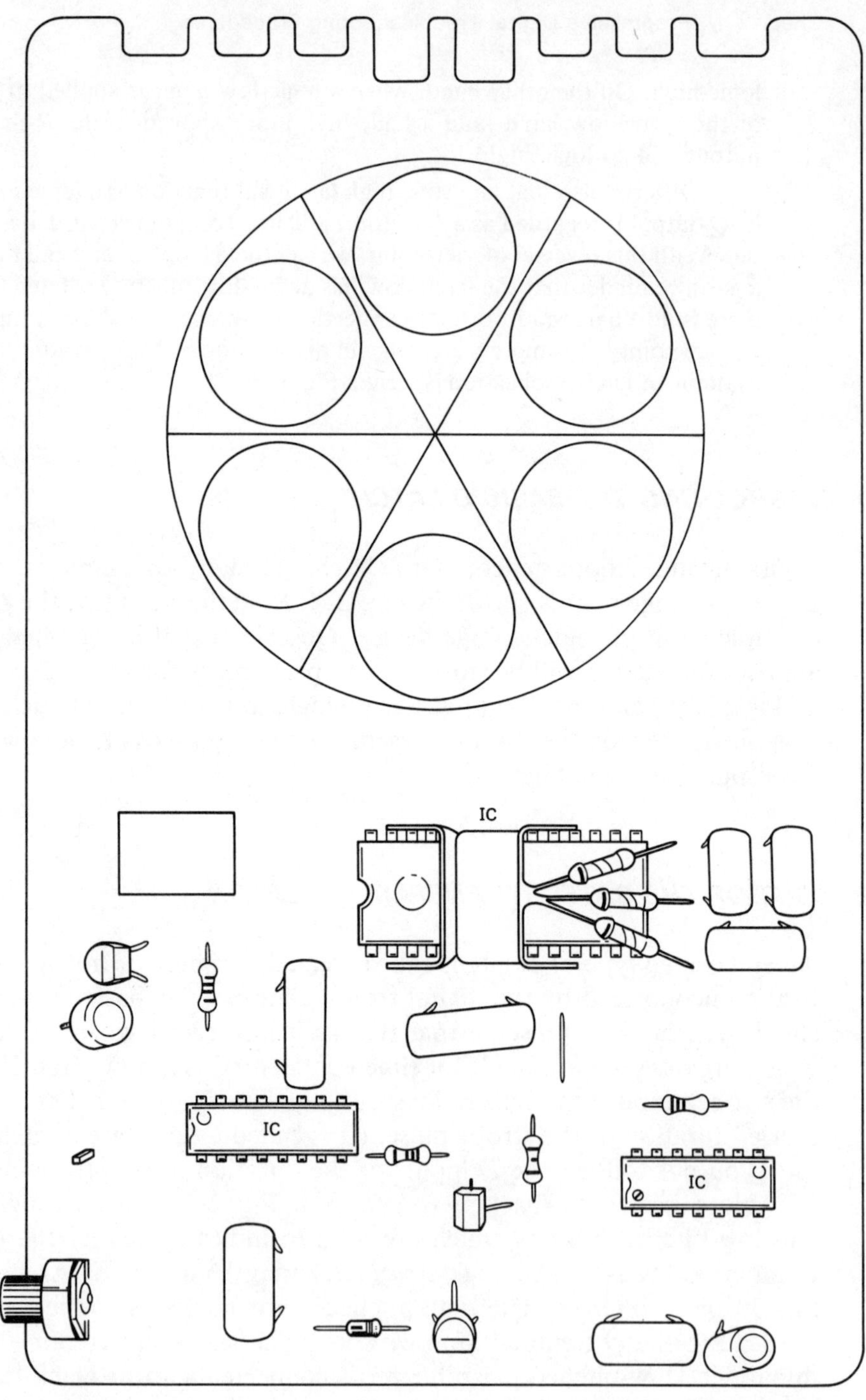

Figure 2–3 Latches are used in digital control circuits such as this motor control module.

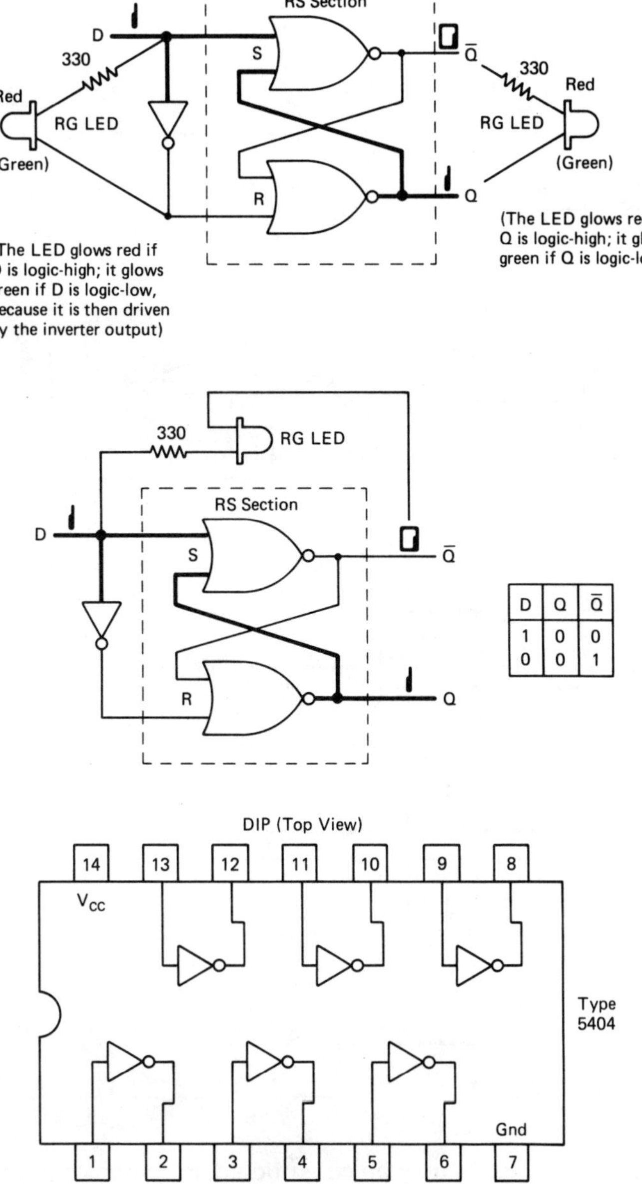

D	Q	$\bar{Q}$
1	0	0
0	0	1

Note: Troubleshooters may refer to a bistable multivibrator as a binary.

Note: Instead of using an *RG* LED at the input of the latch, and another at the output, a single *RG* LED may be connected from the *D* input to the $\bar{Q}$ output. In turn, the LED indicates the transfer states (1/0 or 0/1).

Figure 2–4 Basic *D* latch arrangement and checkout.

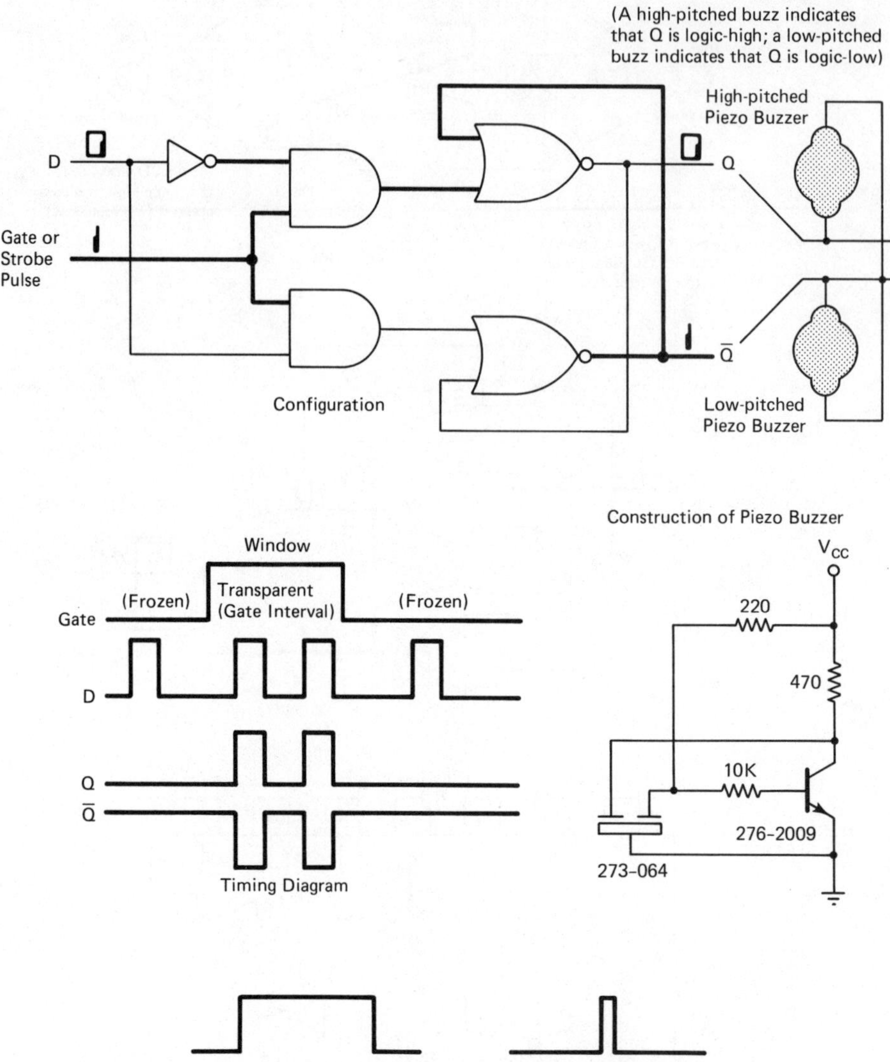

Note: A piezo buzzer may be constructed as shown. Archer (Radio Shack) part numbers are indicated. Note that zero current is drawn if V_{CC} is negative. In turn, two piezo buzzers may be connected back-to-back.

Note 1: When troubleshooting digital equipment without service data, *measure* V_{CC} *first.* It is discouraging to waste time and effort looking for a defective device or component, when the malfunction is simply due to a deteriorated power supply.

Figure 2–5 Basic gated (transparent) latch.

Note 2: A piezo buzzer used in troubleshooting digital equipment typically has an output frequency of 4.7 kHz. This is a sufficiently high frequency that acoustic standing waves are prominent in the sound field. Accordingly, if the piezo buzzer seems to be silent, it is advisable for the troubleshooter to move back or forward slightly, to determine whether he may have been listening in a "dead spot."

Figure 2–5 (*cont.*)

AND-OR GATE OPERATION

The AND gate is a basic type of gate; the OR gate is a basic type of gate. Observe that the AND-OR gate depicted in Fig. 2–6 is not a basic type of gate. In a strict sense, the AND-OR gate is a combinatorial-logic arrangement wherein basic gates are interconnected to provide a specified logic function. The AND-OR gate output will go logic-high if its inputs *A* and *B* are simultaneously driven logic-high, or, if its inputs *C* and *D* are simultaneously driven logic-high. The AND-OR gate output will also go logic-high if all four of its inputs are simultaneously driven logic-high.

Note that an *RG* (tri-color) LED may be connected in parallel with an inverter for operation as a quick checker of single-ended sources, such as an AND-OR gate. In other words, the shunt inverter functions as a digital paraphase inverter and converts a single-ended signal into a double-ended signal.

AND-OR-INVERT GATE OPERATION

Another gate arrangement called the AND-OR-INVERT gate is often encountered by the digital troubleshooter. As seen in Fig. 2–7, the AND-OR-INVERT gate is essentially an AND-OR gate followed by an inverter. Observe also the two-tone single-ended logic-level tester shown in Fig. 2–7. The red lead of one piezo buzzer is connected to the inverter input, and the black lead is connected to ground. In turn, the red lead of the other piezo buzzer is connected to the output of the inverter, and the black lead is connected to the inverter input. *This is the third step in troubleshooting digital equipment, using time-saver shortcuts.*

An ohmmeter quick check for TTL IC gate packages is shown in Fig. 2–8. This is an out-of-circuit test. As previously noted, most device faults

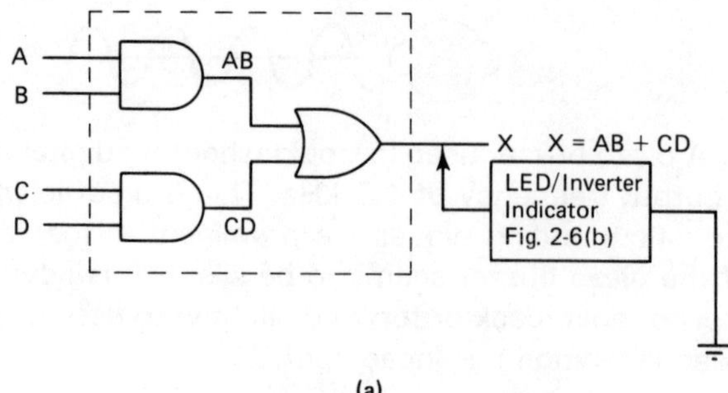

(a)

A	B	C	D	X
0	0	0	0	0
0	0	0	1	0
0	0	1	0	0
0	0	1	1	1
0	1	0	0	0
0	1	0	1	0
0	1	1	0	0
0	1	1	1	1
1	0	0	0	0
1	0	0	1	0
1	0	1	0	0
1	0	1	1	1
1	1	0	0	1
1	1	0	1	1
1	1	1	0	1
1	1	1	1	1

Note 1: More than two AND gates may be used to drive an OR gate in some digital systems.

Note 2: IC packages usually have type numbers marked on the packages. However, the numbers may sometimes be illegible. A few IC packages will be encountered which are unmarked. In-circuit identification of IC packages is accomplished to best advantage with a logic comparator, as detailed subsequently.

Note 3: Troubleshooters will encounter AND-OR gates configured from NAND gates (NAND implementation), as detailed in Chapter 3.

Figure 2–6 (a) Arrangement of an AND-OR gate. (b) Digital paraphase inverter for indicating logic-high/logic-low output states from AND-OR gate.

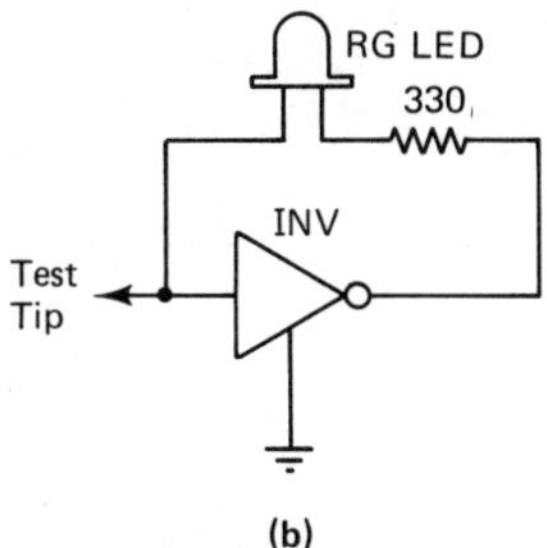

(b)

Note: This is a two-color logic-high/logic low quick checker for single-ended outputs (such as AND-OR gates) in TTL circuitry. Since the tri-color *RG* LED requires double-ended drive, an inverter is used in this tester to change a single-ended input into a double-ended output. The inverter functions as a digital paraphase inverter. It may be one section of an Archer (Radio Shack) 276-1802 hex inverter IC package. The quick checker operates as follows:

1. If the test tip is ''floating,'' the RG LED is dark.
2. If the test tip is applied to a logic-high point in a circuit, the LED glows red.
3. If the test tip is applied to a logic-low point in a circuit, the LED glows green.
4. If the test tip is applied to a point in a circuit which is outputting a high-frequency pulse train, the LED glows yellow. (If the repetition rate is greater than approximately 20 Hz, persistence of vision causes the red and green lights to look like a yellow light.)

Figure 2–6 *(cont.)*

are catastrophic; short-circuits and open-circuits are the most common failures. A short-circuit (zero resistance indication) in an ohmmeter quick check is a sure sign that the IC is defective. An open-circuit indication is usually another sure sign that the IC is defective. However, as noted in the previous chapter, there are a few exceptions. As an illustration, the 7420 dual 4-input NAND gate has NC pins 3 and 11, and these pins normally give an open-circuit indication.

JUNCTION CHARACTERISTICS

A highly practical experiment is shown in Fig. 2–9. This arrangement indicates the test voltage that is applied across a semiconductor junction by

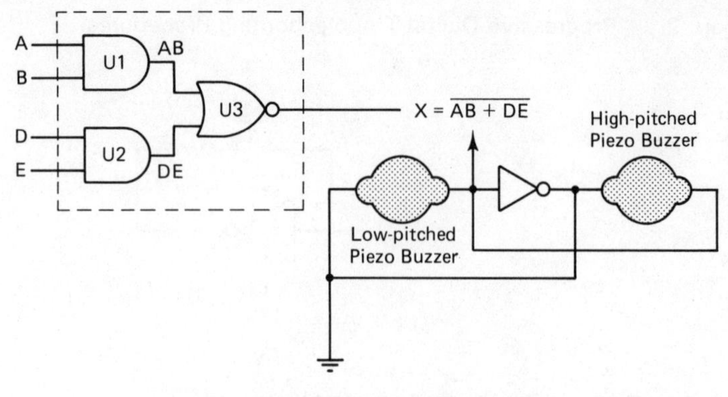

A	B	D	E	X
0	0	0	0	1
0	0	0	1	1
0	0	1	0	1
0	0	1	1	0
0	1	0	0	1
0	1	0	1	1
0	1	1	0	1
0	1	1	1	0
1	0	0	0	1
1	0	0	1	1
1	0	1	0	1
1	0	1	1	0
1	1	0	0	0
1	1	0	1	0
1	1	1	0	0
1	1	1	1	0

Note 1: More than two AND gates may be used to drive the NOR gate in some digital systems.

Note 2: When a large lnumber of inputs are required to an AND gate, an EXPANDABLEAND gate is commonly used, as explained subsequently.

Note 3: An AND-OR-INVERT gate is also called an AOI gate.

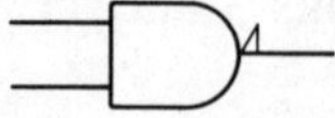

Note 4: Troubleshooters who may be working on obsolescent digital logic equipment will occasionally find a NAND gate symbolized with a "triangle" instead of a "bubble" at the output. These two symbols mean the same thing and denote that the device is a NAND gate.

Figure 2-7 Arrangement of an AND-OR-INVERT gate.

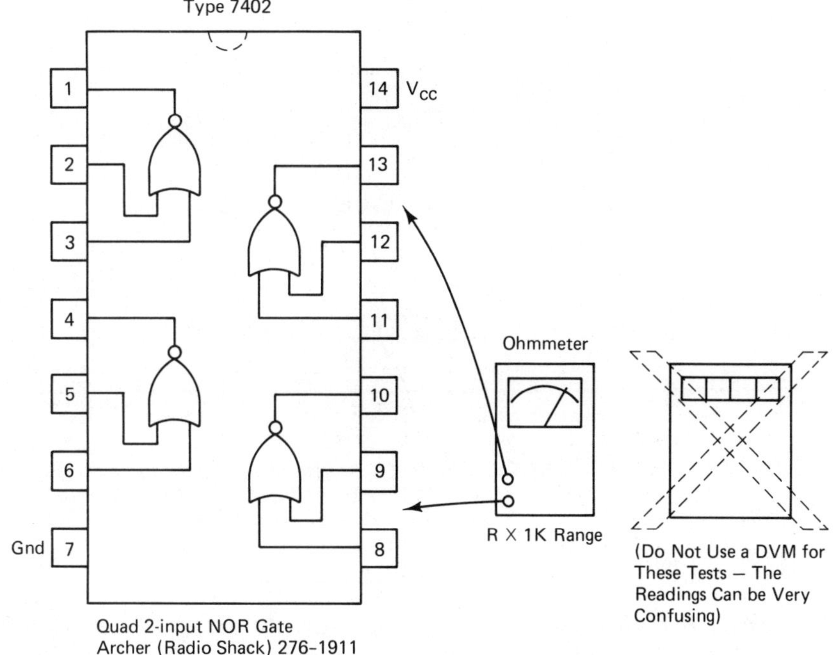

With Pin 7 Negative:

Infinite resistance to all pins (except pin 14).
(Otherwise, IC is defective).

28 kilohms to pin 14 (in this exapmle).
(If infinite or zero resistance, IC is defective).

With Pin 14 Negative:

Infinite resistance to all pins (except pin 7).
(Otherwise, IC is defective).

28 kilohms to pin 7 (in this example).
(If infinite or zero resistance, IC is defective).

With Pin 7 Positive:

Resistance to all pins in the 4-to-10 kilohms range.
(If infinite or zero resistance, IC is defective).

With Pin 14 Positive:

Resistance to all pins in the 15-to-40 kilohms range.
(If infinite or zero resistance, IC is defective.)

Note: The resistance ranges noted in this example are typical. Other ICs may have other resistance ranges. *In any case, with pin 7 or pin 14 negative, infinite resistance must be measured to all pins (except pin 14 or pin 7). Also with pin 7 or pin 14 positive, finite resistance must be measured to all pins.*

This is an out-of circuit test. A companion in-circuit test with V_{CC} applied to the IC is detailed in Chapter 3.

Figure 2–8 How a normally operating NOR gate IC package “looks” to an ohmmeter.

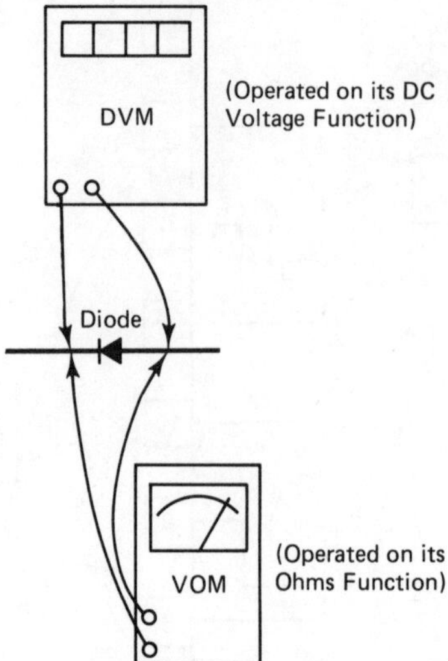

Note: A DVM has a very high input resistance and essentially indicates the true test voltage applied across the diode by the ohmmeter. Observe that the value of test voltage applied by a particular ohmmeter will depend on the resistance range to which the ohmmeter is set.

Figure 2–9 Various ohmmeters apply different test voltages across semiconductor junctions, and indicate different resistance values.

an ohmmeter. It shows that a widely different resistance value may be indicated by the same ohmmeter when operated on various ranges. It also shows that widely different resistance values may be indicated by different ohmmeters, although operated on the same range.*

For example, if this experiment is made with a typical 1000 ohms-per-volt VOM, a 20,000 ohms-per-volt VOM, and a 50,000 ohms-per-volt VOM, the following resistance readings and applied test voltages are representative:

1000 ohms/volt VOM	*20,000 ohms/volt VOM*	*50,000 ohms/volt VOM*
Applied test voltage = 0.24 V	Applied test voltage = 0.16 V	Applied test voltage = 0.19 V
Indicated resistance = 500 Ω	Indicated resistance = 3,000 Ω	Indicated resistance = 1,600 Ω

*A constant-current ohmmeter may be constructed, as subsequently explained.

OHMMETER QUICK CHECK OF D LATCH

The pinout for a typical commercial TTL latch package was shown in Fig. 2–1. Insofar as ohmmeter quick checks are concerned, the same principles apply as in ohmmeter checks of gate packages. Latch packages follow the 7/11 and 8/12 ground rules. For example, the 5400 latch package depicted in Fig. 2–1 has a pin-8 ground. The 7475 4-bit latch has pin-12 ground.

BASIC LATCHES IN 14-PIN IC PACKAGES

Troubleshooters will find very few latches in 14-pin IC packages. The chief exception is the 54/7477 quad *D*-type latch, which provides Q outputs but lacks $\bar{Q}$ outputs.

BASIC LATCHES IN 16-PIN IC PACKAGES

Most latches are provided in 16-pin IC packages. Examples are a 4-bit bistable latch (*D*-type), a dual 4-bit addressable latch, a quad $\bar{S}$-$\bar{R}$ latch, and an 8-bit addressable latch.

GATES IN 16-PIN IC PACKAGES

Some gate combinations are encountered in 16-pin IC packages. For example, the quad EXCLUSIVE-OR/NOR arrangement depicted in Fig. 2–10 is provided in a 16-pin package. Another example is a 13-input NAND gate. Note that AND-OR gates, AND-OR-INVERT gates, and the expandable AND-OR and AND-OR-INVERT gates are provided in 14-pin IC packages.

THE 10/12/14 GROUND RULE

Digital troubleshooters are chiefly concerned with 14-pin and 16-pin IC packages. However, 20-pin, 24-pin, and 28-pin packages will occasionally be encountered. Here, the 10/12/14 ground rule applies; in other words, if the IC package has 20 pins, pin 10 will be the ground terminal. If the pack-

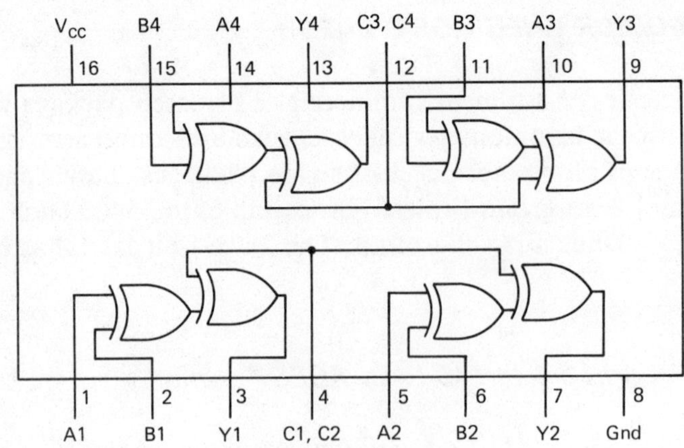

Truth Table

Inputs			Output
A	B	C	Y
0	0	0	0
0	1	0	1
1	0	0	1
1	1	0	0
0	0	1	1
0	1	1	0

$Y = A + B + C$ or, $Y = A\bar{B}\bar{C} + \bar{A}B\bar{C} + ABC$

Note: These gates can be implemented from NAND gates. Any gate can be implemented from NAND gates. Any gate can also be implemented from NOR gates. However, no gates can be implemented from AND gates or OR gates.

De Morgan's Theorem

$$\bar{A}\bar{B}\bar{C} = \overline{A + B + C} \qquad \bar{A} + \bar{B} + \bar{C} = \overline{ABC}$$

Figure 2–10 Quad EXCLUSIVE-OR/NOR gate IC package.

age has 24 pins, pin 12 will be the ground terminal. If the package has 28 pins, pin 14 will be the ground terminal. V_{CC} will always be the last pin.

Observe also that NC (no-connection) pins are present on a few 20, 24, and 28-pin packages. Accordingly, if a quick test indicates that a pin is open-circuited, it should not be concluded that a fault is present until a cross-check is made on the solder side of the PC board to determine whether or not a PC conductor is connected to the open-circuited pin.

"FALSE ALARM" NODES

It is a general rule that all input and output terminals on TTL gates present infinite resistance to the positive test lead of an ohmmeter. However, there is an occasional exception that the troubleshooter should keep in mind. For example, expandable AOI gates are encountered, as exemplified in Fig. 2–11. An expander is connected to an expandable gate in order to increase the number of inputs.

The "tricky" aspect of this arrangement from the troubleshooter's viewpoint is the fact that the expander inputs on the expandable gate do not present infinite resistance to the positive test lead of an ohmmeter. In turn, an ohmmeter quick check of these nodes will result in a "false alarm." The bottom line is that ohmmeter quick checks are not completely reliable in all cases—in doubtful situations, follow-up tests are required.

COMMON FAULTS

When preliminary tests show that an open circuit is present, the fault may be either inside of the IC, or it may be located in the associated PC circuitry. Open circuits outside of an IC are typically caused by cold-soldered joints, cracked PC conductors, poor pin contact in an IC socket, or a defective plated-through hole in the PC board. IC packages with a comparatively large number of pins are sometimes inserted into an IC socket with one pin bent at right angles under the package. On rare occasion, a pin breaks between the socket and the IC package—the break can easily be overlooked until the IC is unplugged from its socket. (See also Fig. 2–12).

Bottom Line: One of the "secrets" of efficient use of shortcuts is the hands-on experience gained by making the tests and experiments described in the previous pages.

INTERMITTENT MONITORING

Digital troubleshooting without service data is particularly aggravated by intermittent operation of circuitry. Fortunately, intermittent problems are not routinely encountered. However, when the troubleshooter must contend with intermittent circuitry, a logic-low probe such as shown in Fig. 2–13 can

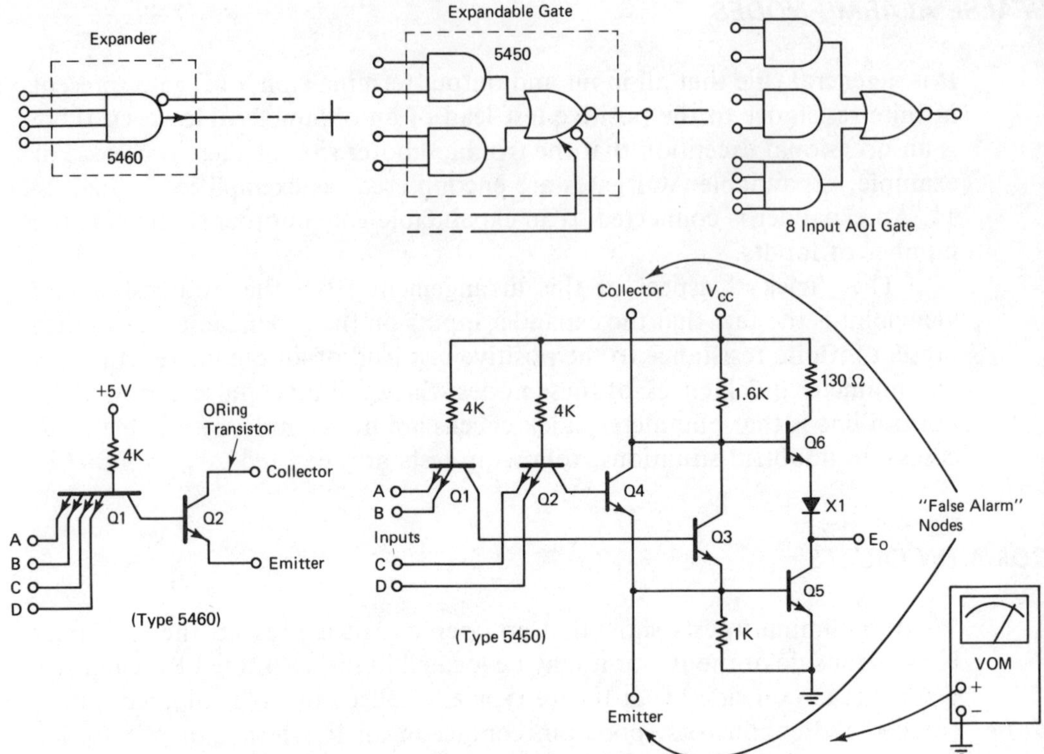

Note 1: Inputs *A,B,C,D* pass no current when a positive test voltage is applied. Output E_O passes no current when a positive test voltage is applied. However, the base of *Q*5 draws current when a positive test voltage is applied. Similarly, the 1 K resistor draws current. The base of *Q*6 and the 1.6 K resistor draw current. In turn, the expander inputs on the expandable gate are ohmmeter quick-check "false-alarm" nodes.

Note 2: A digital paraphase inverter indicator may be used to check the circuit action of expanders and expandable gates.

Figure 2–11 Expander provides additional inputs to AND-OR-INVERT gate.

be very helpful. It frees the troubleshooter by monitoring the circuit and giving a visual and/or audible indication of signal transition from a logic-high to an erroneous logic-low state—even for a fraction of a second.

If a recording voltmeter (as previously described) is included in the intermittent monitoring arrangement, a permanent record of faulting is provided. In turn, the troubleshooter can review the time factor(s) that are

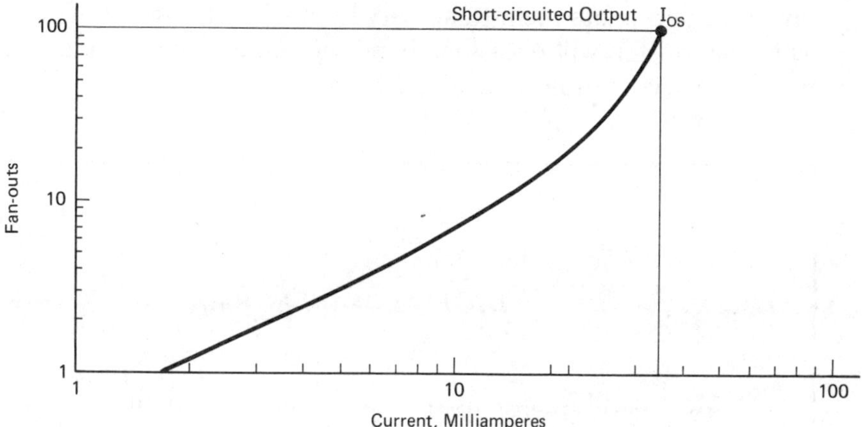

Note: The normal output short-circuit current for a TTL device is subject to a fairly wide tolerance. For example, a NAND gate may have an I_{OS} from 18 to 55 mA. A short-circuited output is similar to a fan-out of 100, approximately.

Figure 2–12 Fan-out current demands and short-circuit current demand.

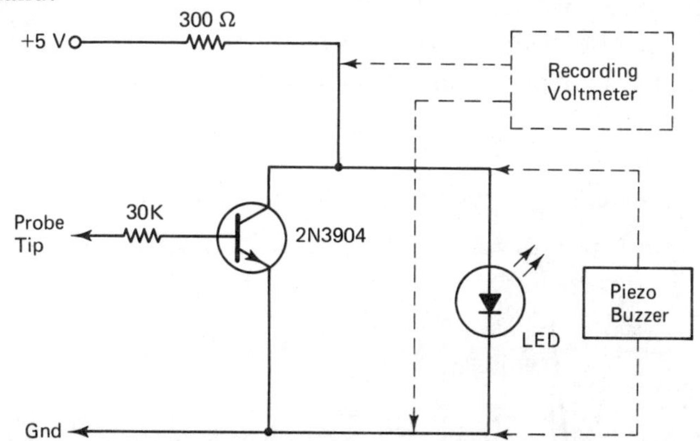

Note 1: The LED may need to be selected for complete darkness when the probe tip is logic-high. If the probe tip is logic-low, the LED glows (it also glows if the probe tip is "floating"). The piezo buzzer provides audible logic-low indication; it may be a Mallory 5C628.

Note 2: This intermittent monitor provides visual indication. A recording voltmeter may be included to provide a record of the time (or times) that the input signal faulted, and the length of time over which the fault(s) occurred. A piezo buzzer may also be included for real-time audible indication of fault occurrence.

Figure 2–13 Logic-low probe, used for intermittent monitoring.

involved, and relate them to activity in other sections of the digital system. This relationship will occasionally provide clues concerning the cause of an intermittent condition. (See also Chart 2-2.)

Chart 2-2

DIGITAL WAVEFORMS

Most oscilloscopes used in digital troubleshooting procedures are comparatively sophisticated. Some high-performance oscilloscopes are combined time-frequency domain and data-domain display instruments. Data-domain instruments, generally called logic state analyzers, are employed to monitor binary digits (bits), digital words, digital addresses, and digital instructions and to display the information as data sequences. The display is in binary form of 1s and 0s arranged in columns and rows on the CRT screen.

A time-domain display of a digital clock signal, with typical distortions, is depicted in the following illustration.

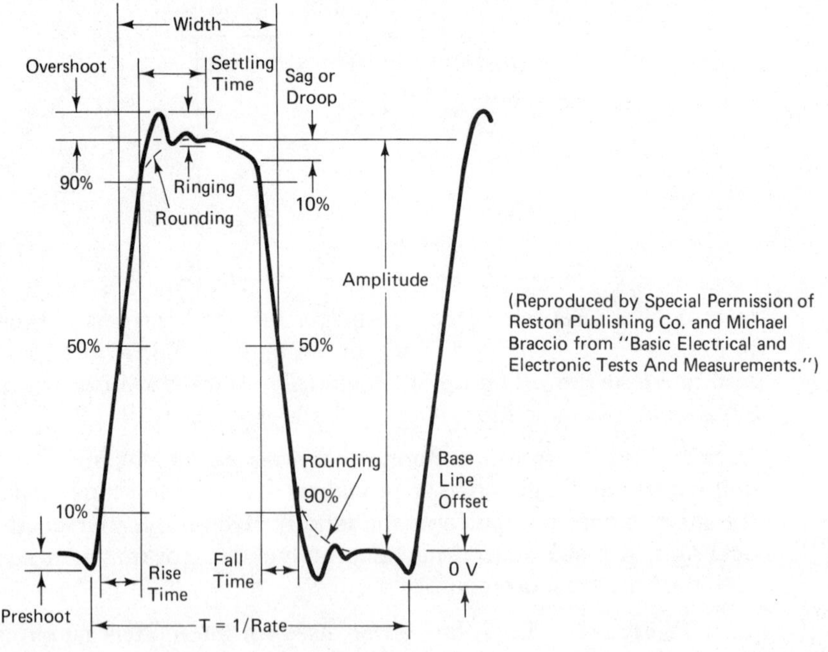

(Reproduced by Special Permission of Reston Publishing Co. and Michael Braccio from "Basic Electrical and Electronic Tests And Measurements.")

By way of comparison, a data-domain display of digital events in a data stream is exemplified next. This data-domain display is also called a window of digital events.

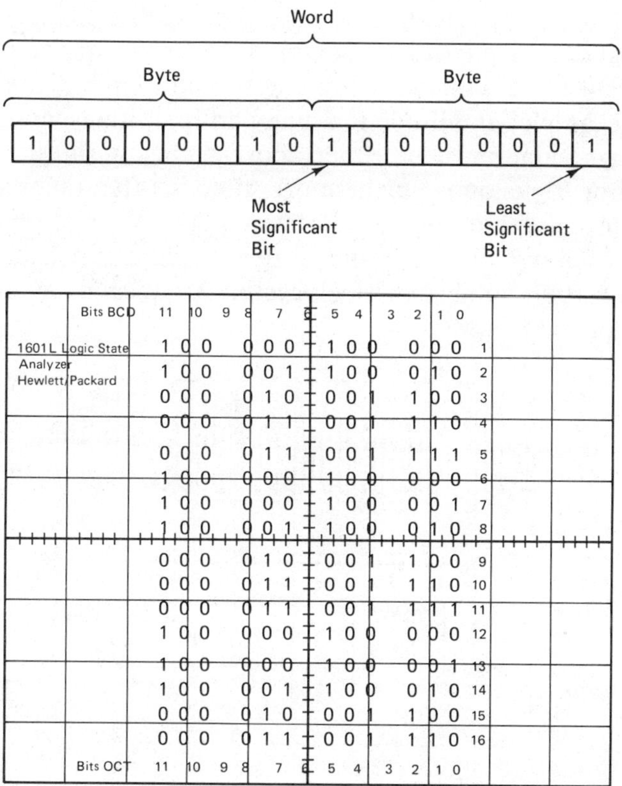

Electrical and functional analyses are not separable, but each is used to complement the other. As an illustration, only when word flow is incorrect as determined from a functional display need the operator concern himself with the voltage conditions that produced the digital words. Even when word-flow errors require electrical analysis, the number of signal nodes in the vicinity of the error complicates the use of an oscilloscope.

It is helpful to define the oscilloscope functions of probing, triggering, and display in terms of words versus event or sequence, or words versus time, rather than in terms of volts versus time. The traditional analog picture of absolute versus sweep time provides a careful analysis of electrical parameters. This is because the important information—amplitude versus time—is carried by the waveform. This method can help to decipher noise, ringing, spikes, constant dc voltage levels, voltage swings, and so on. Furthermore, it is the analysis domain in which the majority of operators are most experienced and have the greatest confidence.

Digital information is often nonrepetitive. Extremely long (and rapid) data sequences are common. Moreover, parameters that are significant for analog analysis are less important in digital measurements. For example, amplitude is usually an important parameter in that the voltage value must be either above or below threshold values (logic-high or logic-low). Furthermore, time is often unimportant in an absolute sense, although time becomes critical when related to the clock rate of a system. As an illustration, a pulse may normally be delayed by a certain number of clock cycles. A typical delay of 20 clock cycles is depicted below.

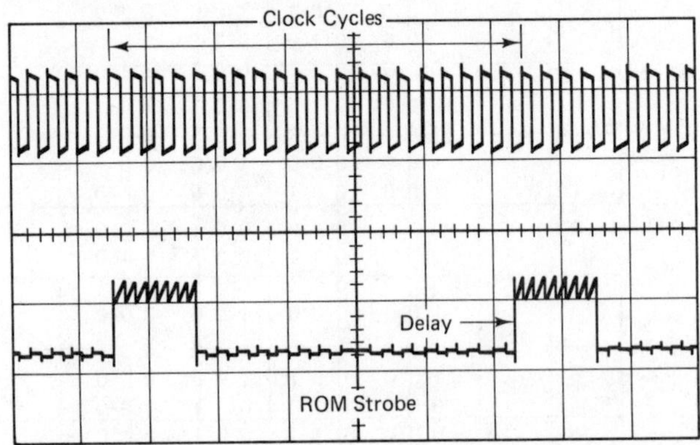

In other words, a functional measurement consists of an observation of digital information (logic-high or logic-low) versus system (clock) time. In turn, a hierarchy of logic-state test and measurement levels can be established. Each of these levels provides only the information necessary for that particular level of digital checkout. It follows that to effectively cope with digital circuit activity, a logic state analyzer must meet several basic requirements:

1. Data must be read and displayed in binary format for ease of reading and with no interpretation required.
2. A sufficient number of inputs are necessary, so that the entire data word can be monitored at the same time.
3. A trigger point is required that is related to a unique data word within a sequence.

4. Digital delay is needed to position the display window to the desired point in time from the reference (trigger word).
5. Digital storage is needed to retain single-shot events along with negative time (the data leading up to a desired trigger point).

Digital signals are almost invariably multiline and are accordingly difficult to interpret from a volts-versus-time display (the operator is concerned only with logic states versus system time). A typical logic state analyzer meets this requirement by displaying digital words that are 32, 16, or 12 bits wide versus system time in a tabular display that is easy to comprehend when the operator is examining functional relationships.

As illustrated above, tabular displays are in terms of logic highs (1s) and logic lows (0s) versus the clock signal. Triggering is accomplished by employment of trigger-word switches that permit selection of a unique trigger point. Note also that the display may be moved in system time from the trigger point by utilizing digital delay in either a positive or a negative direction. An illustration of setup time for a device is shown in the following.

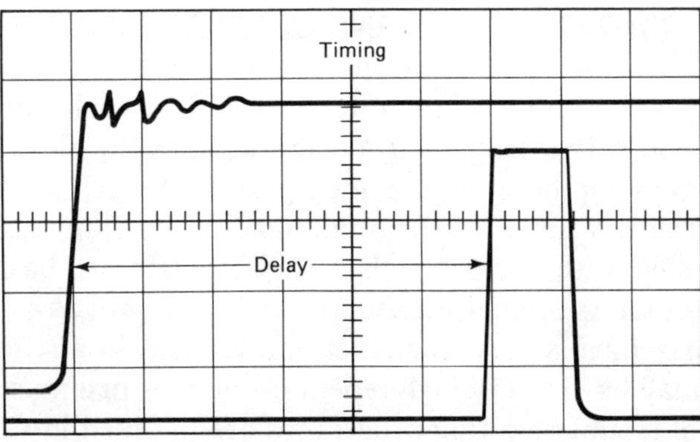

3

Comparison and Quasi-Comparison Quick Tests

TROUBLESHOOTING FLIP-FLOPS AND CLOCKS

Digital troubleshooters are concerned with asynchronous devices such as latches, and with synchronous devices such as flip-flops. (The basic *D* latch is not a flip-flop because it is not clocked.) A latch responds to a test pulse at any arbitrary time. On the other hand, a flip-flop responds to a test pulse only in step with a clock pulse. Clock pulses are basically square waves (troubleshooters often use a square-wave generator as a clock subber). Clock pulses may have a very low repetition rate, as in a scanner-monitor radio, or they may have a very high repetition rate, as in a microcomputer. Follow-up digital troubleshooting procedures are outlined in Chart 3–1.*

Clock pulses ideally have a rectangular waveshape. Troubleshooters will find that clock pulses may become distorted in various ways, as they proceed through elaborate digital networks (the distortion is not necessarily a trouble symptom). Fig. 3–1 shows a basic flip-flop configuration—this is a clocked *RS* bistable multivibrator arrangement. Observe that the latch function is necessarily locked in step with the clock input. This is an active-low ar-

*Manual clocking of digital networks can greatly facilitate preliminary troubleshooting procedures.

Chart 3–1

FOLLOW-UP DIGITAL TROUBLESHOOTING PROCEDURES (WITHOUT SERVICE DATA)

1. Clock line identification: The clock line is the PC conductor which exhibits digital activity at all times and under all conditions of system operation. (There is no standardization regarding clock input pins on IC packages).
2. When making comparison tests on similar units of digital equipment and clocking both units from a clock subber, remember that long clock lines have a very low internal resistance. The clock subber must provide sufficient output energy to drive this low value of load resistance.
3. In "tough dog" situations, it is often helpful to stop the clock, and to clock the system manually by means of a test lead from the clock line which is connected alternately to V_{cc} and to ground.
4. When manually clocking a digital system, it is desirable to have a number of paraphase digital indicators with micro-clip leads available. These indicators can be connected from various nodes to ground, with a resulting "light map" of system operation from one clock pulse to the next. (Light-mapping techniques are explained later.)
5. If in doubt whether data is flowing into, or out of, an IC pin, apply a lab-type DVM at separated points along the data line. With data pulses flowing in the line, the direction of data flow will be from the positive point to the negative point along the conductor. (This topic is subsequently explained in greater detail.)
6. Remember that wire-AND circuitry produces "false alarms" on various quick tests. (Look for an external pull-up resistor near the associated ICs on the PC board.)
7. Although nearly all digital circuitry is direct-coupled, there is an occasional exception which the troubleshooter should keep in mind. Older designs of digital systems with monostable multivibrators employed circuitry with time-constant capacitors. In turn,

the data processing would "crash off in the weeds" if the clock frequency was reduced. However, newer designs of digital systems employ one-pulse circuits such as shown below, instead of RC monostable multivibrators. In turn, the troubleshooter may reduce the clock frequency as much as desired, without affecting data-processing action.

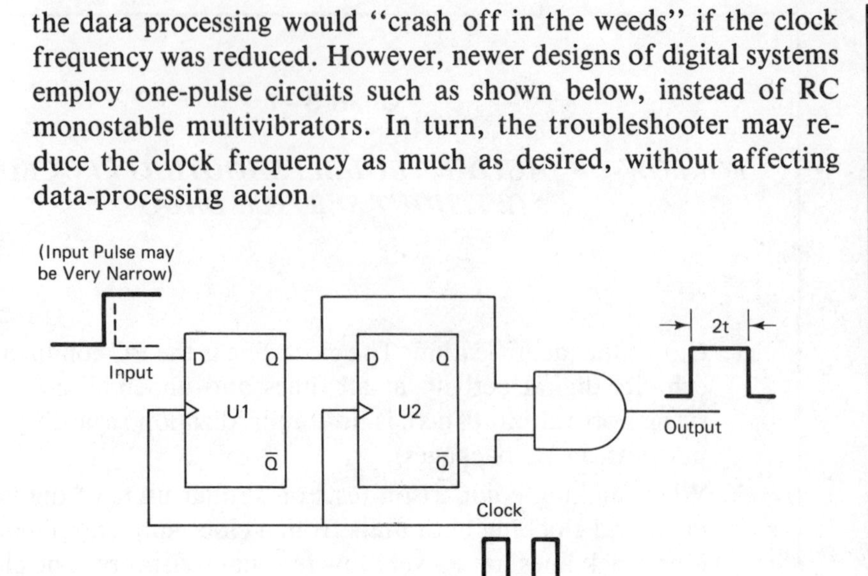

rangement; in other words, the R and S outputs can be complemented only while the clock is logic-high.

A commercial D-type flip-flop is illustrated in Fig. 3–2. Digital troubleshooters need to be familiar with this form of logic circuitry. This is the type 9H74 flip-flop. Observe that the configuration has clear and preset inputs. The Clear input enables the flip-flop to be reset at any time ($Q = 0$) by pulsing the Clear or Reset input logic-low. Similarly, the Preset input permits the flip-flop to be set at any time ($Q = 1$) by pulsing the Preset or set input logic-low.

Observe that although the flip-flop is clocked and that its Q and $\bar{Q}$ outputs can respond to D input pulses only in step with the clock, the Preset (S_D) and Clear (R_D) inputs are asynchronous. In other words, the flip-flop can be preset or cleared at any arbitrary time, without regard to the clock pulses. Note also that the truth table for the flip-flop is stated with respect to bit time before the clock pulse, and with respect to bit time after the clock pulse.

This logic configuration is called a D-type edge-triggered flip-flop. As detailed subsequently, edge-triggering differs from level triggering. Level triggering corresponds to a logic-high level of the clock pulse. On the other

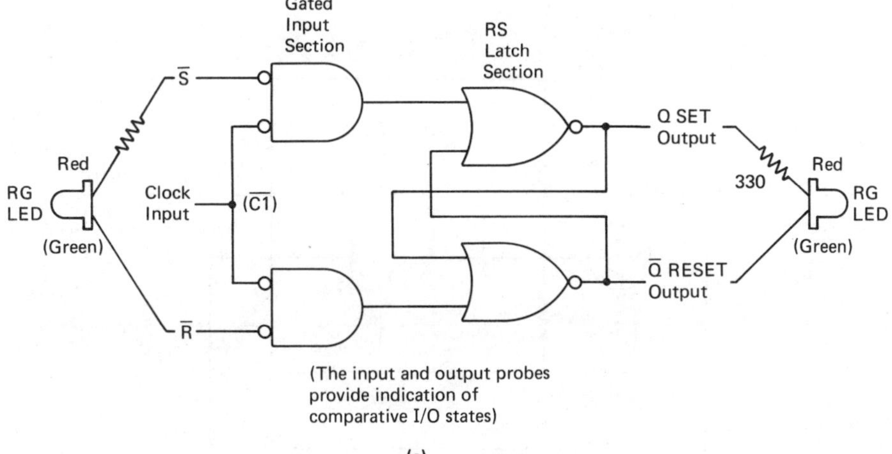

(a)

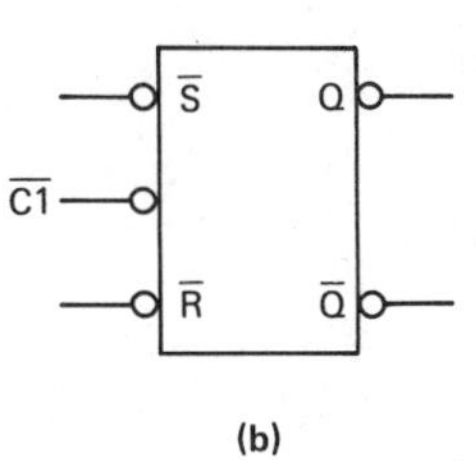

(b)

Input			Command	Output	
$\overline{S}$	$\overline{R}$	$\overline{C1}$		Q	$\overline{Q}$
0	0	0	(Invalid)		
0	0	1		NC	NC
0	1	0	Set	1	0
0	1	1		NC	NC
1	0	0	Reset	0	1
1	0	1		NC	NC
1	1	0	Remember	NC	NC
1	1	1		NC	NC

(c)

Note 1: **The two-color probe will have a yellow glow if the clock frequency exceeds the persistence of vision (approximately 20 Hz). This is because a yellow light is produced when red and green lights are mixed.**

Note 2: **Input terminals will ''look'' logic-high if permitted to float.**

Note 3: **The inversion bars over *R, S,* and *C*1 denote that these inputs are active-low.**

Figure 3–1 Basic clocked RS flip-flop arrangement. (a) Logic diagram. (b) Flip-flop symbol. (c) Truth table. (d) Clocks. (e) Determination of whether a probe is indicating a dc level or a pulse train.

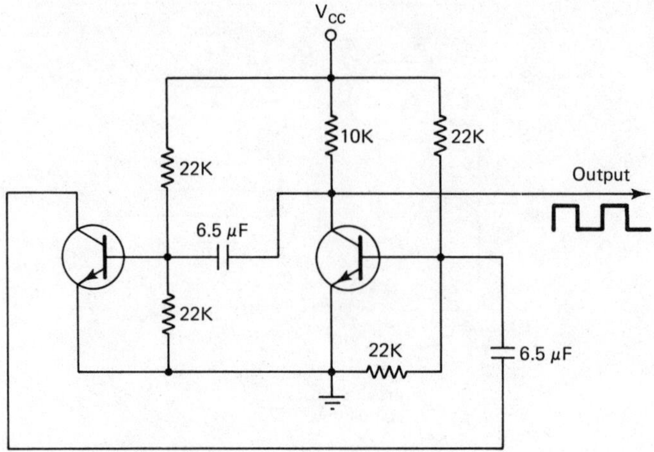

Clock Configuration for a Typical Scanner-monitor Radio.

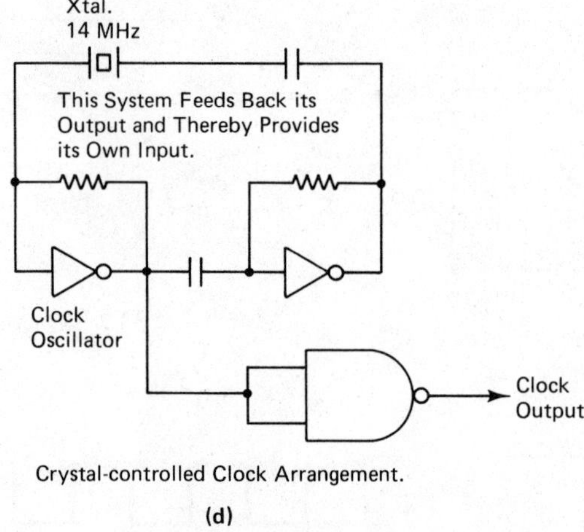

Crystal-controlled Clock Arrangement.

(d)

Note 1: Although clock circuits often contain capacitors, the remainder of a digital system is usually direct-coupled throughout. Accordingly, a digital system will ordinarily operate normally at any reduced clock speed.

Note 2: A clock configuration is typically an astable multivibrator. It generates a square-wave output. Clocks that operate at higher frequencies are usually crystal-controlled. A clock in a scanner-monitor radio typically operates at 4 Hz. On the other hand, a clock in a video game may operate at 20 MHz. A clock in a microcomputer may operate at 2 or 3 MHz.

Note 3: When making comparison tests, digital troubleshooters often use a square-wave generator as a clock subber.

Figure 3-1 (*cont.*)

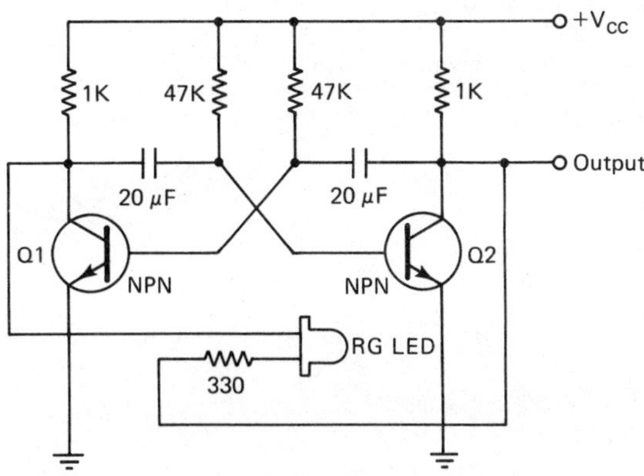

Experimenter's Clock (Slow Pulse Generator)

(d) cont.

Note 4: The easiest way to check clock frequency is with a frequency counter. Alternatively, an oscilloscope may be used.

Note 5: An *RG* LED glows red when *Q*1 is logic-high, and glows green when *Q*2 is logic-high, for example. It is basically a two-color LED. However, troubleshooters often call it a three-color LED because it glows yellow if the clock frequency exceeds persistence of vision.

Figure 3–1 *(cont.)*

hand, edge triggering corresponds solely to the rising edge (or the trailing edge) of the clock pulse. In the case of the 9H74 flip-flop, information at the *D* input is transferred to the *Q* output on the positive-going edge of the clock pulse. In turn, while the clock waveform proceeds along its logic-high level or along its logic-low level, the *D* input is locked out. The only instant at which the *D* input is not locked out is during the rapid rise of the positive-going edge of the clock pulse.*

We will find that other types of edge-triggered flip-flops transfer information at the *D* input to the *Q* output on the negative-going edge of the clock pulse. The advantage of an edge-triggered flip-flop is that the data line can feed a number of flip-flops simultaneously without confusion, provided only that the pertinent information for each flip-flop is correctly timed on the data line.

*Edge-triggering, level-triggering and other flip-flop details must be carefully observed in selection of replacement IC flip-flop packages.

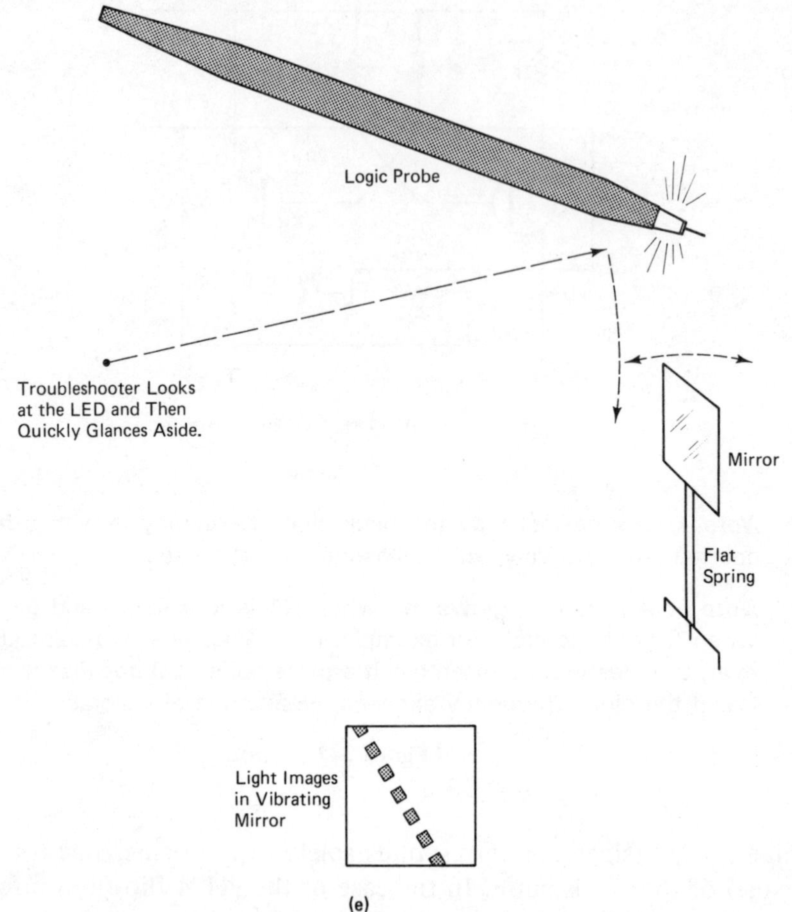

(e)

Note: If the troubleshooter sees a series of light images when he glances aside, he knows that the probe is indicating a pulse train. Or, if he sees only one light image, he knows that the probe is indicating a dc voltage level, or that the pulse train has such a high frequency that it cannot be resolved in this test.

A better test can be made with a vibrating mirror. The troubleshooter looks at the reflection of the LED in the vibrating mirror. If he sees a train of ''fireballs'' in the mirror, he knows that the probe is indicating a pulse train.

Figure 3–1 (*cont.*)

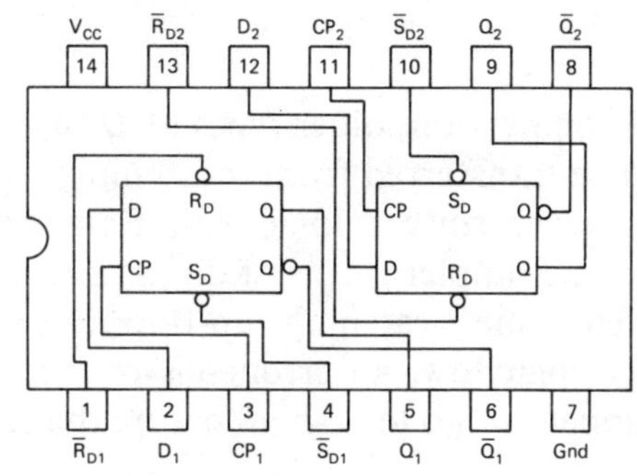

Logic Diagram (Each Flip-flop)

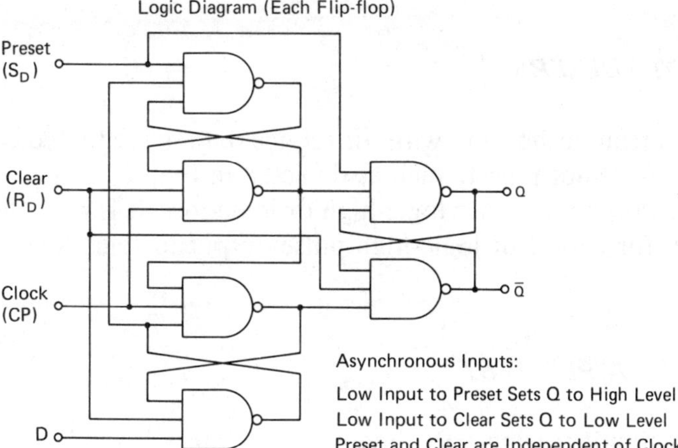

Truth Table (Each Flip-flop)

t_n	t_{n+1}	
Input D	Output Q	Output $\overline{Q}$
L	L	H
H	H	L

H = High Level, L = Low Level

Notes: t_n = Bit Time Before Clock Pulse.
t_{n+1} = Bit Time After Clock Pulse.

Note 1: To check comparative responses of similar (although not necessarily identical) flip-flops, connect corresponding inputs in parallel, and check the outputs with LED indicators.

Note 2: This is a dual flip-flop package inasmuch as the IC contains two flip-flops. Each flip-flop has separate Presets, Clears, and Clocks. The troubleshooter will also encounter flip-flops that do not have separate Presets and/or Clears and/or Clocks.

Figure 3–2 A commercial *D*-type flip-flop.

By way of comparison, older types of *D* flip-flops were level-triggered, and data would be transferred from the *D* input to the *Q* output at any time while the clock waveform was logic-high. Some level-triggered *D* flip-flops are still in use in the simpler forms of digital circuitry. A level-triggered flip-flop is also called a one's-catching flip-flop; this means that if the flip-flop were first driven logic-low, an erroneous output would result in case that the data line happens to go logic-high before the clock waveform goes logic-low.

SINGLE-SHOT PULSERS

Digital troubleshooting without service data is facilitated by the availability of a single-shot pulser, such as depicted in Fig. 3–3. This pulser can drive a *D* flip-flop input either logic-high or logic-low. It is provided with an SPDT switch for choice of logic-high pulse output or logic-low pulse output.

CONTINUITY "WIPE" TEST

When circuit-tracing a PC board, a continuity "wipe" test such as depicted in Fig. 3–4 can speed up the procedure. The "beeper" in the low-power ohmmeter indicates when the test prods are being applied to the same interconnect. In the case of involved networks, it is advisable for the troubleshooter to make a sketch of the ICs that are included in the network (such as *U*1 and *U*2 in Fig. 3–4). Then, each time that continuity is indicated between a pair of IC pins, an interconnecting line may be drawn on the sketch. *This is the fifth step in digital troubleshooting without service data, using time-saving shortcuts.*

NAND IMPLEMENTATION

Digital troubleshooters need to recognize that logic circuitry often employs NAND implementation, as exemplified in Fig. 3–5. Thus, two NAND gates may be configured as an AND gate; three NAND gates may be configured as an OR gate; five NAND gates may be configured as an XOR gate; three NAND gates may be configured as an AND-OR gate; four NAND gates may be configured as an AND-OR-INVERT gate.

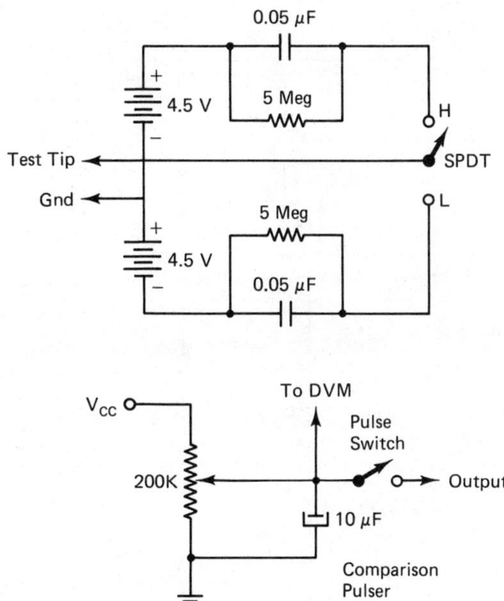

Note 1: This pulser operates by capacitor-charging when the test tip is applied at the input of a device. The 5-mΩ resistors serve to discharge to capacitors in between tests. The input of a device can be alternately pulsed logic-high and logic-low by throwing the SPDT switch back and forth.

Note 2: If it becomes necessary to pulse very low impedance digital circuits, this pulser may not provide sufficient energy to drive the circuit. In such a case, the charging capacitors can be paralleled with larger capacitors in order to store more energy.

Note 3: A comparison pulser is highly informative when troubleshooting marginal malfunctions. The pulser can be adjusted to determine whether a higher peak drive voltage is needed. It also shows the minimum peak voltage that is required for reliable operation.

Figure 3–3 Single-shot digital pulsers.

TRACING CIRCUIT-BOARD WIRING

When troubleshooting digital circuit boards without service data, it is frequently necessary to trace more or less of the wiring. As depicted in Fig. 3–6, the printed-circuit conductors can easily be viewed from the component side of the board if a bright light is placed behind the board. Since the board or card is translucent, light shines through readily.

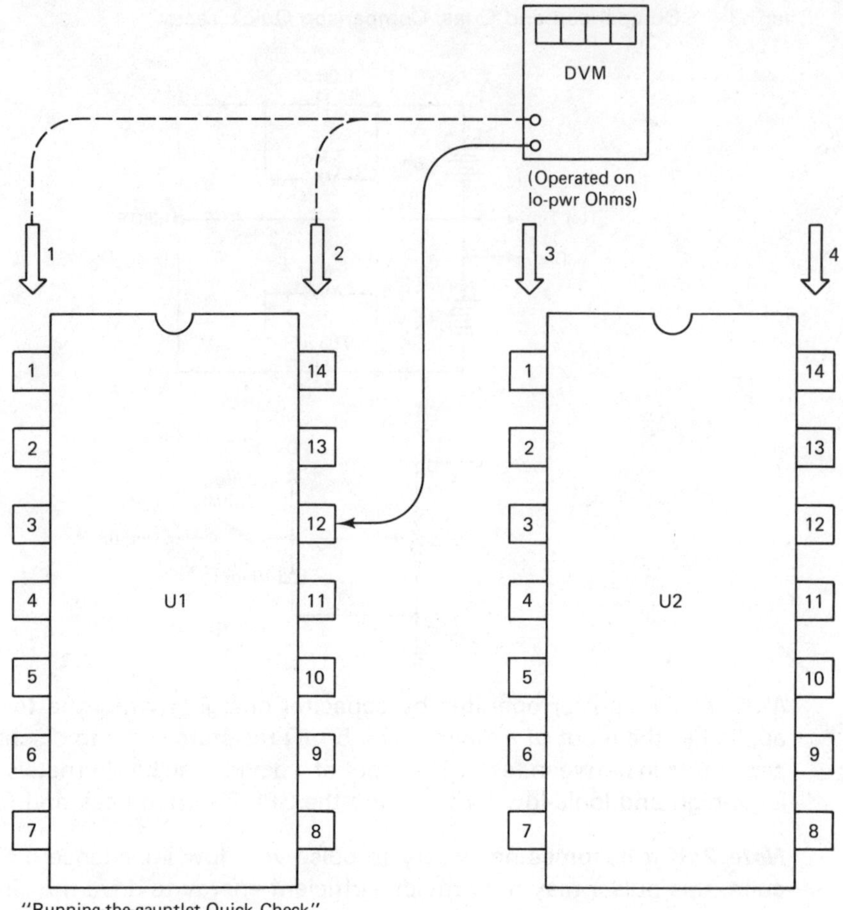

Note 1: This type of test is made with V_{CC} switched off from the circuit board that is being "buzzed out."

Note 2: In this example, the interconnections to terminal 12 of *U*1 are "buzzed out." A DVM with a continuity "beeper" is used. One lead from the DVM is connected to terminal 12 of *U*1. In turn, the other lead from the DVM is "wiped" through terminals 1–7 of *U*1 as indicated at (1). Then, the DVM lead is "wiped" through terminals 8–14 of *U*1 as indicated at (2). The DVM lead is then "wiped" through terminals 1–7 of *U*2 as indicated at (3). Finally, the DVM lead is "wiped" through terminals 8–14 of *U*2 as indicated at (4).

When the continuity "beeper" sounds, the technician knows that the DVM leads are applied at the ends of an interconnection (or to a V_{CC} terminal).

Figure 3–4 A handy "beeper" test for buzzing out PC board conductors.

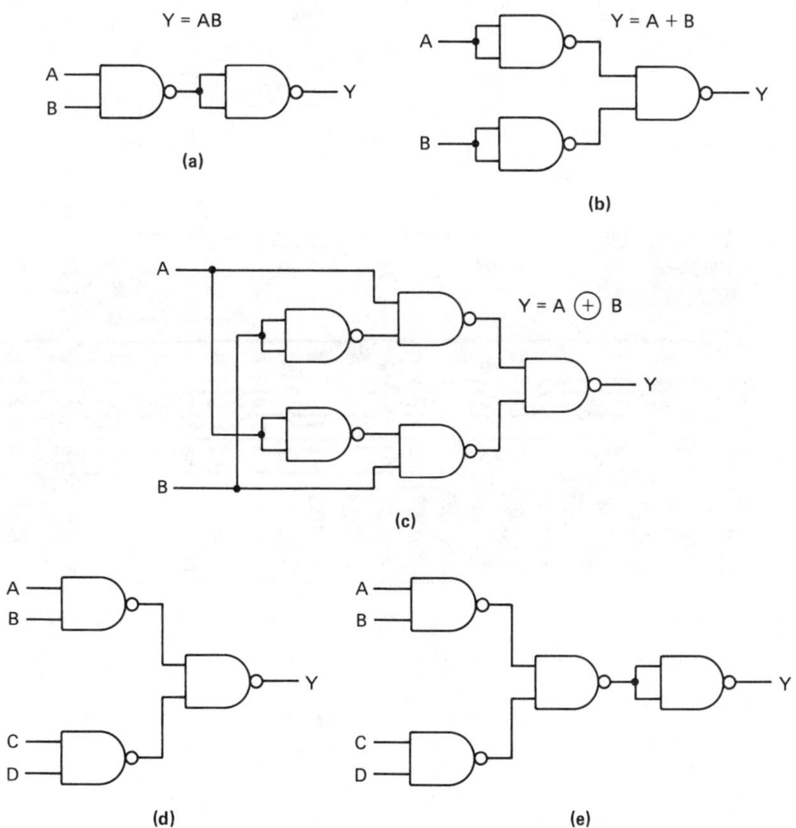

Note: There is also a NOR implementation for any other gate. To derive a NOR implementation, employ the rule that if all three level indicators are changed into their complements, an OR symbol is then replaced by an AND symbol. Conversely, if all three level indicators are changed into their complements, an AND symbol is replaced by an OR symbol. Also, a NOR gate becomes an inverter if both of its inputs are tied together.

Figure 3–5 NAND implementations. (a) AND gate. (b) OR gate. (c) XOR gate. (d) AND-OR gate. (e) AND-OR-INVERT gate.

TROUBLESHOOTING CMOS GATES

Although the digital troubleshooter is chiefly concerned with TTL configurations, CMOS gates will be encountered occasionally. From the troubleshooter's viewpoint, the chief distinction between TTL and CMOS gates is their difference in supply voltages. Also, in the case of CMOS gates, dif-

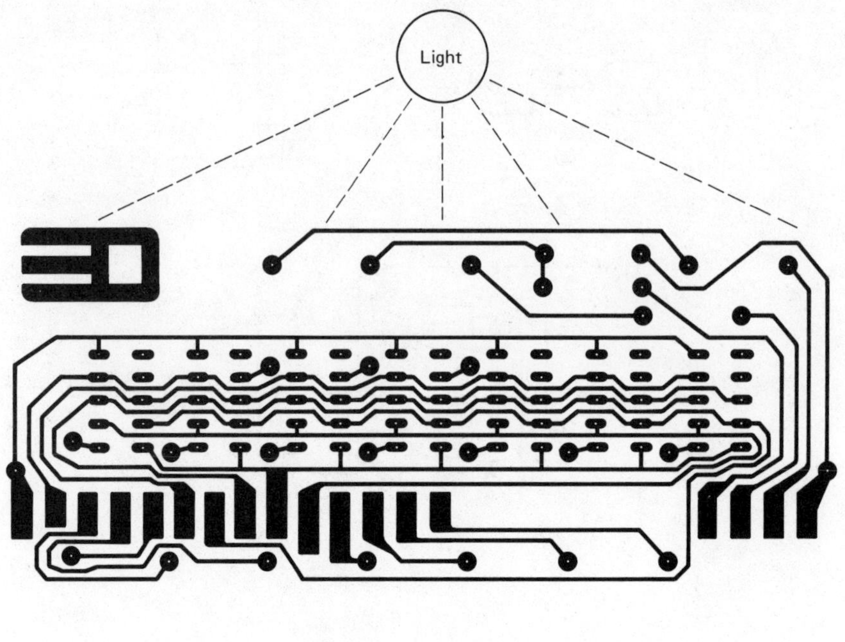

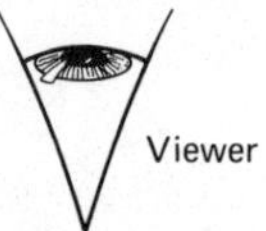

Note: When troubleshooting digital equipment without service data, the technician sometimes needs to determine the direction of data flow along a PC conductor. This knowledge may be required to determine which IC is an input device, and which is an output device. The direction of data flow can be determined while the equipment is processing data by means of a current test made with a lab-type DVM. The DVM test leads are applied at a pair of test points along the conductor to check the IR drop; the polarity of this IR drop indicates the direction of data flow, as explained subsequently.

Figure 3–6 If a bright light is placed behind a digital circuit board, the conductors can be traced from the component side of the board.

ferent values of supply voltage affect the values of logic-high and logic-low threshold voltages.

Troubleshooters will find CMOS gates normally operating from supply voltages in the +3 to +15 V range. From a practical standpoint, CMOS logic-high and logic-low levels are defined as follows: If the supply voltage is in the range from 3 to 10 V, the logic-high state is approximately 70 percent of the supply voltage, and the logic-low state is approximately 30 percent of the supply voltage.

CMOS gates employ the inverter circuit depicted in Fig. 3-7 as their "building block." An enhancement transistor draws zero current until the input terminal is pulsed. CMOS gates have extremely high input and output resistances, and are easily damaged by static electricity. Although many CMOS gates have inputs that are diode protected, comparatively high values of static voltage will nevertheless damage the device. As in the case of TTL gates, all unused CMOS inputs must be grounded or tied to V_{DD}. *No input signal should be applied to a CMOS gate while the power is turned off.*

COMPARATIVE TTL AND CMOS CURRENT DEMANDS

Digital troubleshooters are often concerned with normal current demands and with short-circuit current demands for various gates. CMOS devices have the smallest current demand in normal operation, although the output current increases excessively in case that the output line becomes short-circuited to ground. By way of comparison, a standard TTL gate normally draws as much current as a short-circuited CMOS gate.

Also note that a CMOS gate may draw excessive current and exhibit substantial temperature rise if its drive waveforms have slow rise and slow fall times. In other words, the pull-up transistor is normally nonconducting while the pull-down transistor is conducting (and vice versa). However, in the case of slow rise or slow fall, both the pull-up and the pull-down transistors will conduct simultaneously, resulting in a large increase of current demand.

Observe that there are two TTL families called low-power TTL and Schottky TTL which have less current demand than standard TTL (although they have a higher current demand than CMOS). Notice that not only does CMOS have minimum current demand and power dissipation—it is also a high-density device whereby many more CMOS gates can be formed on a chip than is possible with TTL gates. However, CMOS gates have a disadvantage of comparatively slow speed, compared with TTL.

Checkout of Wired-AND, Wired-OR

Troubleshooters must occasionally cope with wired-AND or wired-OR digital circuitry. These circuits employ open-collector gates instead of totem-pole gates. Outputs of the open-collector gates are tied together so that one

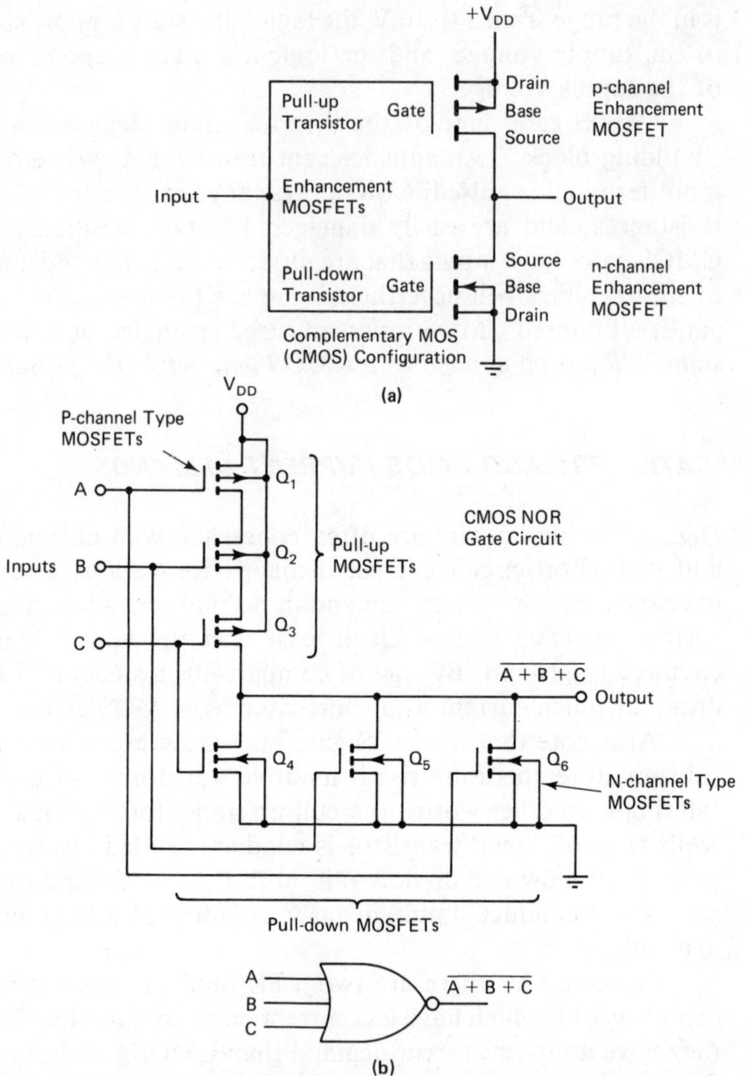

Note 1: **Beginning digital troubleshooters should read the precautions that are printed on packets of CMOS replacement gates, and other CMOS devices.**

Note 2: **CMOS logic circuitry can be checked with a logic pulser and logic probe that are designed to respond to the range of CMOS voltage levels, such as the Hewlett-Packard logic pulser and logic probe.**

Figure 3–7 CMOS inverter is the ''building block'' for various gates. (a) CMOS inverter circuit. (b) CMOS NOR gate arrangement. (c) CMOS NAND gate arrangement.

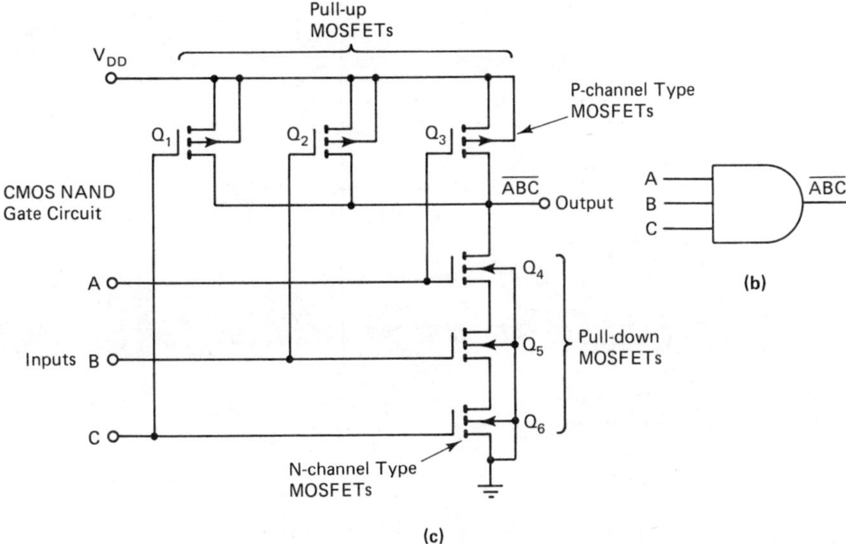

Note 3: If you do not know whether TTL or CMOS gates are in a digital circuit, measure the supply voltage. If V_{CC} is $+5.1$ V, it is reasonable to conclude that the gates are TTL types. On the other hand, if the supply voltage has a higher value or a lower value, it is reasonable to conclude that the gates are CMOS types.

Note 4: The pull-up MOSFETS are P-type devices, whereas the pull-down MOSFETS are N-type devices. Accordingly, when the input signal is logic-high, the pull-up devices conduct, but the pull-down devices are cut off. When the input signal is logic-low, the pull-down devices conduct, but the pull-up devices are cut off. Consequently, the V_{DD} current demand is very small.

Figure 3–7 *(cont.)*

circuit's output can constrain another circuit's output to be in a given state, regardless of their inputs. The tie point is a phantom gate (not a physical gate) which provides wired-AND (or wired-OR) operation.

As seen in Fig. 3–8, gates *U*1, *U*2, *U*3, and *U*4 are "wire-ANDed" together. Under certain conditions, gate *A*'s output can cause gate *B*'s, *C*'s, and *D*'s outputs to be a TTL low, regardless of the *B, C,* and *D* input states. From the troubleshooter's viewpoint, gates *U*2, *U*3, and *U*4 are constrained to operate improperly—at least with respect to the conventional truth table for a NAND gate. The clue to this "normal" malfunction is an asterisk (*) at the output terminal of the gate.

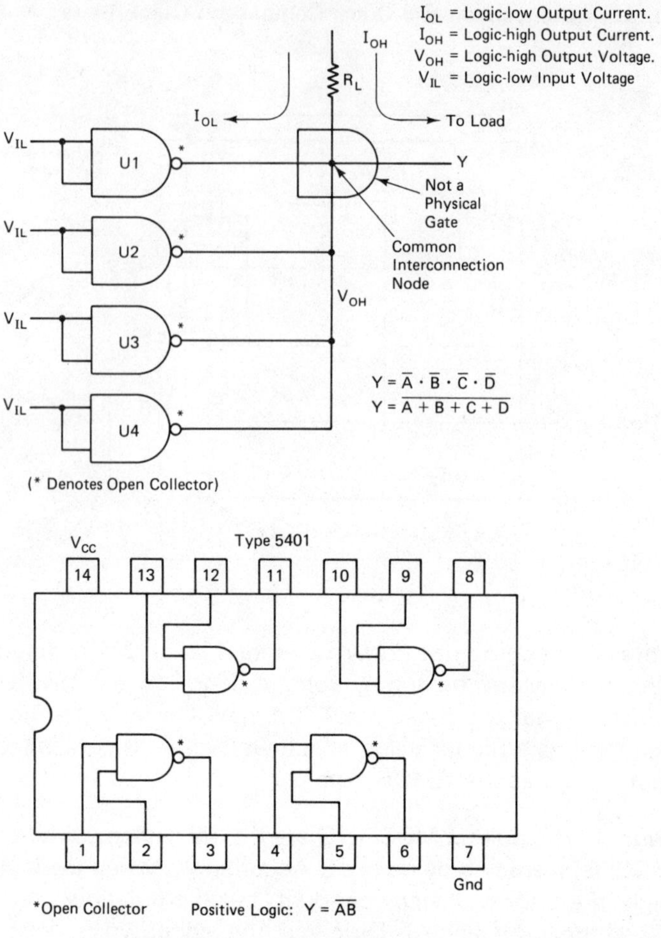

Note 1: When troubleshooting digital equipment without service data, be alert for the presence of a resistor (R_L) on the PC board. The resistor is likely to be associated with a wire-AND circuit.

Note 2: R_L is the common pull-up resistor. It is sometimes implied (not explicitly shown) in logic circuit diagrams.

Note 3: The NAND gates operate as inverters when their inputs are tied together.

Note 4: Wire-AND troubleshooting procedures are often facilitated by current-oriented tests (particularly when the output node of circuit is ''stuck low).'' This troubleshooting procedure employs a logic pulser and a current tracer. Alternatively, a lab-type DVM can be used to analyze the stuck-low fault.

Figure 3–8 Four open-collector NAND gates operating in a wire-AND circuit with a common pull-up resistor.

From a strict technical point of view, although this configuration is commonly called a wired-AND or a wired-OR arrangement, it is actually neither from a functional standpoint. The configuration actually provides negated-AND action, which is the same as OR-NOT action. To obtain AND action, inverters must be included in the *A, B, C,* and *D* inputs. Again, to obtain OR action, an inverter must be included in the output lead.

The so-called wire-AND (wire-OR) configuration is also termed an implied-AND, dot-AND, or phantom-AND arrangement. Note in passing that semiconductor manufacturers occasionally utilize wire-AND circuitry inside of IC packages. Observe the pull-up resistor R_L in Fig. 3-8. *Important: If the pull-up resistor is omitted in a logic diagram, it will nevertheless be present on the digital circuit board. If the pull-up resistor is omitted, the wire-AND configuration is inoperative.*

Either a wire-AND symbol or a wire-OR symbol may be shown in logic circuit diagrams. This choice depends upon the viewpoint of the circuit designer. If he is considering some form of AND action, he will use the wire-AND symbol. On the other hand, if he is considering some form of OR action, he will use the wire-OR symbol. Of course, the circuit action is the same, regardless of the symbolism that is utilized. (See Fig. 3-9.)

An example of wire-OR circuitry inside of an IC package is shown in Fig. 3-10. Wire-AND (wire-OR) circuitry has been extensively used in computer systems; however, as will be explained subsequently, three-state logic has largely supplanted wire-AND circuitry in microcomputers. From a technical standpoint, a wire-AND configuration with a common pull-up resistor has the disadvantage of comparatively slow operating speed.

INVERTER IMPLEMENTATION

We have seen that NAND implementation is common, and that NOR implementation has been employed in the past. Wire-AND and wired-OR is obsolescent, but will be encountered on occasion by the digital troubleshooter. It follows from foregoing discussion that any form of logic circuitry may be implemented from open-collector inverters. This is a principle of basic interest in digital circuit theory, and it also has occasional applications in construction of digital test equipment.

With reference to Fig. 3-11, observe that the simplest type of open-collector inverter is an NPN transistor. When two transistors are configured as shown in part (b), negated-AND action is obtained. Negated-AND action is the same as NOR action, as depicted in Fig. 3-12(a). It is evident that if this NOR gate is followed by an inverter, OR-gate action results. Again, we

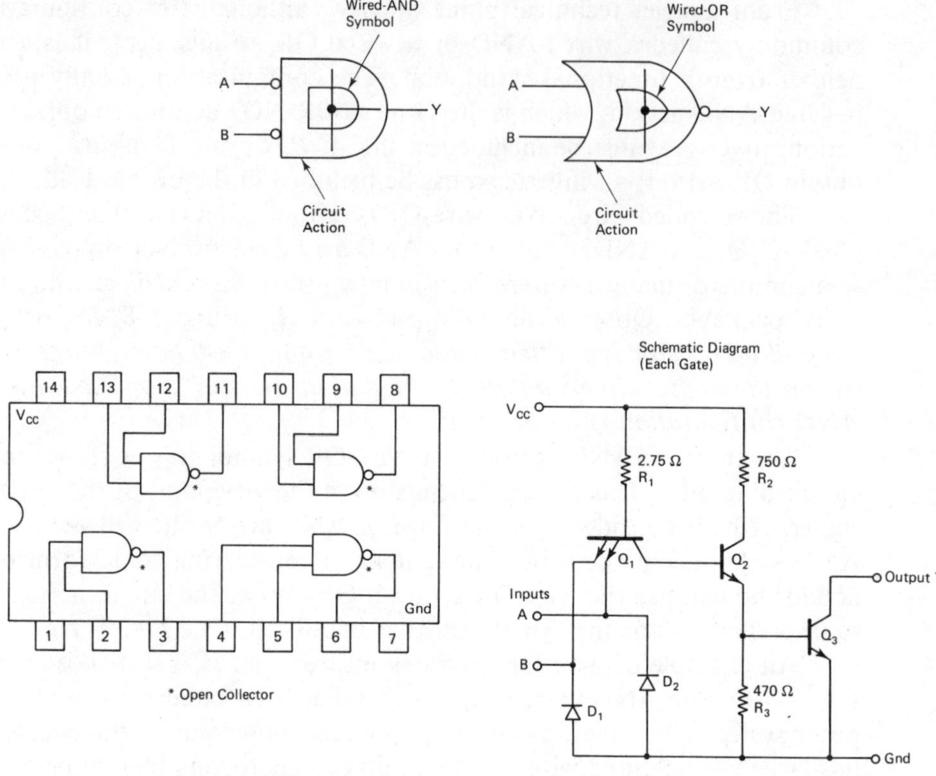

Note 1: Wire-AND (wire-OR) circuitry is very difficult to troubleshoot on the basis of voltage or resistance tests. However, the circuitry can be checked out readily by means of current-oriented tests, as detailed subsequently.

Note 2: Since open-collector inverters can be configured to provide wire-AND, wire-OR, wire-NAND, wire-NOR, wire-XOR, and wire-XNOR gates, all logic circuits can be designed using only open-collector inverters with common pull-up resistors. However, this would not be economical, nor would high-speed digital circuit action be obtained. The fact that this design direction could be taken is nevertheless of considerable theoretical importance.

Note 3: A NAND gate functions as an inverter when its input terminals are tied together.

Figure 3–9 Wire-AND and wire-OR symbols comprise a dot connection inside of an AND gate or an OR gate.

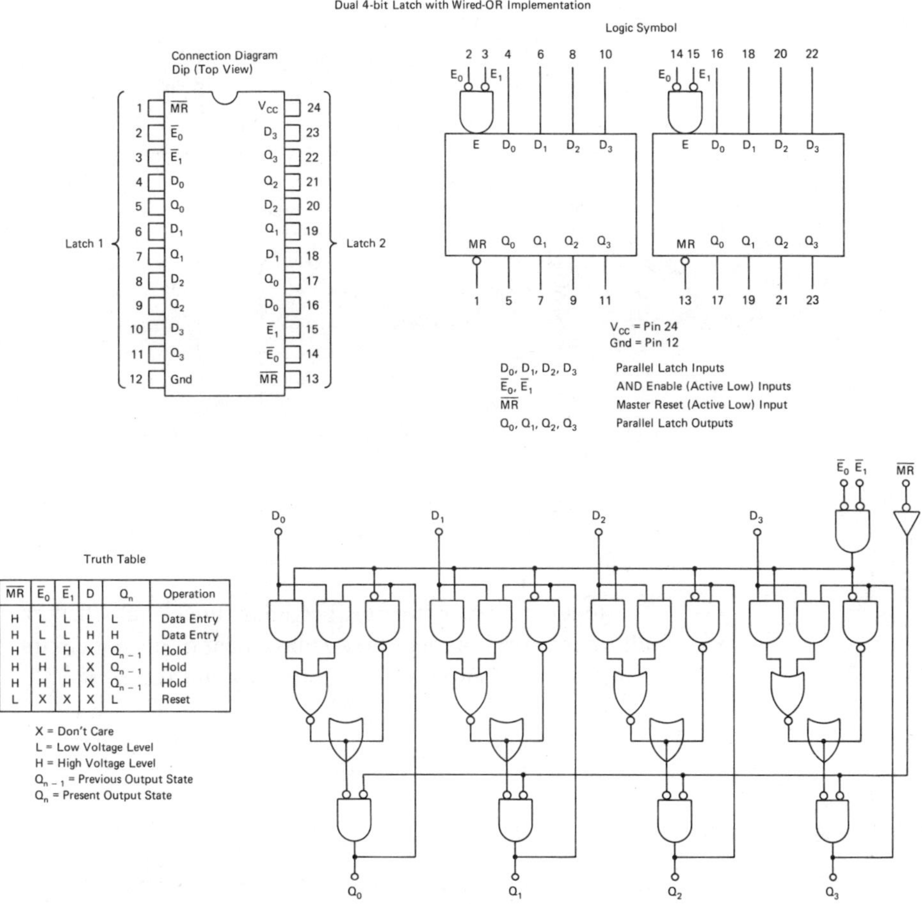

$\overline{MR}$	$\overline{E}_0$	$\overline{E}_1$	D	Q_n	Operation
H	L	L	L	L	Data Entry
H	L	L	H	H	Data Entry
H	L	H	X	Q_{n-1}	Hold
H	H	L	X	Q_{n-1}	Hold
H	H	H	X	Q_{n-1}	Hold
L	X	X	X	L	Reset

Functional Description—Latch Operation: Data can be entered into the latch when both of the enable inputs are LOW. As long as this logic condition exists, the output of the latch will follow the input. If either of the enable inputs goes HIGH, the data present in the latch at that time is held in the latch and is no longer affected by data input.

The master reset overrides all other input conditions and forces the outputs of all the latches LOW when a LOW signal is applied to the master reset input.

Figure 3–10 Example of an IC containing a 4-bit latch which employs internal wire-OR logic circuitry.

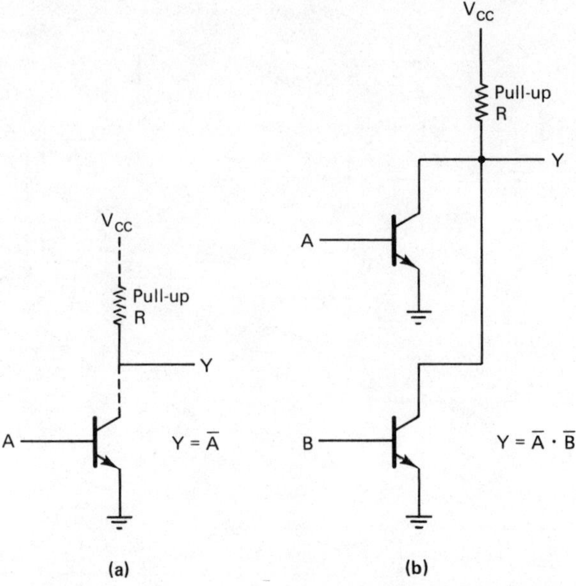

Figure 3–11 Open-collector inverter becomes functional when an external pull-up resistor is added. (a) Arrangement of output circuit for inverter action. (b) Output circuit configuration for wire-AND action.

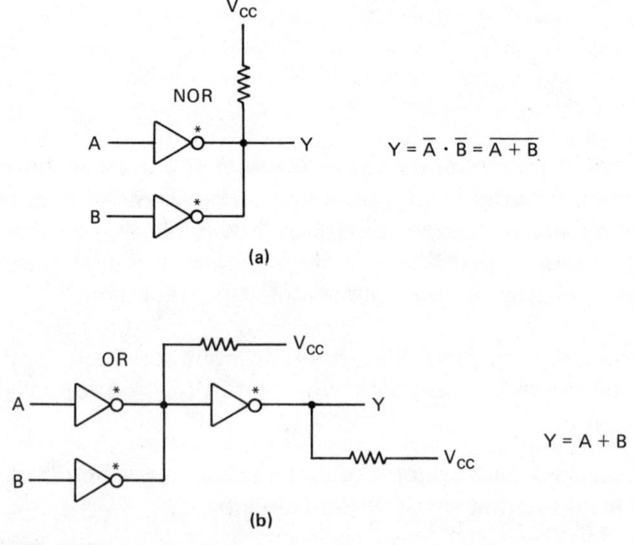

Figure 3–12 Implementation of gates with open-collector inverters. (a) NOR configuration. (b) OR configuration.

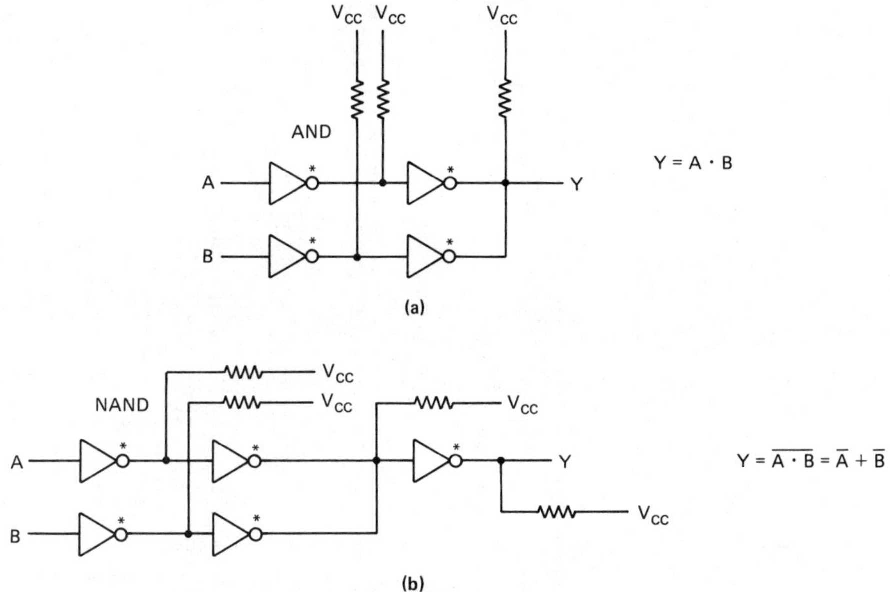

Figure 3–13 Implementation of gates with open-collector inverters. (a) AND gate. (b) NAND gate.

observe that AND-gate action and NAND-gate action are provided by the inverter circuits shown in Fig. 3–13.

PRACTICAL NOTE

When troubleshooting IC gate packages, we expect to find that a gate has two or more inputs, and one output. However, an occasional exception to this principle will be encountered. For example, as shown in Fig. 3–14, a 9014 quad 2-input XOR/XNOR gate has two outputs. These outputs are the complements of each other, and are customarily denoted Z and $\bar{Z}$.

The advantage of this type of gate with two outputs is in its simplification of comparison circuitry, code conversion, and parity generation/checking, as explained in greater detail subsequently. The 9014 also finds application in specialized digital test equipment.

Bottom Line: Wire-AND/OR digital circuitry can be a troubleshooter's curse unless he understands the time-saving shortcuts that can be used.

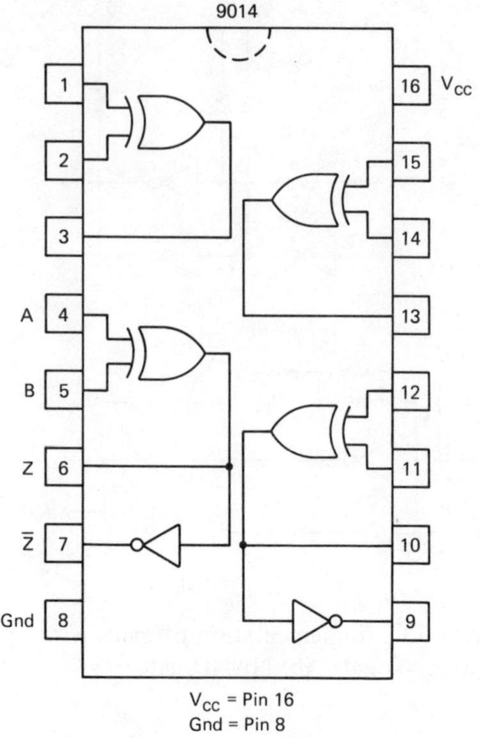

Figure 3–14 Example of gate with two outputs.

PRINCIPLES OF DATA TRACING

Considerable time can often be saved when "sizing up" a logic board by employing the data-tracing principle. In other words, we check various nodes with an LED indicator to determine:

1. Is the same data stream present at the two nodes under test?
2. Is the same data stream present under some conditions of operation only?
3. Is the same data stream never present under any condition of operation?

For example, an eight-channel multiplexer is illustrated in Fig. 3–15. A multiplexer is a device for obtaining simultaneous transmission of two or more digital signals over a common transmission channel. In this example,

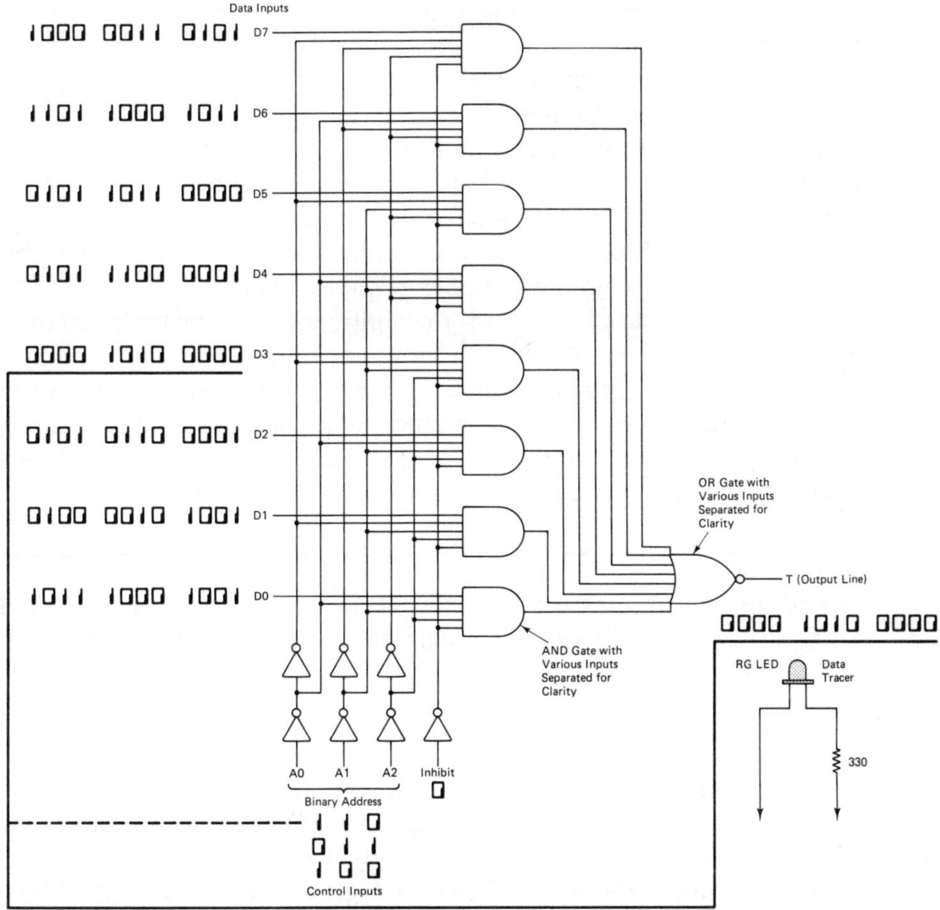

Basis of Test: If two nodes have the same data flow, the LED will remain dark. However, if the LED flashes red or green, the two nodes do not have the same data flow.

Figure 3–15 An eight-channel multiplexer arrangement with eight data inputs, one data output, and three control inputs (Courtesy, Hewlett-Packard.)

the information on data line *D*3 is being multiplexed into the output line *T*. Observe that the Inhibit line is logic-low, and that the binary address 110 is being applied to control inputs $A_0A_1A_2$. In turn, all inputs to the AND gate connected to data line *D*3 are logic-high; on the other hand, not all of the inputs to any other AND gate are logic-high at this time. Accordingly, the information on data line $D3$ is channeled through the multiplexer and into the output line T.

Basically, multiplexing is a processing of digital signals from multiple sources into a lesser number of outputs. As an illustration, other widely used multiplexer arrangements can multiplex 4 or 16 lines into one output line; word multiplexers can be used, for example, to multiplex three 4-bit-wide parallel words into a single 4-bit-wide output. This is just another way of saying that a multiplexer may have several output lines, although by definition it will have a greater number of input lines.

In the example of Fig. 3–15, the multiplexer may be inhibited (disabled) at any time by driving the Inhibit input logic-high. The binary address may be changed at any time to channel some other data lines through the multiplexer. If the Inhibit line is maintained logic-low, and a binary address such as 110 is continuously applied, the multiplexer is said to function as a data selector. (These terms are often used interchangeably).

> **Bottom Line:** We can use a simple LED data tracer such as shown in Fig. 3–15 to quickly "size up" a logic board by means of the data-tracing principle.

Next, with reference to Fig. 3–14, note that each XOR gate is provided with complementary outputs. In other words, pins 6 and 7 provide Z and $\bar{Z}$ data streams, respectively. Accordingly, when we are "sizing up" a logic board, and apply the tester shown in Fig. 3–15 to the Z and $\bar{Z}$ lines, the LED's will remain glowing continuously without interruption, regardless of the information that is in the data streams (the information on the Z line is necessarily the complement of the information on the $\bar{Z}$ line). Even if the data ceases for a time in the streams, the LEDs will remain glowing inasmuch as pins 6 and 7 will rest in complementary states.

LED quick checkers are not infallible, inasmuch as they impose appreciable loading on the circuit under test. This is just another way of saying that if a gate is fanned out almost to its limit, an LED quick check might be "the straw that breaks the camel's back," and in turn give a false indication. Accordingly, in case of doubt, the troubleshooter should check the possibility of a marginal fan-out condition from the node under test. (Follow the line and observe how many times it may branch off to other IC inputs.)

As a practical example of complementary data streams, digital systems often employ two-phase clocks. The Ø1 clock is complementary to the Ø2 clock. Accordingly, when an LED quick checker is applied between a Ø1 clock line and a Ø2 clock line, the LEDs remain glowing continuously, regardless of the presence or absence of data input to the system. A single-phase clock line can be "spotted" without difficulty also, inasmuch as it is active at all times, regardless of any condition of system operation.

4

Voltage-Based, Current-Based, and Resistance-Based Troubleshooting Procedures

TROUBLESHOOTING DIGITAL COUNTERS

Digital troubleshooters need a practical working knowledge of counter action. An outline of application notes on voltage, current, and resistance based troubleshooting procedures for counter circuits (and related circuitry) is tabulated in Chart 4–1. Counters are based on the elementary discrete toggle latch depicted in Fig. 4–1.* This is an *RST* latch, with positive feedback from the Q output to the R input, and from the $\bar{Q}$ output to the S input. This feedback conditions the R and S inputs for toggle action. One pulse is outputted for each two pulses that are inputted.

A toggle latch functions as a storage device, as a simple counter, and as a divider. It is called a modulus *2* (mod-2) counter. Toggle latches can be cascaded as required to count up to any desired number (or to store any desired number). Latches can be preset or cleared when required, as explained subsequently.

Experiment: Construct the *RST* latch shown in Fig. 4–1. The devices and components may be assembled on a plug-in experimenter's socket such as the Radio Shack 276-174, or equivalent. A 6-V power supply or lantern battery may be

*Digital troubleshooters encounter discrete logic circuitry only in simple digital equipment, or in obsolescent equipment.

Chart 4–1

APPLICATION NOTES ON VOLTAGE, CURRENT, AND RESISTANCE BASED TROUBLESHOOTING PROCEDURES

1. A "voltage" source has comparatively low internal resistance and provides an essentially constant voltage across a high-resistance load or a low-resistance load. A commercial logic pulser is an example of a "voltage" source.
2. A "current" source has comparatively high internal resistance and provides an essentially constant current through a considerable range of load resistance. A constant-current ohmmeter is an example of a "current"-based tester.
3. Semiconductor junctions have very low forward resistance and very high reverse resistance. An ohmmeter "wipe" test is an example of a resistance-based troubleshooting procedure.
4. A silicon semiconductor junction does not "turn on" until more than 0.5 V of forward bias is applied. Low-power ohmmeters apply less than 0.5 V across the points under test; this is an example of a voltage-based troubleshooting procedure.
5. Most resistance-based troubleshooting procedures are concerned with static resistance, such as the forward or reverse resistance of a semiconductor junction.
6. Some resistance-based troubleshooting procedures are concerned with dynamic resistance, such as the internal resistance of a gate (its fan-out capability). For example, a negative fan-out test is an example of a dynamic resistance-based test.
7. Dynamic resistance cannot be measured with an ohmmeter, because it can be measured only as a voltage/current ratio in an active circuit. However, a specialized dynamic resistance ohmmeter can be constructed, as will be explained.
8. A negative-fanout test is made by injecting a constant current into an output load circuit and noting the resulting change in load voltage with the load logic-low. Thus, this test is based on the dynamic resistance of the circuit using a constant-current based troubleshooting procedure.

9. As a practical note, nodes (IC pins) can be connected to test instruments by means of microclips. Troubleshooters often prefer to use proto clips instead of microclips, because the terminals on top of the proto clip are more easily accessed. For example, Jimpak® PC-14 or PC-16 proto clip is suitable for 14-pin or 16-pin IC packages.

used for V_{CC}. Voltage measurements at the Q and $\bar{Q}$ outputs will show that the logic-high state is approximately 5 V, and that the logic-low state is 0.12 V, approximately.

Check the *RST* latch operation as follows: Connect a test lead to the *T* terminal and touch the end of the lead to V_{CC}. In turn, a positive charge is placed on the coupling capacitors. Next, touch the end of the test lead to ground. In turn, the coupling capacitors are discharged. Discharge of the coupling capacitors generates a negative edge trigger pulse which causes the latch to toggle. This toggling action is shown by a dc voltmeter connected to the Q output terminal (or to the $\bar{Q}$ output terminal).

Observe that toggling occurs each time that the coupling capacitors are discharged from a charged condition. To set or reset the latch without regard to the *T* input, the base of *Q*1 (or the base of *Q*2) may be pulsed. For example, a 0.05 μF capacitor serves as a practical one-shot pulse generator. Charge the capacitor by touching its leads between V_{CC} and Ground. Then touch the positive lead of the capacitor to the base of *Q*2, with the negative lead grounded. If *Q*2 was previously nonconducting, it will be driven into the conducting state; if *Q*2 was previously conducting, it will remain in the conducting state.

Repeat the foregoing test, touching the positive lead of the charged capacitor to the base of *Q*1. The test pulse will drive *Q*1 into conduction. Note that mechanical switching action can cause "bounce" trouble symptoms. "Bounce" will cause an *RST* latch to change its output state several times in very rapid succession. Accordingly, troubleshooters will find debouncing circuits included in various digital systems that employ mechanical switches, such as microcomputers. Note that if the *RST* latch is driven from a square-wave generator, or from a pulse generator, "bounce" will not occur.

CHECKOUT OF 4-BIT ASYNCHRONOUS BINARY UP COUNTER

Troubleshooters will recognize that all counters are related to the basic serial (ripple) up counter shown in Fig. 4–2. It is a serial counter inasmuch as Latch 1 triggers Latch 2, Latch 2 triggers Latch 3, and so on. It is a ripple

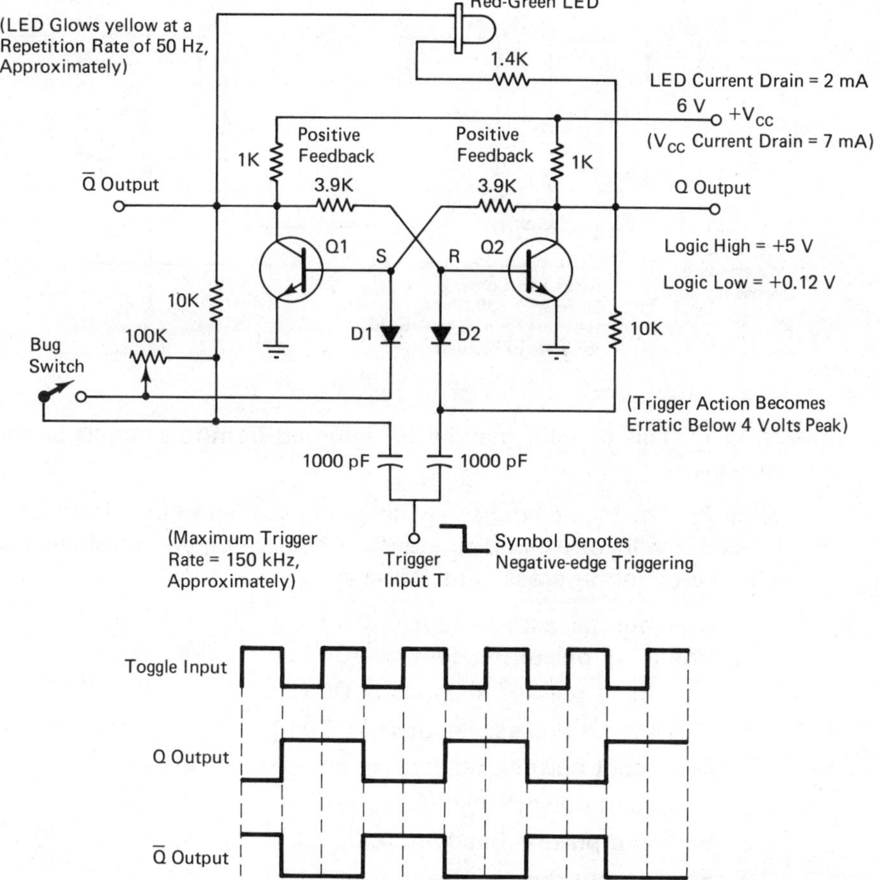

Note 1: This is an example of discrete transistor logic. The transistors are connected in an *RS* latch circuit with feedback from $\bar{Q}$ to *R*, and from *Q* to *S*. Toggling is provided by the two coupling capacitors and the two steering diodes. The steering diodes are arranged for negative-edge triggering. The diodes operate in combination with the transistor collector-base voltages to momentarily drive the conducting transistor into nonconduction. This circuit action that drives the nonconducting transistor is driven into conduction, and the *RST* latch locks up. It remains locked until another trigger pulse arrives, whereupon the latch again reverses its output states.

Note 2: The bug switch enables the experimenter to observe the faulted circuit action due to a deteriorated steering diode.

Figure 4–1 A basic form of the RST or toggle latch. (a) Timing diagram. (b) Configuration.

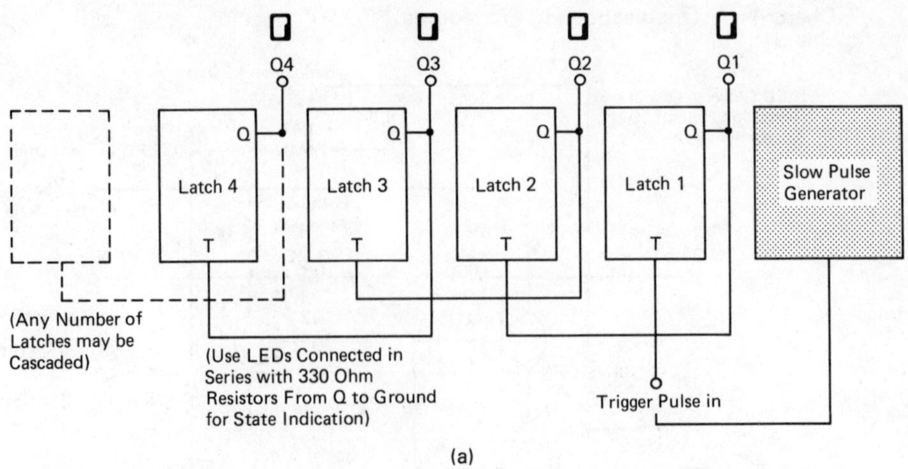

(a)

Note 1: This counter may be constructed from four toggle latches such as shown in Fig. 4–1.

Note 2: The toggle latches are shown cascaded from right to left, so that readouts will correspond to standard binary-number notation. For example, the count progresses as follows:

1. No input pulses: Readout = 0000
2. One input pulse: Readout = 0001
3. Two input pulses: Readout = 0010
4. Three input pulses: Readout = 0011
5. Four input pulses: Readout = 0100
6. Five input pulses: Readout = 0101
7. Six input pulses: Readout = 0110
8. Seven input pulses: Readout = 0111
9. Eight input pulses: Readout = 1000
10. Nine input pulses: Readout = 1001
11. Ten input pulses: Readout = 1010
12. Eleven input pulses: Readout = 1011
13. Twelve input pulses: Readout = 1100
14. Thirteen input pulses: Readout = 1101
15. Fourteen input pulses: Readout = 1110
16. Fifteen input pulses: Readout = 1111
17. Sixteen input pulses: Readout = 0000

This is an "up" counter, because the readout increases with each trigger.

(counter "overflows" and returns to zero.)

Note 3: When the counter overflows, the start-up "garbage" will have been run out, and the readout will then be 0000.

Figure 4–2 (a) Arrangement of asynchronous 4-bit binary counter. (b) A program counter "orchestrates" data processing in a microcomputer.

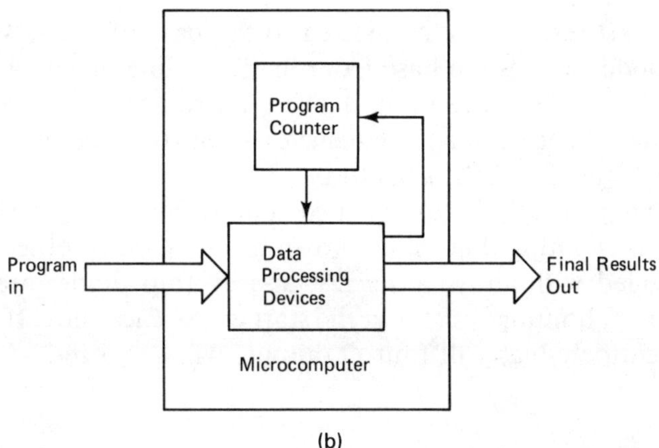

(b)

Note 4: Specialized counters used in microcomputers, digital clocks, and scanner-monitor radios operate with a recirculating data bit. They are variously called shift counters, ring counters, timing-slot generators, or recirculating shift registers, as detailed subsequently.

Figure 4–2 *(cont.)*

counter because the carry bit from Latch 1 ''ripples'' through Latch 2, Latch 3, etc. Consider the situation wherein seven input trigger pulses have been applied. The *Q* output count is then 0111. On application of the next pulse, *Q*1 goes to 0 and a carry bit passes into Latch 2. Next, *Q*2 goes to 0 and a carry bit passes into Latch 3. Then, *Q*3 goes to 0 and a carry bit passes into Latch 4. Finally, *Q*4 goes to 1, and the readout becomes 1000.

RUNNING OUT THE ''GARBAGE''

When you start the counter depicted in Fig. 4–2, it is unlikely that the readout will be 0000. This unpredictable starting count is in random tolerances on transistors, diodes, resistors, and capacitors. Accordingly, the first requirement is to clear the counter and make it read 0000. One way, although not the most elegant, is to run the ''garbage'' out by applying a sufficient number of pulses to make the counter ''overflow'' and read zero. To briefly recap, ''clear'' denotes return of the counter readout to 0000. ''Reset'' denotes return of the counter readout to some predetermined value.

CLEARING OUT THE ''GARBAGE''

With reference to Fig. 4–1, the latch is reset (*Q* is driven to a logic-low state) by momentarily pulsing the base of *Q*2 positive. For this purpose, a diode and a 1-kΩ resistor are connected to the base of *Q*2 as shown in Fig. 4–3. (The diode isolates the base from the Reset line, unless a positive voltage is applied to the Reset terminal.) The 1-kΩ resistor serves as a current limiter to prevent device damage. Whenever a positive pulse is applied to the Reset terminal, the *Q* output goes to 0.

This principle is extended to four latches using a single 1-kΩ resistor and four isolating diodes, as shown in Fig. 4–3. In other words, each latch is provided with an isolating *D*3, and all four diodes are connected to the single 1-kΩ limiting resistor at the start of the Clear line. If the Clear terminal is driven logic-high, all four *Q* outputs will go to zero.

CHECKOUT OF ASYNCHRONOUS 4-BIT DOWN COUNTER

The foregoing principles can also be used to construct an asynchronous 4-bit down counter, as shown in Fig. 4–4. The only difference in the circuitry is in driving the toggle from $\bar{Q}$ instead of *Q*. The down-counter readout starts at 1111 and decreases to 0000, after which it recycles. Note that if a Clear line is employed, the isolating diodes are connected to the *Q*1 transistor bases. In turn, if the Clear terminal is driven logic-high, all four *Q* outputs will go to 1.

EXPERIMENTAL IC ASYNCHRONOUS UP COUNTER

Observe next the integrated-circuit asynchronous up-counter depicted in Fig. 4–5. This arrangement uses *JK* latches which are an elaboration of the basic *RS* latch. (Although commonly called a *JK* flip-flop, the 7476 IC functions as a latch in this counter configuration.) The logic circuit is shown in Fig. 4–8. Observe all of the *J* and *K* terminals are tied to V_{CC}—this provides toggling action. The input terminal of the counter is commonly called the clock input (although this counter is unclocked).

Note that the Preset and Clear inputs are all connected to V_{CC} in the arrangement of Fig. 4–5. In other words, this simple configuration does not provide for clearing the "garbage" out from the counter. To obtain a 0000 readout, a sufficient number of input pulses must be applied to cause the counter to overflow. In spite of its simplicity, this counter arrangement is of great practical importance—its basic features will be encountered frequently by the digital troubleshooter.

Consider next the oscilloscopic display of *Q* waveforms shown in Fig. 4–6; these represent normal operation of the *JK* latch. Although the wav-

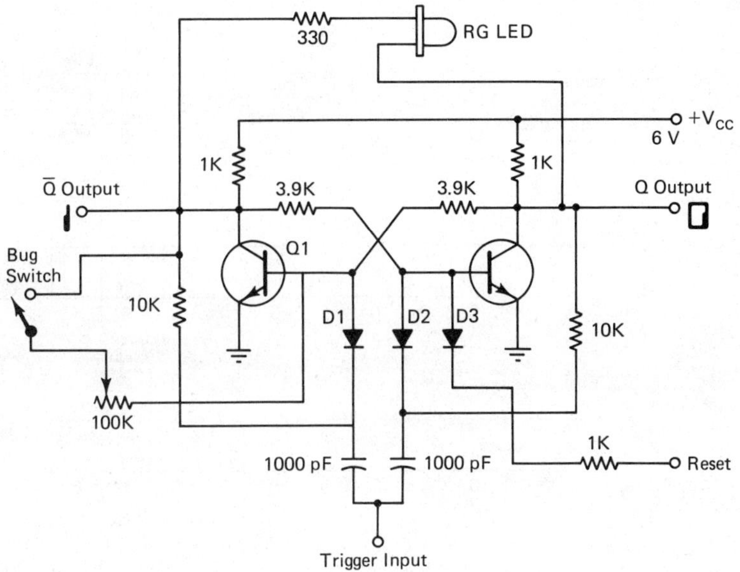

Note: Application of a logic-high level to the Reset terminal drives the Q output logic-low ($Q = 0$).

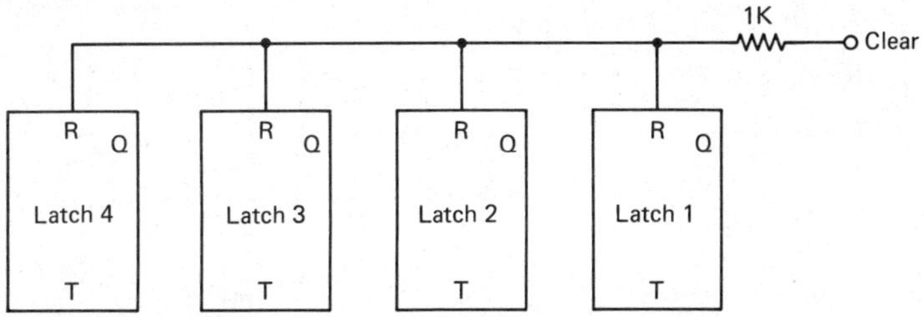

Note 1: The Clear line clears the "garbage" out of the counter and resets each *RST* latch to $Q = 0$. Each latch employs an isolating resistor $D3$; only one 1-kΩ current-limiting resistor is used in the Clear line. Application of a logic-high level to the Clear terminal forces the count to 0000.

Note 2: When checking digital circuit operation with an LED and series resistor, it is helpful to operate the LED at a high brightness level, to facilitate distinction between a dc voltage source and a pulse voltage source, as explained in Chapter 9. A conventional LED will provide higher level brightness if it is backed by a small concave mirror. An LED with a Fresnel lens such as the Archer (Radio Shack) 276-033 provides higher brightness than a conventional LED.

Note 3: The bug switch provides for observation of faulted circuit action due to transistor deterioration.

Figure 4–3 Provision of a Clear (Reset) line for a single *RST* latch, and for a 4-bit counter. Illustration of up-counter operation.

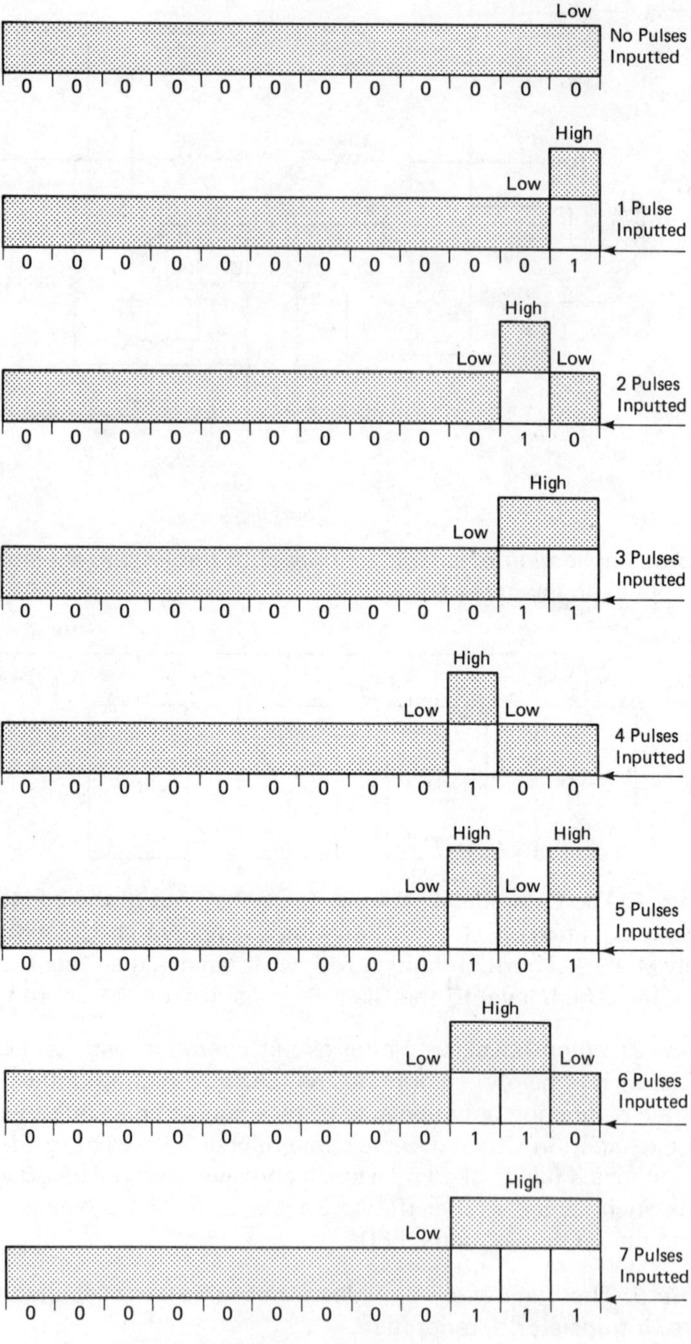

Figure 4–3 *(cont.)*

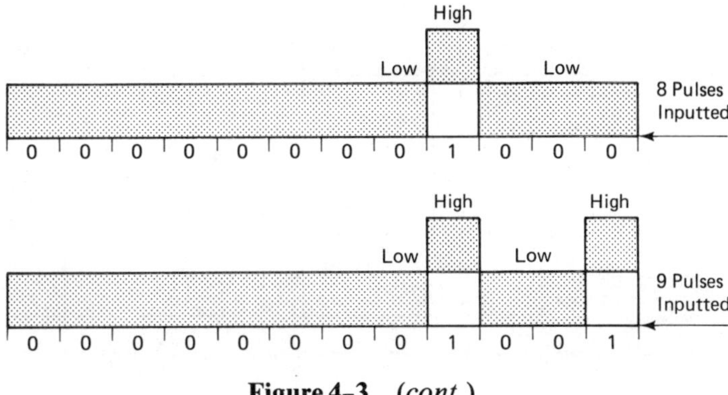

Figure 4–3 (*cont.*)

eforms are not ideal, the waveshapes are within normal tolerance. A logic diagram for the *JK* latch is shown in Fig. 4–7. Note that this configuration consists of a master section and a slave section. The latch is edge-triggered. Data is entered into the master section on the leading edge of the trigger pulse, and the data is transferred from the master section to the slave section on the trailing edge of the trigger pulse.

Toggling is provided by the *JK* latch as a result of its master-slave circuit action. This action is essentially the same as provided by the coupling capacitors and steering diodes in the arrangement of Fig. 4–1. However, the master-slave implementation of the toggle latch is preferred in modern technology because it does not require coupling capacitors. Virtually all present-day digital circuitry is direct-coupled throughout.

It follows from previous discussion that the up-counter arrangement depicted in Fig. 4–5 can be changed into a down-counter arrangement by driving the toggle input terminals from the $\bar{Q}$ outputs, instead of the Q outputs. Note also that, if desired, the Preset and Clear lines need not be tied to V_{CC}, but may be wired into switching circuits for presetting or clearing the counter at will.

TROUBLESHOOTING WITH THE LOGIC COMPARATOR

Troubleshooting digital equipment without service data is greatly facilitated by the availability of a logic comparator. With reference to Fig. 4–8, a logic comparator serves as both a troubleshooting device and as an identification

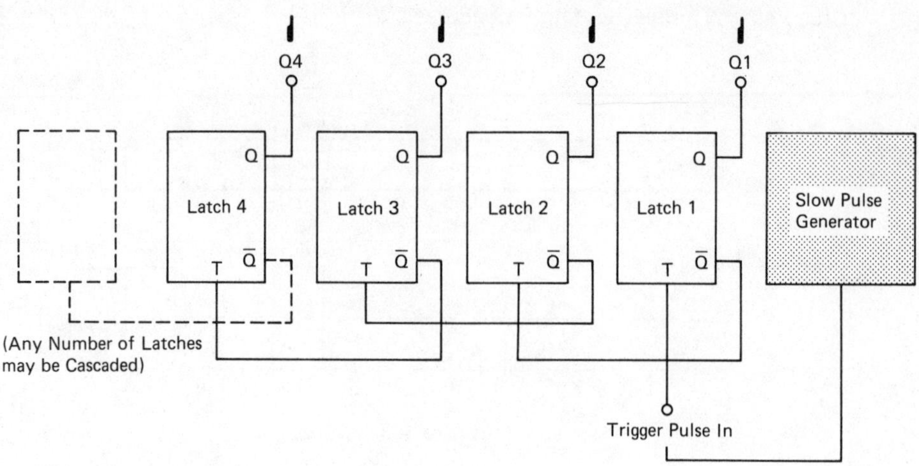

Note 1: The toggle latches are shown cascaded from right to left, so that readouts will correspond to standard binary number notation. For example, the count decrements as follows:

1. No input pulses: Readout = 1111
2. One input pulse: Readout = 1110
3. Two input pulses: Readout = 1101
4. Three input pulses: Readout = 1100
5. Four input pulses: Readout = 1011
6. Five input pulses: Readout = 1010
7. Six input pulses: Readout = 1001
8. Seven input pulses: Readout = 1000
9. Eight input pulses: Readout = 0111
10. Nine input pulses: Readout = 0110
11. Ten input pulses: Readout = 0101
12. Eleven input pulses: Readout = 0100
13. Twelve input pulses: Readout = 0011
14. Thirteen input pulses: Readout = 0010
15. Fourteen input pulses: Readout = 0001
16. Fifteen input pulses: Readout = 0000
17. Sixteen input pulses: Readout = 1111

This is a "down" counter, because the readout decreases with each trigger

Note 2: To clear out the "garbage" from the down counter, a Clear line is employed, similar to the arrangement shown in Fig. 4–3. However, the Q1 transitors are driven by the Clear pulse, in order to reset the counter to 1111.

Figure 4-4 Asynchronous binary down counter configured from *RST* latches.

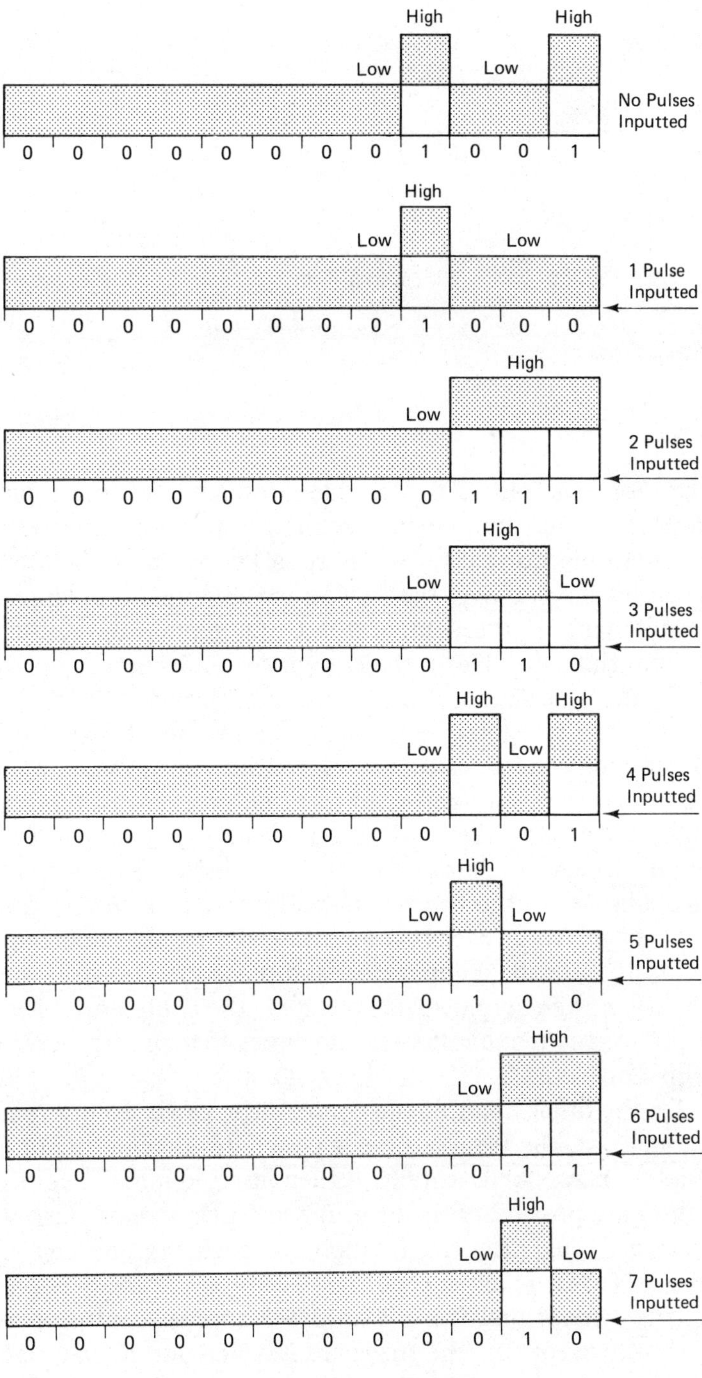

Figure 4–4 (*cont.*)

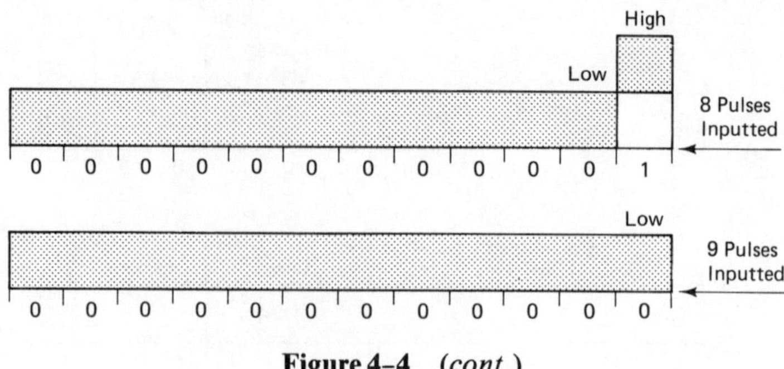

Figure 4–4 (*cont.*)

device. For example, if the troubleshooter knows that a particular IC is a 7476 dual *JK* flip-flop, he can use the comparator to determine whether the IC is operating normally. He inserts a known good 7476 into the comparator, places the pod over the 7476 in the equipment under test, and activates the equipment. If the comparator LEDs remain dark, the troubleshooter concludes that the 7476 in the equipment under test is operating normally.

On the other hand, if one or more of the comparator LEDs is observed to glow in the foregoing test, the troubleshooter recognizes that the 7476 in the equipment under test is not responding in the same manner as the reference IC in the comparator. In turn, he proceeds to replace the faulty IC.

Next, consider the application of the logic comparator as an identification device. As an illustration, the troubleshooter may not be sure whether a particular IC in the equipment under test is a *JK* flip-flop or a NAND-gate package. To answer this question, the troubleshooter may insert a *JK* flip-flop IC into the comparator, and place the pod over the IC under test. Then, when the equipment is activated, the comparator LEDs may remain dark. If so, the troubleshooter concludes that the IC under test is indeed a *JK* flip-flop.

On the other hand, if the comparator LEDs do not remain dark in the foregoing test, the troubleshooter concludes that the IC under test is not a *JK* flip-flop. A NAND-gate IC is then inserted into the comparator to check the alternate possibility. In turn, if the LEDs remain dark when the equipment is activated, the troubleshooter concludes that the IC under test is indeed a NAND-gate package.

It is evident that there is an off-chance of confusion in identification tests, in the event that the supposed NAND-gate IC package contains a de-

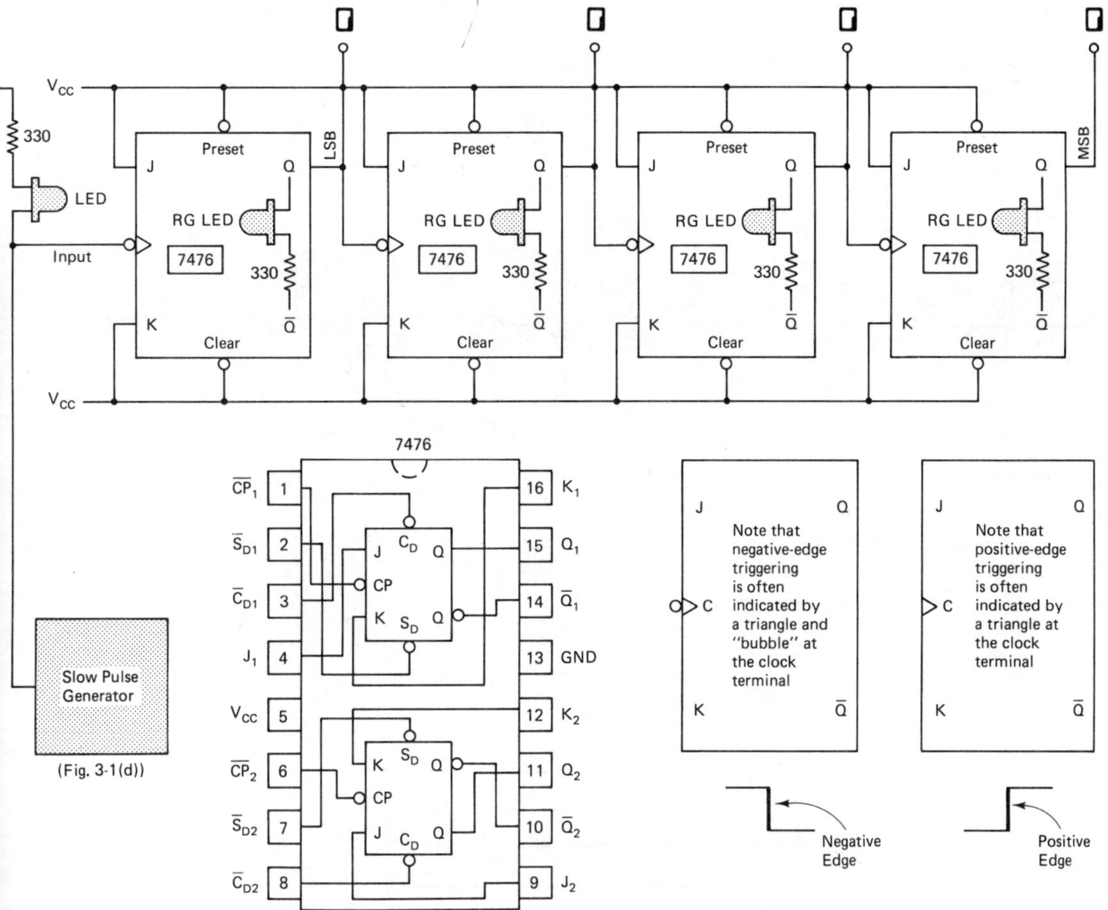

Note 1: The LEDs provide a basic "light map" of digital circuit operation.

Note 2: This up-counter is configured from two 7476 dual *JK* flip-flop packages. These are 16-pin IC packages, as shown in the pinout diagram. The counter may be easily assembled on an experimenter's socket, such as the Archer (Radio Shack) 276-174.

Note 3: If permitted to float, the input terminal will "look" logic-high. If returned to ground through a 200-Ω resistor, the input terminal will rest at a logic-low state, and can be driven high with a logic pulser; or, it can be connected to V_{CC}.

Figure 4–5 Experimental asynchronous up-counter arrangement.

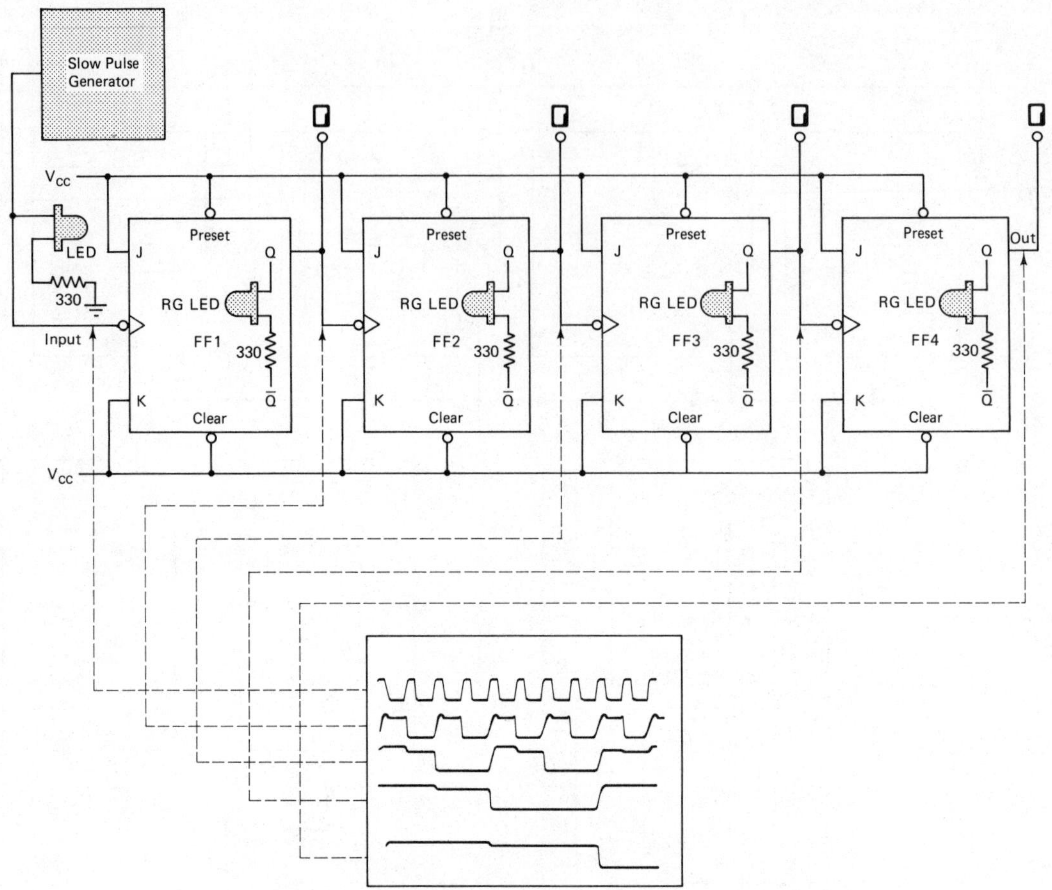

Figure 4-6 Normal 4-bit counter output waveforms.

fective gate, for example. In such a case, the comparator will indicate that the IC under test is not functioning in the same manner as the reference IC, and the troubleshooter would then conclude that the IC under test is not a NAND-gate IC package. If this situation happens to occur, a "tough-dog" problem surfaces. This problem can be eventually resolved, of course, by elimination. In other words, after the equipment under test is completely mapped-out, it will become evident that the IC package under test is indeed a NAND-gate package, but that it contains a defective gate.

This is the sixth step in troubleshooting digital equipment without service data, using time-saving shortcuts.

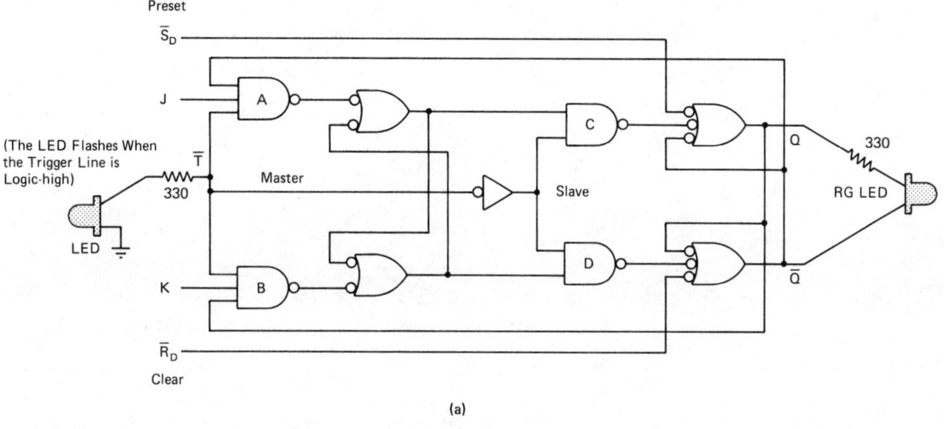

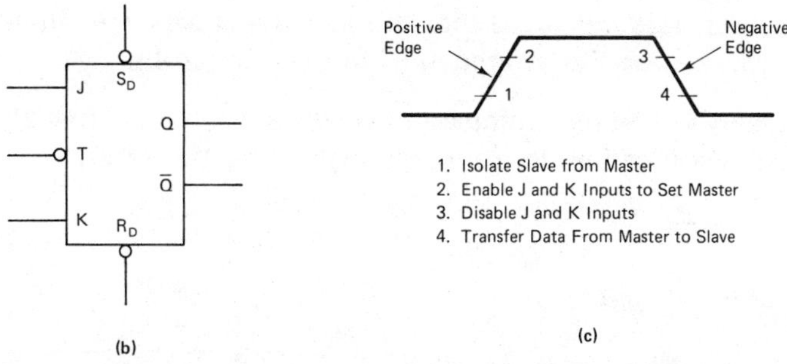

Note 1: The preset line is active-low, and is indicated by $\overline{S}_D$ in the diagram. However, the preset line is indicated by S_D in the logic symbol. An inversion bar is not placed over S_D in the logic symbol because the indication is inside the symbol (beyond the inversion "bubble"). In other words, a 0 on the preset line becomes a 1 past the "bubble." Similarly, the toggle or clock terminal is indicated by $\overline{T}$ in the diagram, but is indicated by T in the logic symbol.

Note 2: When the J and K inputs are tied to V_{CC} and the Clear and Preset terminals are tied to V_{CC} the latch toggles each time that a trigger pulse is applied to the $\overline{T}$ terminal.

Figure 4–7 *JK* latch configuration. (a) Logic diagram. (b) Logic symbol. (c) Data transfer timing diagram.

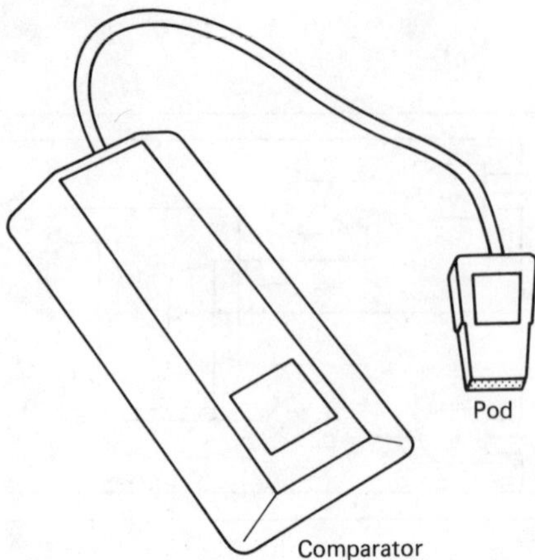

Note: The logic comparator is applied by inserting a chosen IC into the comparator and placing the pod over a corresponding IC in the equipment under test. The equipment is then activated in various ways, and the comparator LEDs are observed. If any LED glows during the activation of the equipment, it is indicated that the IC under test is not functioning in the same manner as the reference IC in the comparator.

Figure 4–8 Logic comparator is both a troubleshooting device and an identification device. (Courtesy, Hewlett-Packard.)

DIODE SWITCH EXPERIMENT

Diodes are used as electronic switches in various digital circuits. The experiment shown in Fig. 4–9 provides a good understanding of diode switching action. This is a simple form of AND gate comprising a pair of diodes and a resistor. A piezo buzzer is used as an output indicator. Observe that when inputs *A* and *B* are both logic-high, the output is also logic-high, and the piezo buzzer sounds. However, if either input *A* or *B* (or both) are logic-low, the piezo buzzer is silent.

NEGATIVE FAN-OUT TEST

Although this one is not really "new," it is of sufficient practical importance that it should be mentioned here. As shown in Fig. 4–10 a negative-fanout tester can be arranged to check suspected marginal thresholds in fanned-out nodes. The switch positions from 1 to 5 provide negative fan-outs from 1

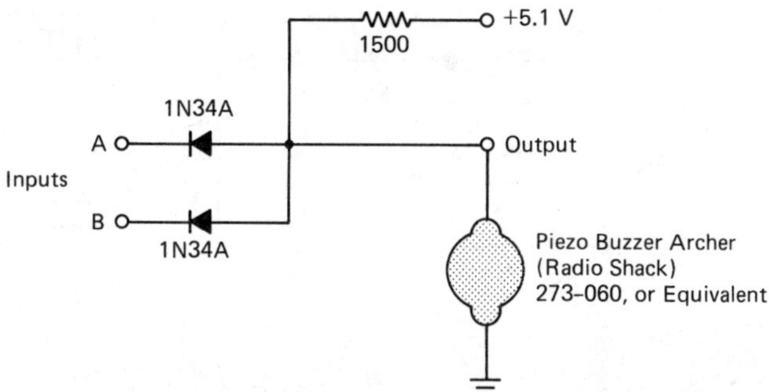

Note: Diode switching action is employed in this simple AND circuit. When both inputs are logic-high, the "switches" are open, and current from the resistor flows through the piezo buzzer to ground. However, if either one or both of the inputs are logic-low, the current from the resistor is switched directly to ground and the piezo buzzer is silent.

Figure 4–9 Diode switch experiment.

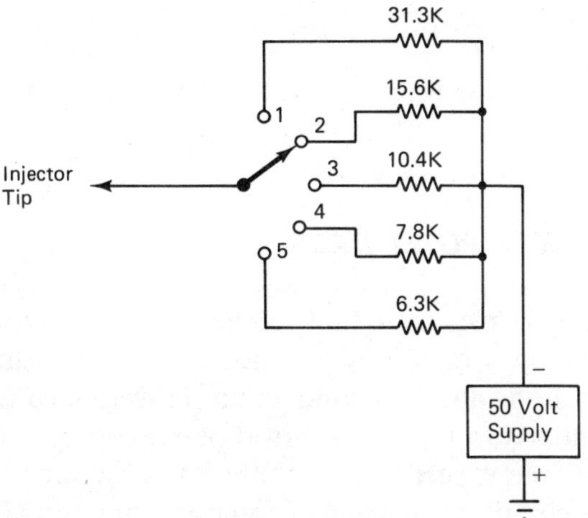

Figure 4–10 A negative fan-out tester.

to 5 unit loads (ULs) when the "constant-current" flow from the injector tip is applied to the output node of a TTL device in its logic-low state.

An output load may be held logic-low during test by stopping the clock, or equivalent expedient. Note that negative fan-out does not remove a load (or loads) in the sense of disconnection; it effectively reduces the load that is imposed on the driver. In turn, the troubleshooter can determine whether

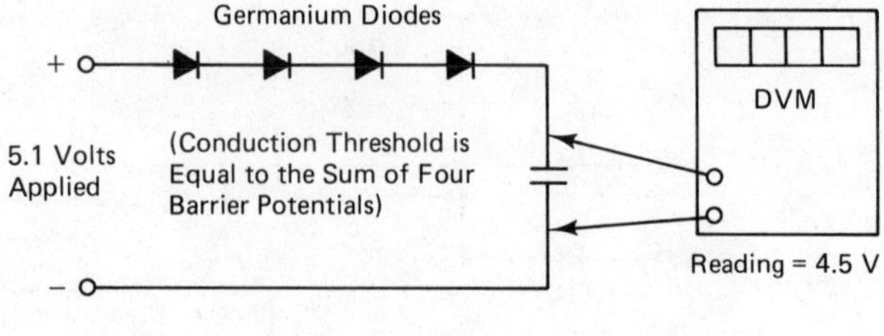

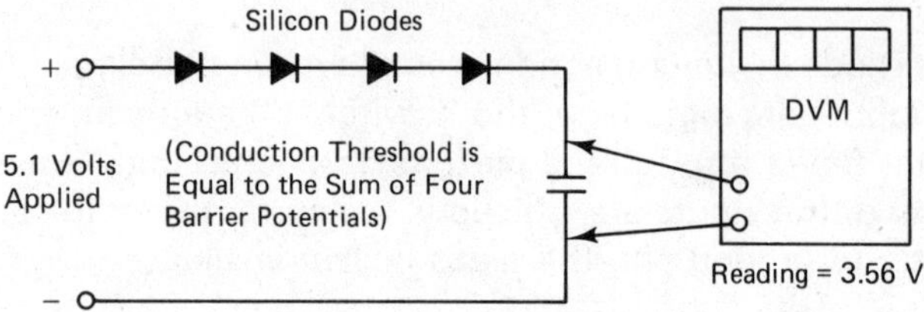

Figure 4–11 Demonstration experiment for cumulative barrier potentials.

an erratic trouble symptom is being caused by a marginal threshold condition.

CUMULATIVE BARRIER POTENTIALS

Troubleshooters recognize that semiconductor diodes have a certain barrier potential, below which they will not conduct. As an illustration, a typical germanium diode does not conduct until the applied forward voltage exceeds 0.25 V. Similarly, a typical silicon diode does not conduct until the applied forward voltage exceeds 0.4 V. When two or more diodes are series connected in a circuit, their barrier potentials are cumulative, as shown by the experiment in Fig. 4–11. Accordingly, the sum of their barrier potentials establishes the threshold of forward conduction.

> **Bottom Line:** Practical experience plus a few time-saving quick checkers and shortcuts are an unbeatable combination.

5

Digital Mapout Troubleshooting Procedures

TROUBLESHOOTING SYNCHRONOUS COUNTERS

Digital troubleshooters are routinely concerned with various types of synchronous counters. Some counter-related problems fall into the "tough-dog" category. An introductory listing of "tough-dog" troubleshooting procedures is given in Chart 5–1. As noted previously, a synchronous counter differs from an asynchronous counter in that the former is clocked. Most synchronous counters employ *JK* flip-flops. Observe (Fig. 5–1) that clock pulses are applied to the input gating section, which is followed by a master latch and a slave latch. The *J* and *K* inputs are basically regarded as signal inputs; however, they are often tied to V_{CC} in counter circuitry.*

Consideration of the *JK* truth table will show that if the *J* and *K* inputs in Fig. 5–1 are tied to V_{CC}, the flip-flop will then toggle in response to clock pulses. As a practical note, if the *J* and *K* inputs are floating, they "look" logic-high and toggling occurs as if they were tied to V_{CC}. (Floating *J* and *K* inputs are regarded as undesirable design practice, although the circuitry is thereby simplified.)

*Preliminary mapout procedures are sometimes facilitated by use of the quick comparison checker described at the end of this chapter.

Chart 5-1

"TOUGH-DOG" TROUBLESHOOTING PROCEDURES

1. A common type of "tough-dog" troubleshooting is exemplified by erratic functioning of a unit of digital equipment which may have been exposed to high-humidity conditions. In such a case, the troubleshooter should start by operating the equipment continuously for 100 hours in a low-humidity warm environment. If the equipment is still malfunctioning after this drying-out period, quick tests may then be made to localize faulty devices or components.
2. Another type of "tough-dog" troubleshooting can occur when an older design of personal computer unpredictably develops data-processing errors. Experienced troubleshooters monitor the power line with an oscilloscope in this situation, to determine whether line surges are present simultaneously with the processing errors. If so, the computer can then be powered from an automatic line-voltage regulating transformer.
3. When CMOS ICs overheat, and there is no short-circuit fault, remember to check the clock waveform. Slow rise and fall can result in excessive current drain.
4. Complex digital systems may employ many cards (plug-in PC boards). In "tough-dog" situations, check the cards to see whether two or more may have identical circuitry. If so, the troubleshooter can swap similar cards back and forth, to determine whether the trouble symptom changes. In the event that the symptom does change, the troubleshooter has an important "handle" on the problem.

Experiment: Digital troubleshooters will find it helpful to construct and check out the synchronous counter with ripple carry shown in Fig. 5-2. This is a comparatively simple synchronous counter; it has an advantage in that only a single extra two-input AND gate is required for each flip-flop that may be added to the chain (to obtain a greater count). Although this feature contributes to

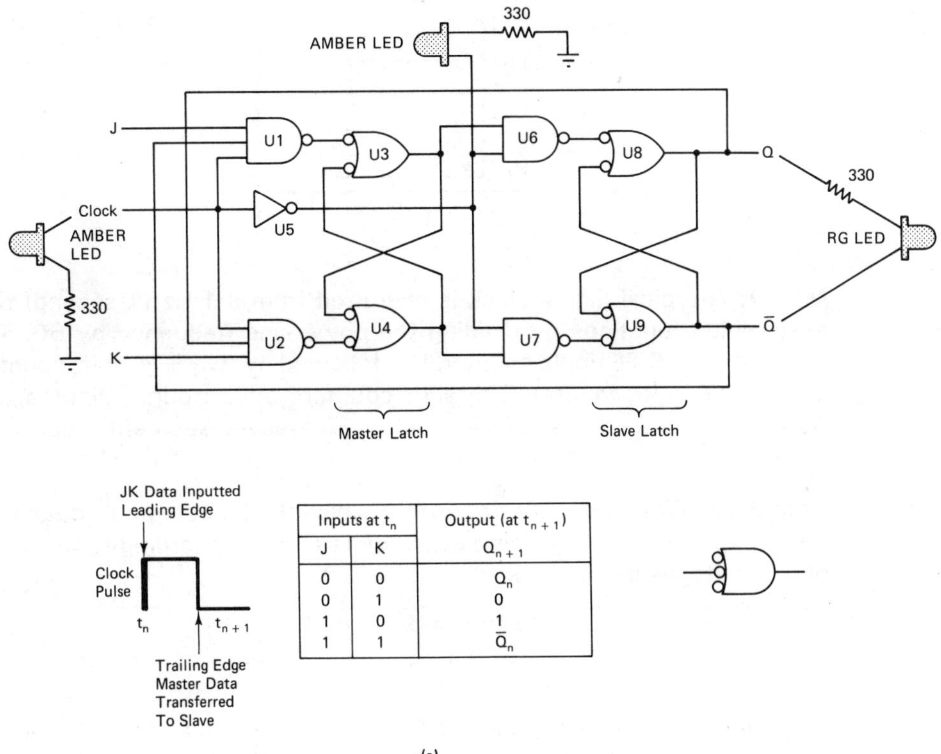

Inputs at t_n		Output (at t_{n+1})
J	K	Q_{n+1}
0	0	Q_n
0	1	0
1	0	1
1	1	$\overline{Q}_n$

Note 1: If Preset and Clear lines are provided, a third input is provided on *U*8 and *U*9 as shown above.

Note 2: Edge triggering (as contrasted to simple level triggering) is accomplished by configuring the gate input circuitry so that the master latch is locked out from the *J* and *K* lines except for a brief instant as the clock goes high. This narrow "window" results from a transient propagation delay that occurs in the gate input circuitry for the master latch. The slave latch is locked out from the master-latch outputs at t_n. However, the gates to the slave latch are opened at t_{n+1}, and the master-latch output data is then transferred into the slave latch.

Note 3: To recall an important fact, the *JK* flip-flop toggles on clock pulses if its *J* and *K* inputs are maintained logic-high. On the other hand, it is "forbidden" to drive the *R* and *S* inputs of an *RS* flip-flop logic-high simultaneously.

Figure 5–1 (a) Typical configuration for a *JK* flip-flop. (Courtesy, Hewlett-Packard.) (b) Synchronous counters are used in digital clocks.

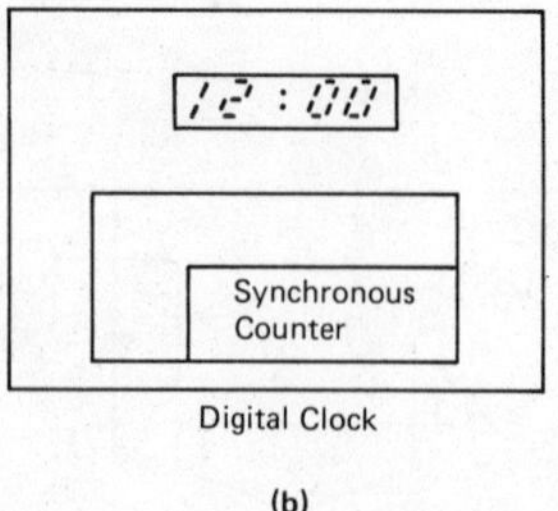

Digital Clock

(b)

Note 1: A typical digital clock is energized from a 1-Hz (1-second) time base which functions by dividing the power-line frequency by 60. This time base is a 6-FF ripple counter. It is followed by two 8-FF shift counters and a 6-FF shift counter. The shift counters drive Hours, Minutes, and Seconds logic sections which in turn drive seven-segment display devices.

Note 2: A shift counter is a network in which the first stage, through logic feedback, produces a certain pattern of 1s and 0s according to the states of other stages in the network.

Figure 5–1 (*cont.*)

circuit simplicity, each added AND gate introduces an additional propagation delay between all of the prior outputs and the *JK* inputs to the flip-flop. The propagation delay for a typical TTL NAND gate is 10 ns.

An advantage of the synchronous counter is that only one gate propagation delay is involved to set up any flip-flop in the chain. However, for each flip-flop that is added, the corresponding AND gate must be provided with one more input. In other words, this involves a trade-off in circuit complexity. Since any counter has a rated maximum operating speed, a synchronous counter with ripple carry introduces more propagation delay between the *JK* input of a flip-flop and all prior outputs than does a "true" synchronous counter, as explained in greater detail subsequently.

This is just another way of saying that as the clock is speeded up, a synchronous counter will arrive at an input frequency limit beyond which the control output will not reach the next flip-flop soon enough to be processed as a valid data input. As noted in Fig. 5–2, the experimenter may check out the counter at slow speed with visual readout provided by LEDs with 330-Ω resistors connected between each *Q* output terminal and ground. It is instructive to make the following tests:

Using a 0.05 μF capacitor charged to V_{CC}, pulse the clock input of the counter shown in Fig. 5–2. The resulting *Q* output from the first flip-flop can be observed by means of an LED and series resistor. Do you obtain the anticipated

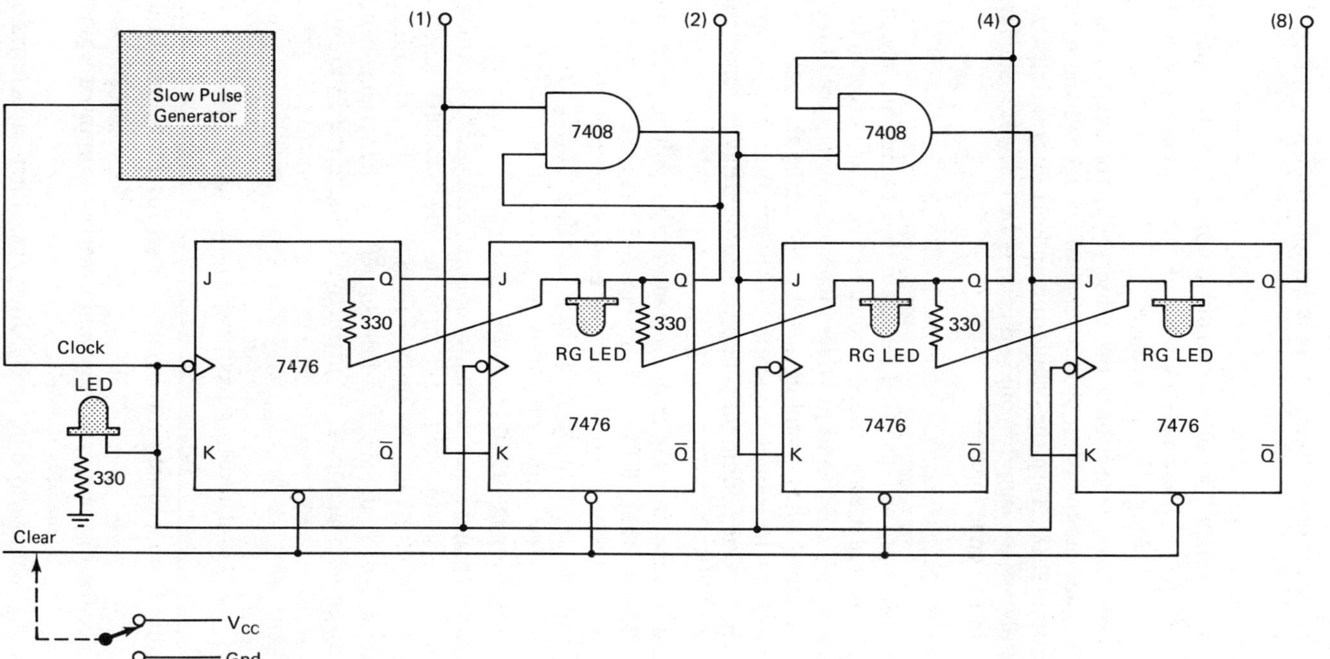

Note 1: The LEDs provide a ''light map'' of ripply carry action in to the clock input pulses.

Note 2: This synchronous counter is driven by clock pulses which are applied simultaneously to all clock inputs of the flip-flops. Clock pulses will be counted as long as the Clear line is logic-high. If the clear line is driven logic-low, the counter is reset to 0000. A square-wave generator can be used as a clock subber in the experiment.

Figure 5–2 An experimental synchronous counter with ripple carry.

response each time in half a dozen tests? (There is a mechanical switch factor involved in this test.)

Next, connect an SPST slide switch between the clock input and ground. When you open and close the switch half a dozen times do you obtain the anticipated response each time? Which of these two methods of manual pulsing would you prefer in practical troubleshooting procedures?

Turn the counter off and then turn it back on. Observe the starting readout that is displayed. Then actuate the Clear switch depicted in Fig. 5-2 and note that the readout becomes 0000. Turn the counter on and off several times. Is the starting readout always the same? Would your reply be true for a similar counter constructed in the same manner?

Experiment: Digital troubleshooters will also find it helpful to construct and check out the synchronous up-counter arrangement shown in Fig. 5-3. This counter is configured to avoid ripply carry, and provides higher operating speed than simpler counters. It obtains higher operating speed by changing the states of all flip-flops simultaneously, instead of sequentially. Any number of flip-flops may be used in the chain. One additional gate propagation delay is added to the set-up time of each MSB.

For each flip-flop that is added into the chain, one more input will be required on each AND gate. Observe that all flip-flops in Fig. 5-3 change state simultaneously inasmuch as they share a common clock edge. Consequently, flip-flop output data is available simultaneously, without any delay owing to ripple carry. Each of the gates receives data from all of the less-significant flip-flops when the more-significant inputs are to be enabled.

Observe next that the foregoing synchronous up-counter arrangement can be rearranged as shown in Fig. 5-4 to provide synchronous down-counter action. The Preset line line is utilized to reset the counter to 1111, and the $\bar{Q}$ terminal of each flip-flop is used to drive the next flip-flop. Otherwise, the arrangement is the same as for up-counter action. It is instructive to make the following tests:

1. Connect LEDs and 330-Ω resistors from each Q output to ground for visual readout.
2. Clock the counter from a square-wave generator set for a low repetition rate.
3. Gradually reduce the V_{CC} supply voltage and observe the point at which counter action starts to become erratic.
4. Using normal V_{CC} voltage, progressively disconnect filter capacitors in the power supply until counter action starts to become erratic. Check the V_{CC} ripple voltage with an oscilloscope.

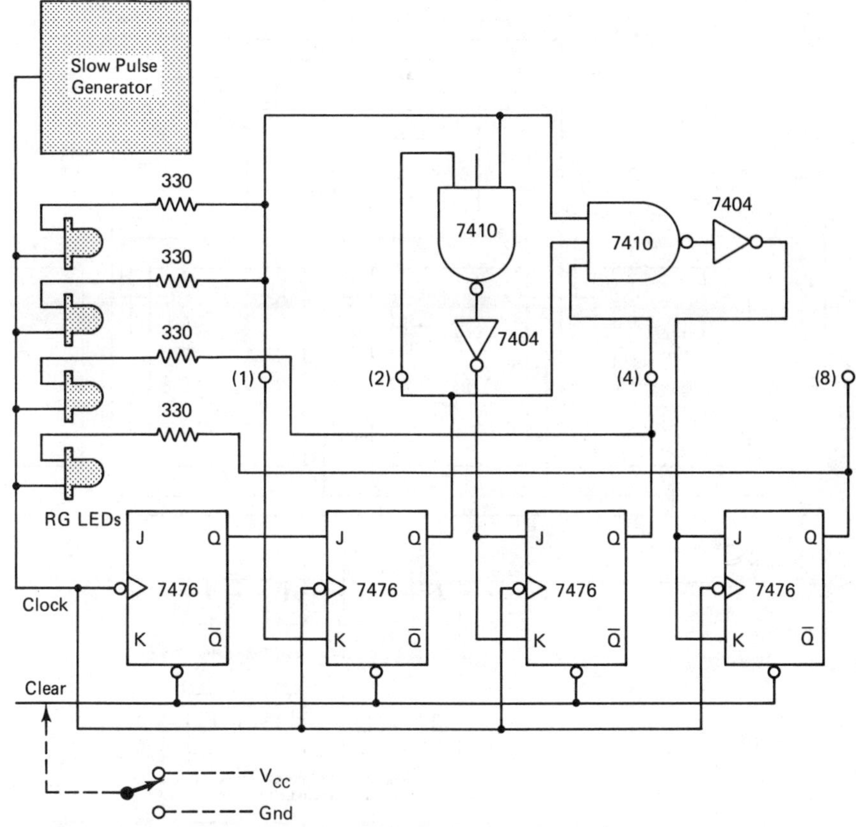

Note 1: The LEDs provide a "light map" of the relation between clock pulses and the *Q* outputs.

Note 2: An eight-bit digital word, such as 10011101, is called a *byte.* A four-bit digital word, such as 1001, is called a *nybble.* Any group of 1s and/or 0s is called a *digital word,* or *bit pattern.* A 1 or a 0 is called a *binary number, binary digit,* or *bit.*

Figure 5–3 Experimental high-speed synchronous up-counter arrangement.

5. Using a well-filtered 5.1-V power supply, touch the tip of a soldering pencil to one of the *JK* flip-flop packages. Heat the IC until counter action starts to become erratic. Measure the temperature of the IC package with a thermocouple probe and DVM.

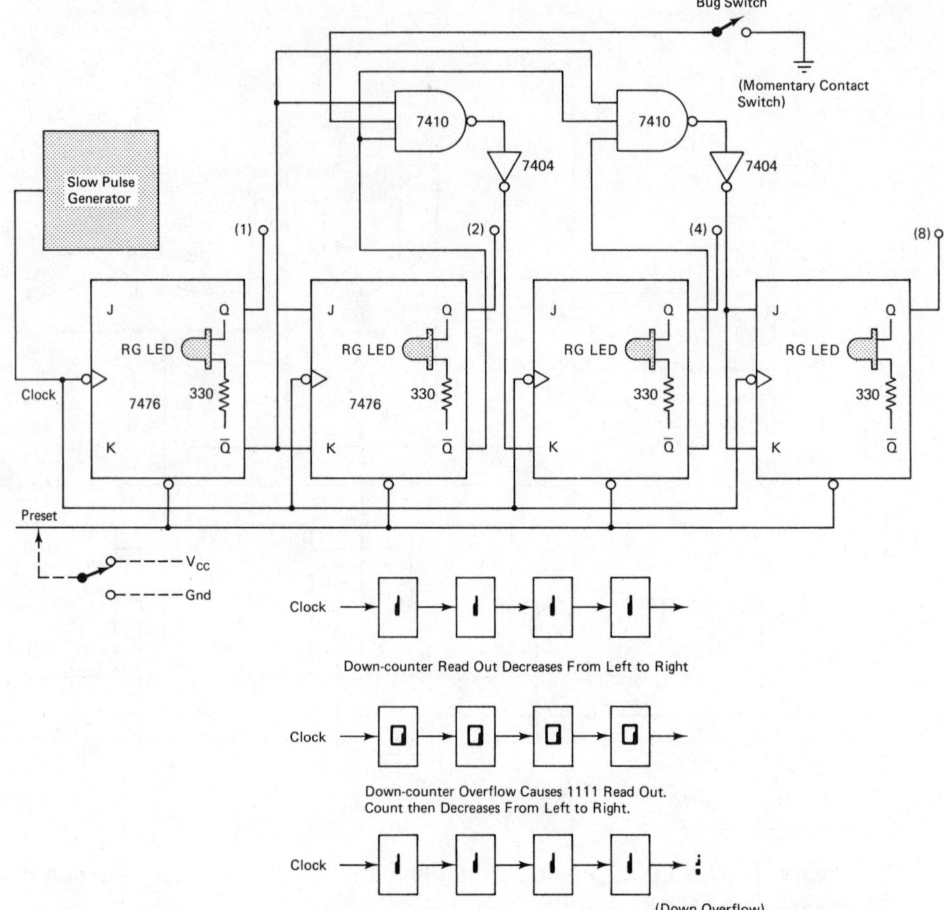

Note: The bug switch enables the troubleshooter to observe the response of the counter to instantaneous shorts to ground at selected times.

Figure 5–4 Experimental high-speed synchronous down-counter arrangement.

PROGRAMMABLE DOWN COUNTER

Digital troubleshooters will also find it helpful to construct and check out the programmable down counter shown in Fig. 5–5. This type of counter is used in the more sophisticated types of commercial digital equipment. This experimental configuration can be easily assembled on an experimenter's

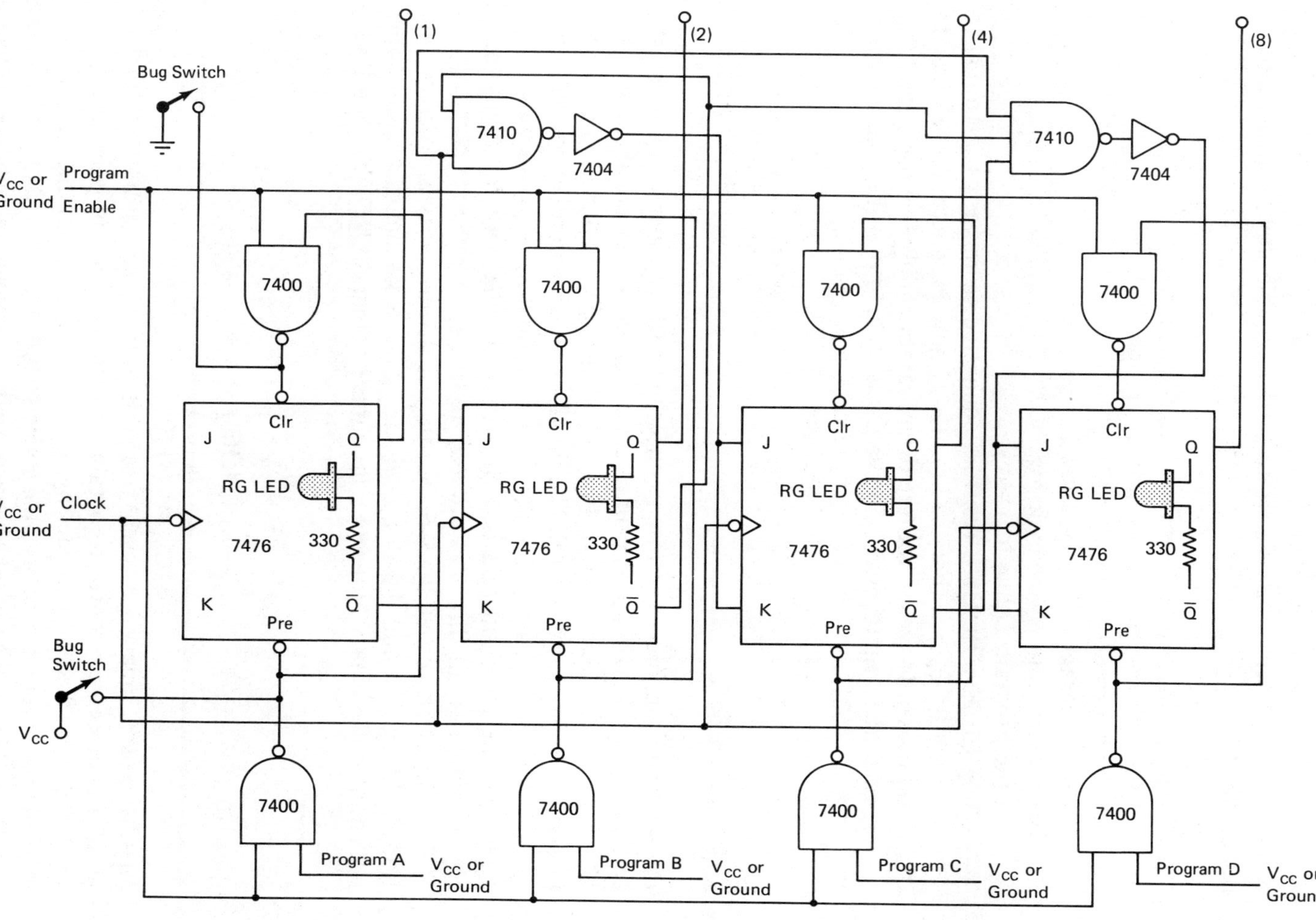

Figure 5–5 An experimental programmable down-counter.

socket. The following tests illustrate programmable down-counter operation.

1. Programming the number nine into the counter:
   Set the program-enable line line logic-high.
   Set the program A line logic-high.
   Set the program B line logic-low.
   Set the program C line logic-low.
   Set the program D line logic-low.
   Using LEDs with 330-Ω resistors connected from each Q terminal to ground. Observe that the counter is now programmed to nine. In other words, when you set the program-enable line logic-low, and clock the counter manually, you will observe that the readout starts from 1001 and counts down on each clock pulse.

FREQUENCY COUNTER

Digital troubleshooters are often concerned with frequency counters. Accordingly, you will find it helpful to construct and check out the divide-by-12 frequency counter shown in Fig. 5–6. This type of counter is employed in digital clocks, for example. The divide-by-12 function is provided by a *JK* flip-flop configuration that divides by two and then divides by six. Observe that the reset function is activated by driving both R_o inputs logic-high, whereupon all flip-flops are reset to 0.

With reference to Fig. 5–6, flip-flop A divides by two, whereas flip-flop B, flip-flop C, and flip-flop D divide by six. Note that this counter does not count up to binary 5; instead, it counts in binary ($Q_DQ_CQ_B$) as follows:

0(000)
1(001)
2(010)
4(100)
6(101)
6(110)

In other words, binary 3 is absent from this counting sequence. Nevertheless, since six pulses are required to obtain one complete output pulse at Q_D, this configuration effectively divides by six. The bottom line is that inasmuch as only the Q_D output is of functional concern, the divide-by-12 processing may employ either conventional or unconventional count sequencing. Unconventional count sequencing minimizes hardware, and is preferred in practice.

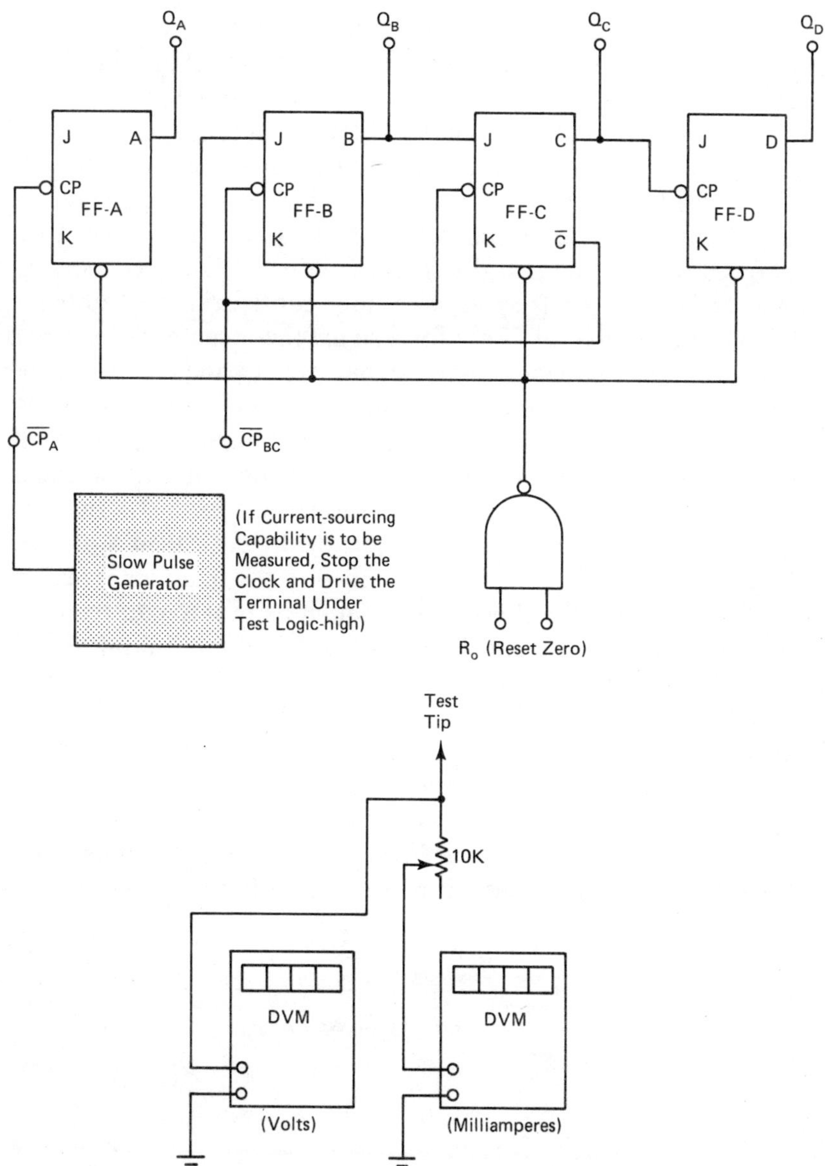

Note: The current-sourcing capability of a *Q* terminal can be easily checked with 10-kΩ resistor and two DVMs. The test tip is applied at a *Q* terminal, and the load resistance is reduced until the logic-high threshold is reached. Then, the current reading shows the current-sourcing capability of the terminal under test.

Figure 5–6 Experimental divide-by-12 frequency counter.

If you use flip-flop A for the input, and use output Q_A as the clock for the divide-by-six section, this configuration will provide divide-by-12 operation. Note that the divider does not count from 0 to 11 in binary sequence (binary six and seven are absent in the counting sequence). Nevertheless, binary 1011 normally outputs after 11 input pulses have been applied.

Consider the gated counter arrangement shown in Fig. 5-7. This arrangement comprises a main gate which is switched on or off as required, followed by a counter. It provides counting action over a desired period of time, and is a basic frequency counter configuration. In other words, if n denotes the number of Hertz for a signal that occurs during a period of time t, the average frequency of the signal over period t is equal to n/t H. If $t = 1$ s, the counter output will equal the signal frequency in Hertz. (Each input pulse corresponds to one complete excursion of the signal.)

In some frequency-counter arrangements, the signal under test is not directly applied to the main gate (shown in Fig. 5-7.) Instead, the signal under test is first applied to a frequency divider called a prescaler. In turn, the output from the prescaler is applied to the main gate, and thence to the final counter. Troubleshooters will find it helpful to construct and check out the prescaler arrangement shown in Fig. 5-8. First, initialize the clock logic-high. Then clear the counter by switching the Clear line from logic-high to logic-low. Proceed by switching the clock line from high-to-low-to-high, and so on. Observe that the output changes state on every twelfth clock pulse.

It is evident that a prescaler is a logic circuit which provides an output signal related to the input signal by some fractional scale factor such as 1/2, 1/8, 1/10, etc. A decade frequency divider is often used as a prescaler; its output frequency is equal to one-tenth of the input frequency.

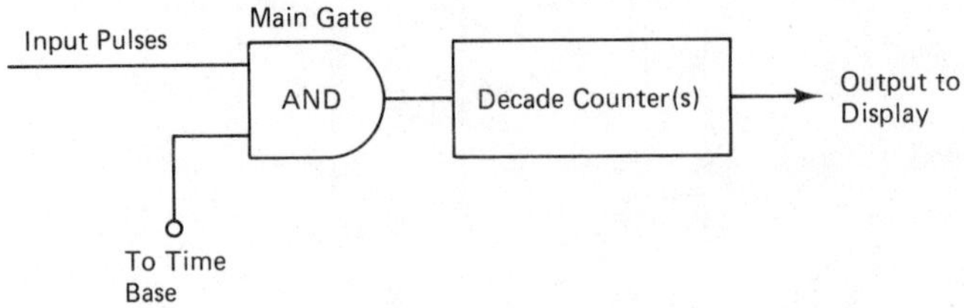

Note: This arrangement is widely used in commercial frequency counters. It is operated in combination with a precision time base so that the counter output will precisely correspond with the frequency of the signal under test.

Figure 5-7 Experimental gated frequency counter arrangement.

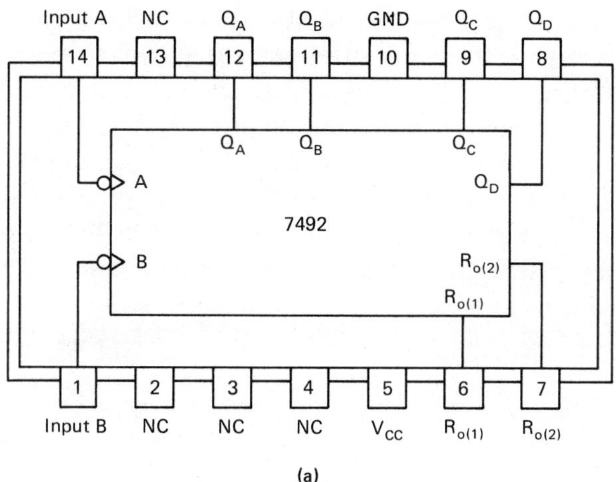

(a)

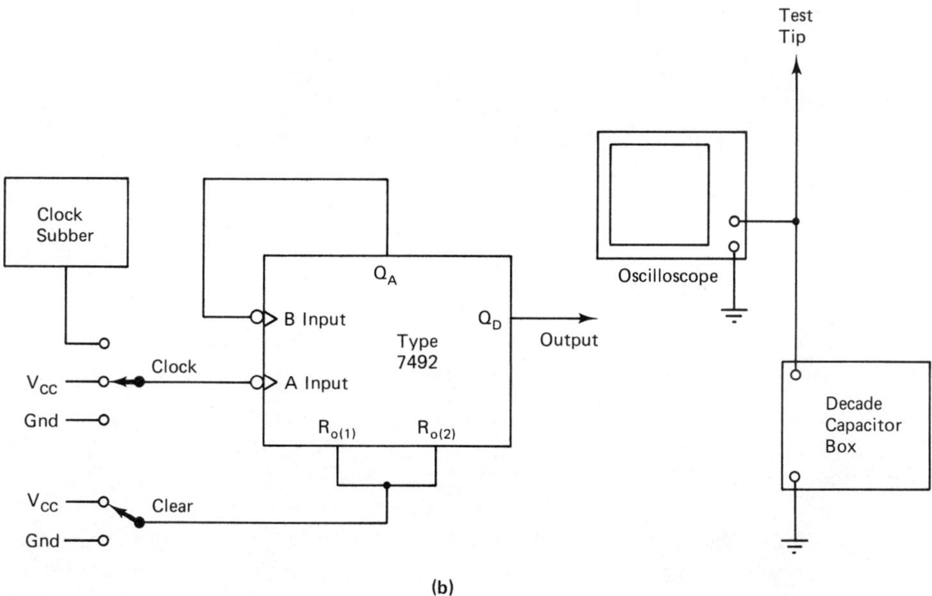

(b)

Note: This experiment shows the effect of a capacitive load on the output of the prescaler. Observe that substantial capacitive loading causes slowed rise and attenuates the output at faster clock rates.

Figure 5–8 Experimental prescaler circuit for frequency counting. (a) Pinout for the 7492 divide-by-12 counter. (b) Prescaler connection diagram.

Digital troubleshooters will also find it helpful to construct and check out the main-gate/prescaler arrangement shown in Fig. 5–9. Note that the prescaler output goes logic-high when the main gate is logic high, and vice versa. In this example, the input frequency is 1,000 kHz, and the output frequency is approximately 8.3 kHz. In turn, a logic probe at the main scaler output will glow continuously while the main gate is logic-high.

The foregoing prescaler arrangement may be driven from a square-wave or pulse generator. Output from the prescaler may be indicated by a logic

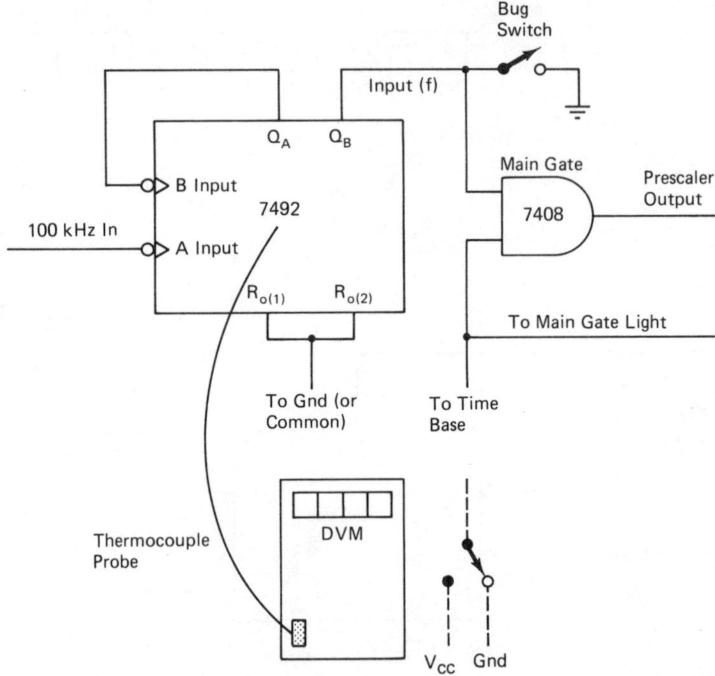

Note: This experiment illustrates the effect of excessive current demand on warm-up time. First, turn the circuit on, and observe the warm-up time that is required to arrive at normal operating temperature. Next, turn the circuit off and let it cool down to room termperature. Then, close the bug switch and again observe the warm-up time that is required to arrive at (and pass) normal operating temperature. Observe that the warm-up time is much shorter with the bug switch closed.

Construction of a 2-channel recording voltmeter is described in Chapter 7, whereby comparison recordings of warm-up times can be made.

Figure 5–9 Experimental arrangement for a prescaler/main-gate configuration.

probe, or by an LED connected in series with a 330-Ω resistor. The time-base function can be simulated by manually switching the time-base line from logic-low to logic-high and vice versa. Observe that the input signal frequency (f) is indicated at the prescaler output when the main gate is logic-high. (The state of the main gate is indicated by an LED; this is called the main gate light in frequency counter circuits.)

DIRECTION OF DATA FLOW

When troubleshooting digital equipment without service data, the technician is often concerned with the question of which direction data flow is going along a PC conductor. This is a key point in identifying output devices and input devices. In the example of Fig. 5-10, an output bus connects a "talker" to a "listener." The talker is an output device, and the listener is an input device. Data flow takes place from the talker to the listener. Troubleshooters state that a talker is a data source, and that a listener is a data destination.

In TTL circuitry, to quickly check which direction data is flowing along a conductor, a lab-type DVM may be applied at a pair of test points along the conductor. With reference to Fig. 5-10, data flows from a totem-pole output to an emitter input. The chief current associated with this data flow is an electron flow from the emitter of $Q1$ along the conductor and thence through $Q3$ to ground. Observe that the direction of data flow is opposite to the direction of electron flow in the conductor.

Accordingly, to check the direction of data flow along a conductor, the troubleshooter applies a sensitive dc voltmeter, such as a lab-type DVM, at a pair of points along the conductor. In turn, he notes which test point is positive with respect to the other test point. He concludes that the data is flowing along the conductor in a direction from the positive test point to the negative test point.

This is the seventh step in troubleshooting digital equipment without service data, using time-saving shortcuts.

BINARY CODED DECIMAL COUNTER

Digital troubleshooters frequently encounter binary coded decimal (BCD) counters. A widely used type of synchronous BCD counter is shown in Fig. 5-11. It is similar to conventional 4-bit up-counters, but has mod-10 operation. This example utilizes the 8421 code, and sequences through 10 states. Note that to maintain FF2 in the 0 state on the next clock pulse following the 1001 state, $\bar{Q}$ of FF4 is connected to gate $U1$. The $\bar{Q}$ output of FF4 is

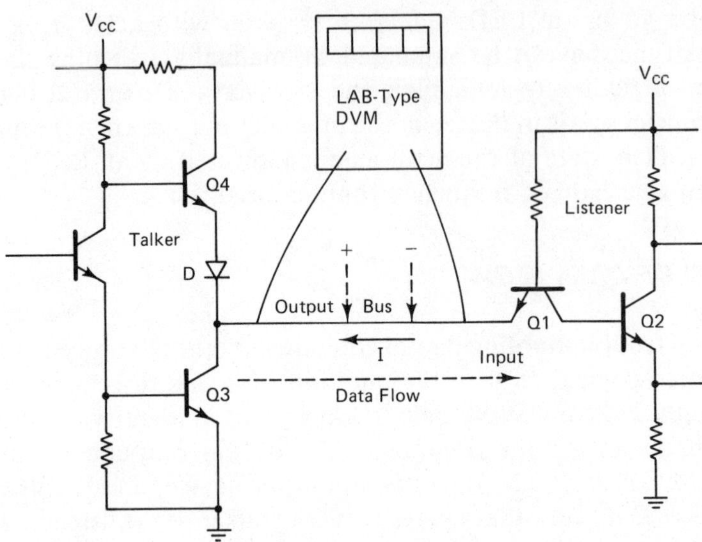

Note: A typical PC conductor has a resistance of 0.047 ohm per inch. A current flow of 1 mA produces an IR drop of 47 microvolts per inch. A lab-type DVM has a resolution of 1 μV.

Note: The *data flow* consists of a succession of current pulses from the listener to the talker. These are dc pulses, and their average value produces an IR voltage drop along the bus wire. At any two points along the bus wire, this IR voltage is more positive at the point nearest to the talker. In turn, the troubleshooter can quickly determine which end of a bus wire is the talker end, and which the listener.

Figure 5–10 Determination of the direction of data flow along a PC conductor.

logic 0; in turn, the *J* and *K* inputs to FF2 are logic 0, and FF2 does not change to logic 1 at the next clock edge.

Furthermore, to reset FF4 to 0, *Q* of FF1 is connected directly to *K* of FF4. Accordingly, the *K* input is alternately high and low during the counts of 1 through 6, but the *J* input of FF4 is continuously low, so that FF4 remains in the 0 state. Observe that at count 7, all inputs to *U*3 become high, and there is a high signal at both *J* and *K* of FF4. Thus, with the next clock edge, (count 8), FF4 toggles to the 1 state. FF4 remains in the 1 state after one more clock pulse (count 9), inasmuch as *Q* of FF1 is now 0, which removes the logic 1 from both *J* and *K* of FF4. At count 9, *Q* of FF1 goes again to 1, and in turn FF4 has a high *K* input although its *J* input is still low. Accordingly, on the next clock edge, FF4 resets back to the 0 state.

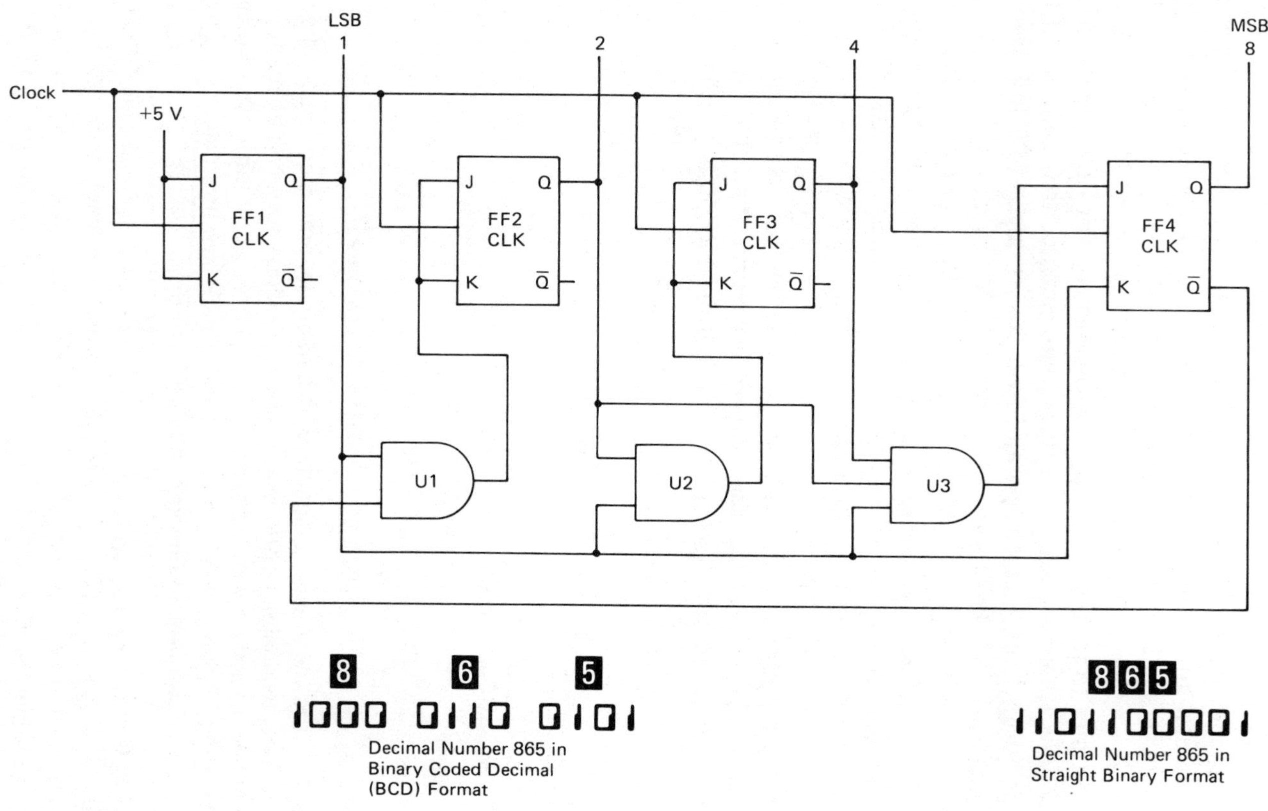

Note: This is a mod-10 counter because it sequences through 10 states.

Figure 5–11 Binary coded decimal counter logic diagram. (Courtesy, Hewlett-Packard.)

SERIES OPERATION OF DEVICES

Devices may be operated in series in order to obtain desired characteristics in logic circuits. For example, we may occasionally see two or three buffers connected in series. This is done when a certain amount of propagation delay is needed at a given point in a network. Consider an AND gate which is subject at times to inputs which are complemented. For example the AND gate may have 1,0 inputs which are then complemented to 0,1 inputs. If the input transitions from high-to-low and from low-to-high are simultaneous, the output from the AND gate will remain logic-low during the complementary transition.

Consider, however, a network in which the input waveforms are not quite simultaneous—one input waveform is slightly delayed with respect to the other waveform. In such a case, the AND gate will produce a glitch (very narrow pulse) during the complementary transition. (See Fig. 5–12.) To prevent glitch generation in this case, one or more buffers may be included in series with the input that has the "early" waveform. Thereby, both input waveforms are made simultaneous, and glitch production is avoided.

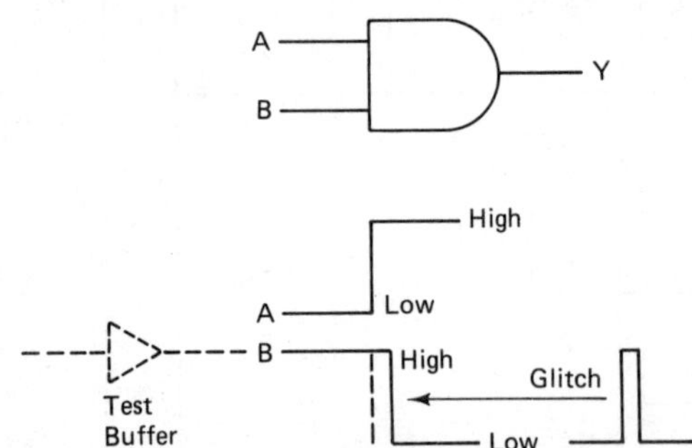

Note 1: If the technician suspects that a marginal timing condition is developing occasional or continuous glitches, a quick check can be made by connecting a buffer in series with the suspected early path. The buffer will introduce a small delay. In turn, if the trouble symptom clears up, the suspicion of a marginal timing condition is verified.

Note 2: It may be observed that any digital network is invariably "full of glitches" inasmuch as it is a switching system. The glitches do not cause false triggering as long as their pulse width (energy content) does not exceed a critical value. A test buffer will not respond to glitches that have widths less than the critical value.

Figure 5–12 Unequal propagation delays can result in glitch generation.

Typical buffers have a propagation delay of 10 ns. It is evident that when digital pulses flow through numerous paths in a logic network, that they acquire more and more propagation delay. Troubleshooters often describe the resulting timing relations as "clock skew." Clock skew is defined as phase shift in a single clock distribution system in a digital network.

HANDY IC COMPARISON QUICK CHECKER

A speedy IC comparison quick checker is depicted in Fig. 5-13. It consists of a pair of proto clips connected to a row of red/green LEDs and 330-Ω resistors. The quick checker is applied by clamping one proto clip over the

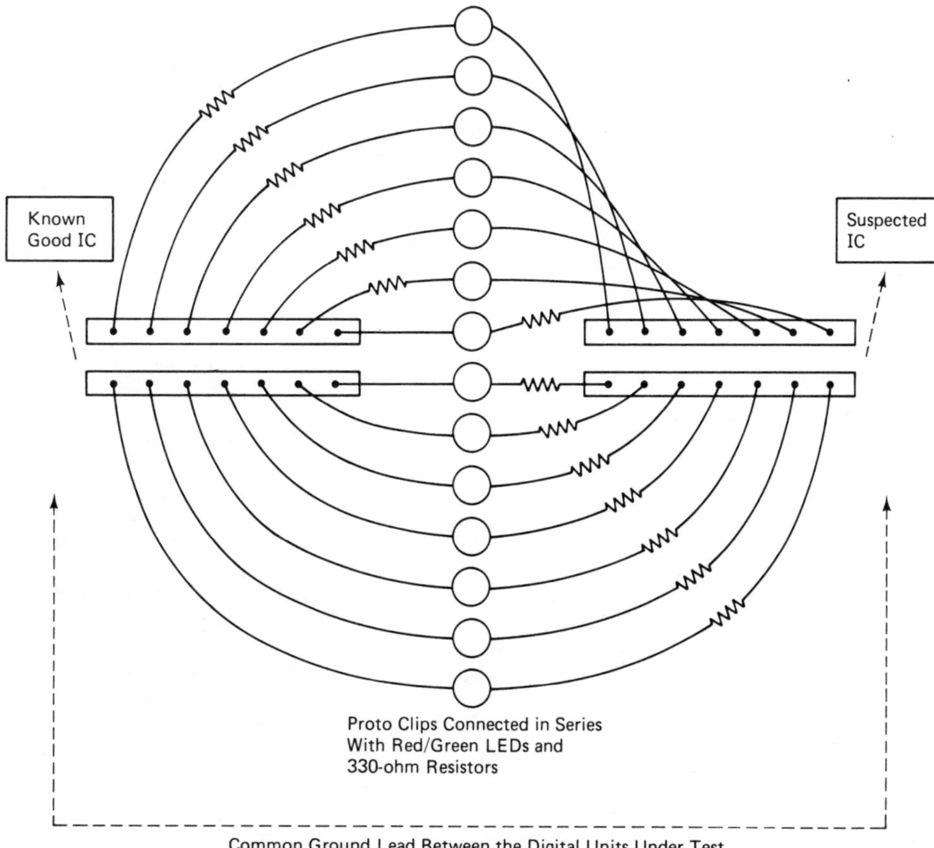

Figure 5-13 A proto-clip/LED quick checker speeds up comparison tests of digital units.

suspected IC in the digital unit under test, and clamping the other proto clip over a corresponding IC in a similar digital unit that is in normal working condition. If any IC glows red or green, it is indicated that the suspected IC is defective.

On the other hand, if all of the ICs remain dark, it is indicated that the suspected IC is ok. This quick check is based on the principle that the ICs in similar digital units will have the same terminal voltages at all times. Of course, the two digital units must be operating under similar conditions. In other words, if the units are clocked, then the troubleshooter must use a substitute clock in lieu of the regular clocks. Or, he can merely disable the regular clocks and make static tests with both clock lines high, or with both clock lines low.

6

Encoder and Decoder Identification and Troubleshooting

DIGITAL STETHOSCOPE

One of the most useful instruments for preliminary troubleshooting of digital equipment without service data is the digital stethoscope, illustrated in Fig. 6–1. This is a localized strayfield (RFI) quick checker. The probe (pickup plate) is placed on an IC package, and an input data signal is applied to the equipment under test. In turn, the probe picks up the RFI produced by the digital activity inside of the IC package, and feeds this RFI signal into a mini-amplifier with a built-in speaker. When the equipment under test is driven with an audio-frequency input data signal, the RFI "signature" of the particular IC package becomes audible from the speaker. It is sometimes helpful to clock the equipment from test from a low-frequency clock subber.

The digital troubleshooter can immediately determine whether an IC package has digital activity, whether it is inactive, and whether its digital "signature" is normal or abnormal (on a comparative basis). In other words, the digital stethoscope is first placed over an IC in the equipment under test, and the resulting digital "signature" is noted. Then, the stethoscope is placed over a corresponding IC in a reference unit of equipment which is in good working condition. If the two "signatures" are the same, the troubleshooter concludes that the stethoscope is not in a trouble area. On the other hand, if the two "signatures" are different, the troubleshooter concludes that the stethoscope is in a trouble area, and he proceeds to make further tests to close in on the fault.

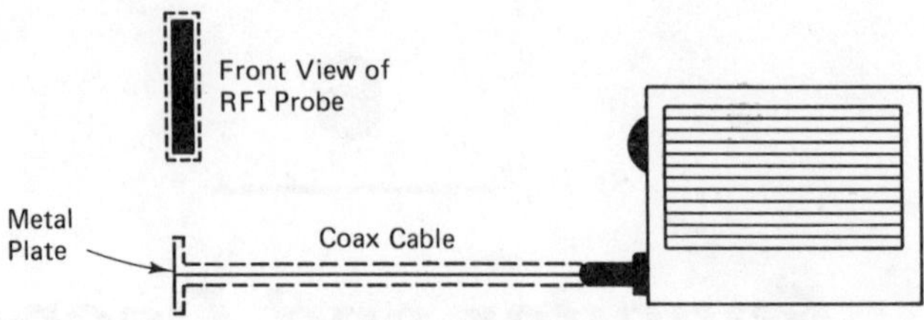

Note 1: The metal plate is the same size as the top of a digital IC package. It functions as a "floating" probe which picks up the RFI signal which is present over the IC package. The plate and cable are shielded to avoid pickup of any RFI except that which is present over the IC under test. The cable connects to a mini-amplifier with built-in speaker such as the Archer (Radio Shack) 277-1008. The great utility of this digital stethoscope is based on the fact that every digital IC package has its unique RFI "signature" when operated at an audio-frequency rate.

Note 2: This digital stethoscope is an audio-frequency tester and the digital equipment to be checked out should be clocked from a square-wave generator or clock subber operated at a low audio frequency. The digital troubleshooter will often find it helpful to patch the mini-amplifier into a tape recorder, so that a permanent record of the test can be made. In turn, the test results can be played back at any time that a comparison test may be needed.

Note 3: In some cases, the digital troubleshooter will find it helpful to input a chosen data sequence from a word generator into the digital equipment under test, instead of inputting a simple pulse or square-wave data signal.

Figure 6–1 Arrangement of the digital stethoscope.

The digital stethoscope owes its diagnostic power to the fact that the troubleshooter can make definitive preliminary tests without concerning himself about the type of IC which is under test, nor what kind of circuit the IC is associated with, nor what sort of data stream is being processed by the IC. *This is the ninth step in troubleshooting digital equipment without service data, using time-saving quick checkers and shortcuts.*

2-TO-4 LINE DECODER

Digital troubleshooters routinely encounter encoders and decoders. An encoder or code converter is a devide that produces coded combinations of outputs from discrete inputs, whereas a decoder is a device that translates

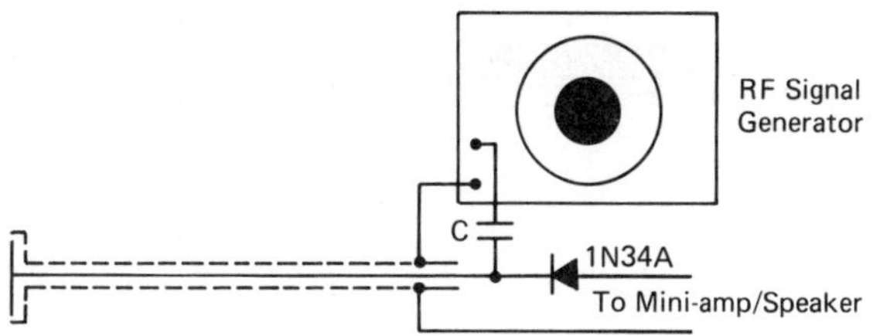

Note 4: This arrangement of the digital stethoscope is utilized for checking for the presence or absence of high-frequency pulse trains in an IC package, and also for measuring the frequency of a pulse train. Observe that a 1N34A diode is connected in series with the output lead from the digital stethoscope, and a "search" voltage is coupled via a small capacitor from an RF signal generator into the diode. If there is a high-frequency pulse train in the digital IC under test, you will hear a "birdie" from the mini-amp/speaker as you tune the RF signal generator to the fundamental frequency of the pulse train. For example, a typical digital pulse train has a frequency of 1 MHz. The mini-amp/speaker cannot indicate the presence of this pulse train unless it is heterodyned with an RF signal in the frequency range from 1,010,000 to 990,000 Hz. The pulse train in this example will zero-beat at 1 MHz. In turn, you can check the operating frequencies in all of the ICs in a digital system.

Figure 6–1 *(cont.)*

a combination of signals into one signal that represents the combination, as illustrated in Fig. 6–2. Since both devices function to change from one digital code to another, there is no sharp dividing line between encoders and decoders.

The 2-to-4 line decoder shown in Fig. 6–3 has two inputs and can output four different state patterns. A different output gate is enabled by each binary state. It follows that a 3-input decoder can have eight different output state patterns, a 4-input decoder can have 16 different output state patterns, and so on. Thus, if n = inputs, the possible outputs are equal to 2^n.

BCD-TO-DECIMAL DECODER

Troubleshooters encounter BCD-to-decimal decoders in various kinds of digital equipment. A typical logic diagram is shown in Fig. 6–4. Since four inputs are available, there are 16 possible output state patterns. However, since the binary decimal code employs only the counts from 0000 to 0001, this arrangement is limited to ten output states. A strobe input is provided

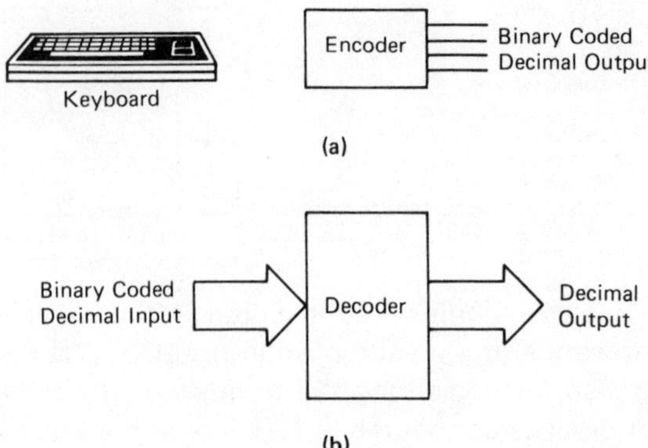

Note: Decoders are used in digital clocks at the outputs of the shift counters to format the Seconds, Minutes, and Hours displays.

Figure 6–2 Examples of encoders and decoders. (a) Encoder for keyboard. (b) Binary-coded-decimal to decimal decoder.

in this example—it functions to enable the decoder for a brief period of time and to lock out the 8421 lines except for the strobe pulse duration.

In the example of Fig. 6–4, an input state pattern 0111 is applied (this is equal to decimal 7). Accordingly, the "7" output lines goes logic-high for the duration of the strobe pulse. Troubleshooters should note that false data rejection capability is not provided in this type of decoder. Stated otherwise, if an "illegal" input code, such as 1010 or 1011 happens to occur, the decoder will respond with an irrelevant output.

It will be seen that if a 1010 input were applied in Fig. 6–4, output lines 2 and 8 would go logic high. (In normal operation only one output goes logic high at any given time.) However, this basic decoder configuration can be elaborated, if desirable, to prevent any gate in the output circuit from going logic-high in consequence of an illegal input state pattern.

BINARY NUMBER DECODER

Binary-to-decimal decoders such as shown in Fig. 6–5 are in very wide use. This arrangement inputs numbers that have been temporarily stored in a shift register. Note that the binary number 00101 is being processed in the example—the number is stored in complementary form at the Q and $\bar{Q}$ outputs of the flip-flops. Both the Q and $\bar{Q}$ outputs are used to drive the AND

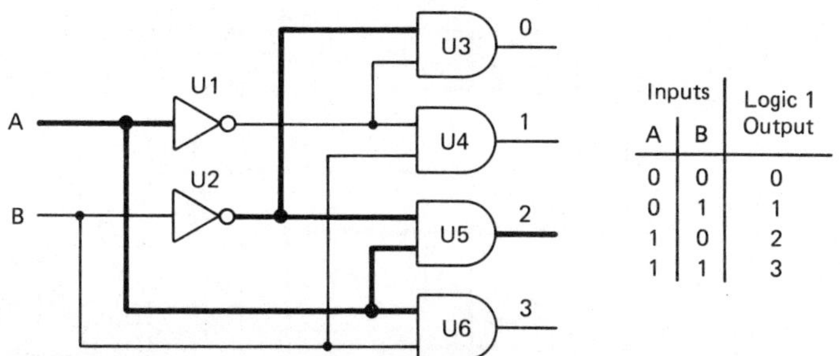

Inputs		Logic 1 Output
A	B	
0	0	0
0	1	1
1	0	2
1	1	3

Note: In this example of decoder action, line 1 is driven logic-high, and line *B* is driven logic low. In turn, both inputs of *U*5 go logic-high and output line 2 goes logic-high.

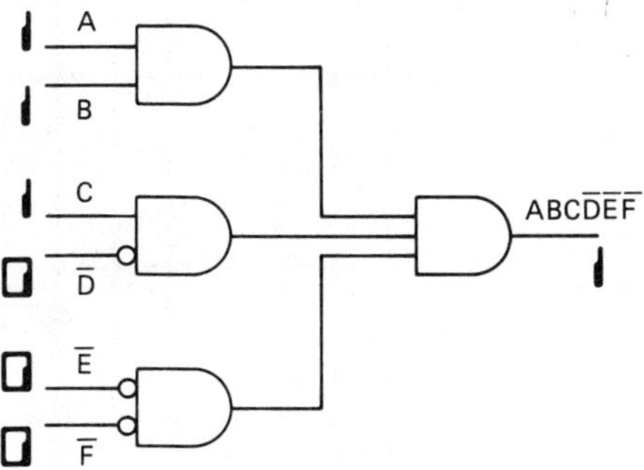

Note: This decoder configuration exemplifies processing of four inputs in order to select a specified single output. Since there are four input lines, there are 64 possible input state combinations. Of these 64 combinations, only $ABC\overline{D}EF$ represents a logic-high output.

Figure 6–3 A 2-to-4 line decoder arrangement. (Courtesy, Hewlett-Packard.)

gates in the decoder output circuit. Inasmuch as 00101 equals decimal 5, the output of gate *U*6 goes logic-high in this example.

The arrangement in Fig. 6–4 also exemplifies provision of supplementary gates—in this case, supplementary gates for decoding the numbers 30 and 31. Observe that if the binary number 11111 (decimal 31) is stored in the shift register, the *Q* output of each flip-flop will be 1. Consequently, gate *U*12 will be driven logic-high. Again, if the binary number 00000 (decimal 0) is stored in the shift register, the $\bar{Q}$ output of each flip-flop is 1. Consequently, gate *U*1 will be driven logic-high.

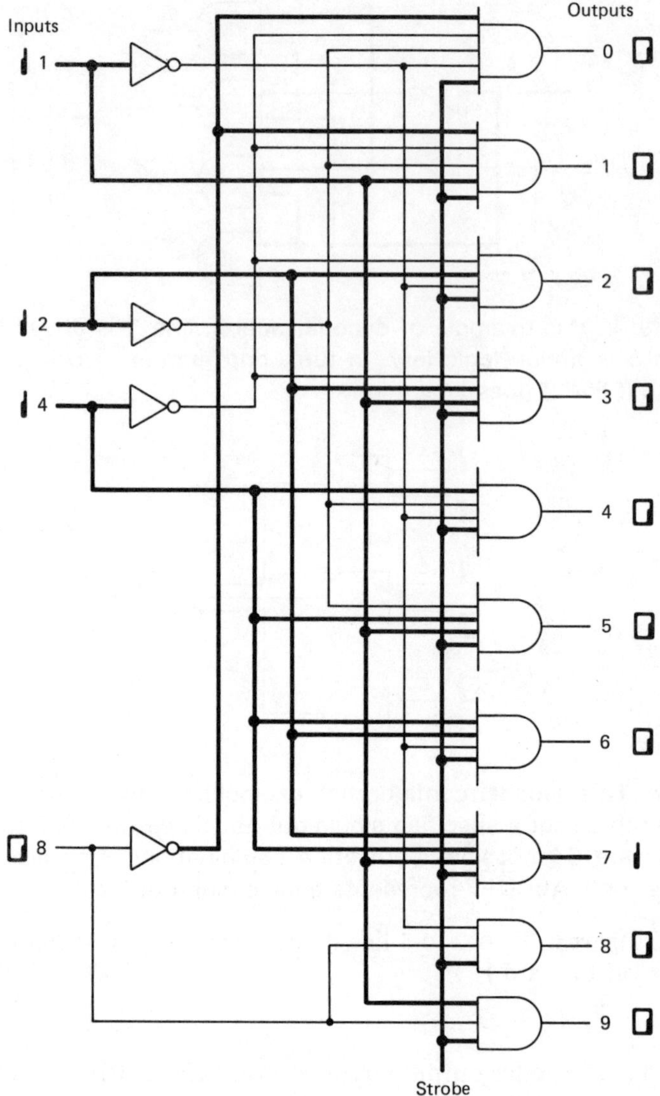

Note: This arrangement is also called a code converter.

Figure 6–4 Example of a BCD to decimal decoder. (Courtesy, Hewlett-Packard.)

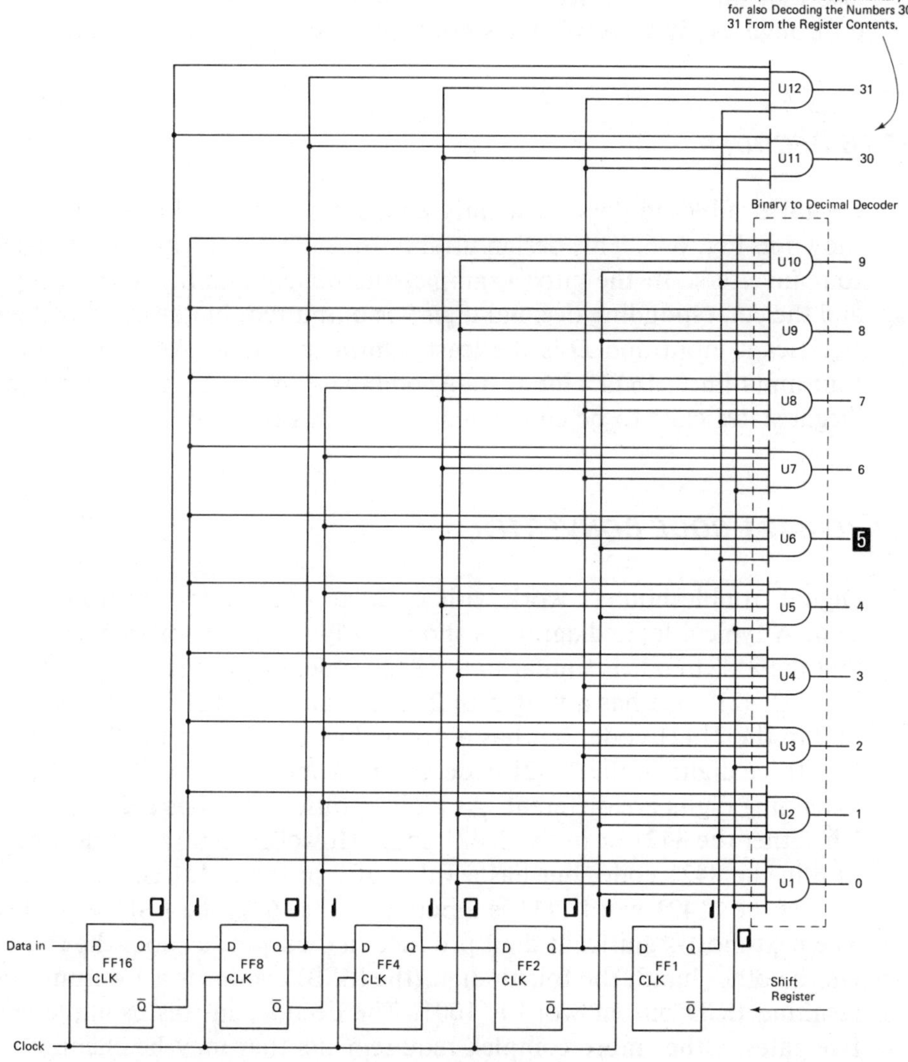

Figure 6–5 Binary-to-decimal number decoder. (Courtesy, Hewlett-Packard.)

The troubleshooter should note that this type of decoder is unresponsive to illegal inputs. Thus, the decoder will output decimal numbers 0 through 9, and will also output decimal numbers 30 or 31. On the other hand, it will be seen that the decoder is unresponsive to an illegal input such as 1100. That is, an 1100 input cannot drive any one gate logic-high. Note also that this type of decoder can be designed to decode all numbers from 0 through 31, if 32 AND gates are employed in the output circuit.

1-OF-16 DECODER

Digital troubleshooters frequently encounter 1-out-of-16 decoders such as shown in Fig. 6–6. This design utilizes 16 AND gates with four buffers and four inverters. In the cited example, the binary number 1001 is inputted, and the corresponding decimal digit 9 is outputted. Note that *A* is the most-significant input and *D* is the least-significant input. Also, since there are four input lines, 16 (2^4) input state combinations can be accommodated. No illegal states exist to be contended with in this configuration.

2′421-TO-8421 CODE CONVERTER

Digital troubleshooters work with 2′421-to-8421 code converters on occasion. A typical logic diagram is shown in Fig. 6–7. It was previously noted that the 2′421 code is similar to the 8421 code, except that the MSB position in the 2′421 code has a weight of 2. For example, 1111 has a decimal value of 9 in the 2′421 code, but has a decimal value of 15 in the 8421 code. The first three digits in the 2′421 code are the same as in the 8421 code, but the remaining digits are different. As an illustration, 111 has a decimal value of 7 in either the 8421 or in the 2′421 code. However, 1000 has a decimal value of 8 in the 8421 code, but has a value of 2 in the 2′421 code.

If the 2′421 word 1111 is inputted in Fig. 6–7, the LSB feeds through. The next-most-significant digit produces a 0 output, as does the third digit. On the other hand, the fourth digit (the MSB) produces a 1 output, and the resulting BCD output word is 1001. The decoder in this example employs five gates. Other more complex requirements that may be encountered require decoders with more gates which are often implemented as read-only memories (ROMs). This topic is detailed subsequently.

KEYBOARD-TO-BCD ENCODER

Troubleshooters encounter keyboard-to-BCD encoders, as exemplified in Fig. 6–8. An encoder, like a decoder, is a code converter; a code converter with a single input and several outputs is generally called an encoder. The

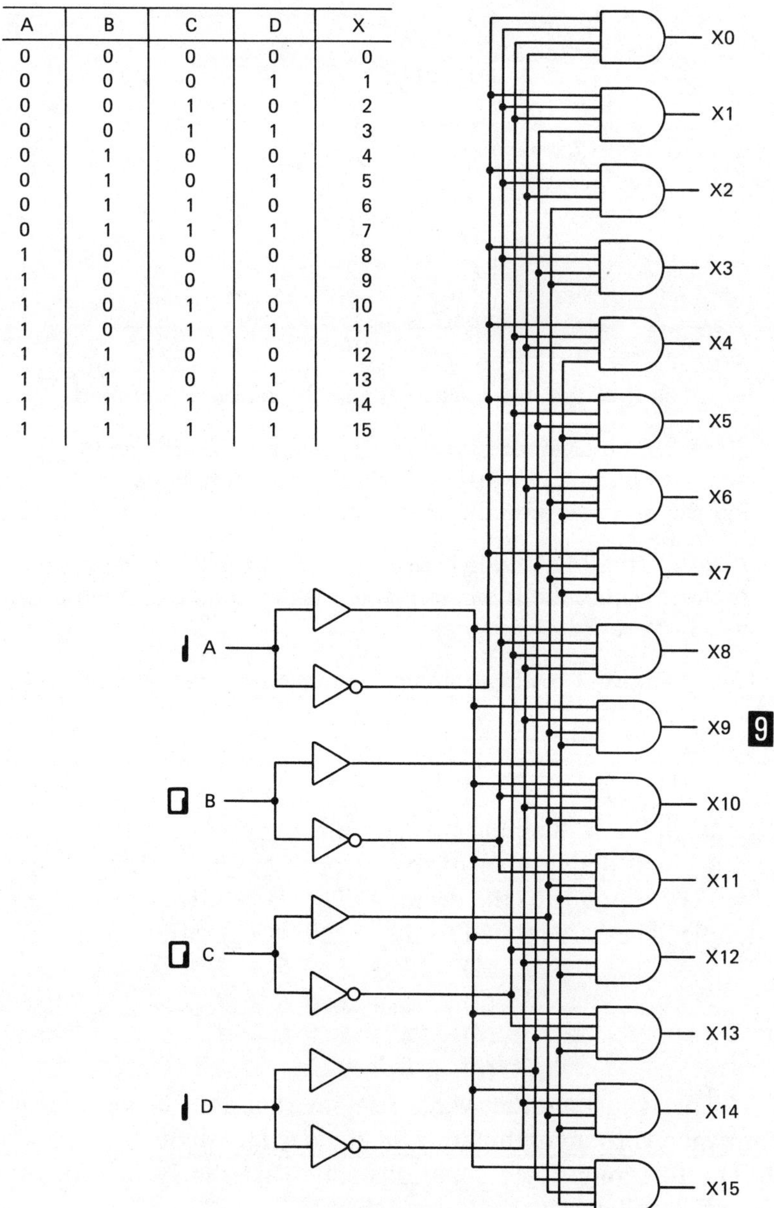

A	B	C	D	X
0	0	0	0	0
0	0	0	1	1
0	0	1	0	2
0	0	1	1	3
0	1	0	0	4
0	1	0	1	5
0	1	1	0	6
0	1	1	1	7
1	0	0	0	8
1	0	0	1	9
1	0	1	0	10
1	0	1	1	11
1	1	0	0	12
1	1	0	1	13
1	1	1	0	14
1	1	1	1	15

Figure 6–6 A 1-of-16 decoder arrangement.

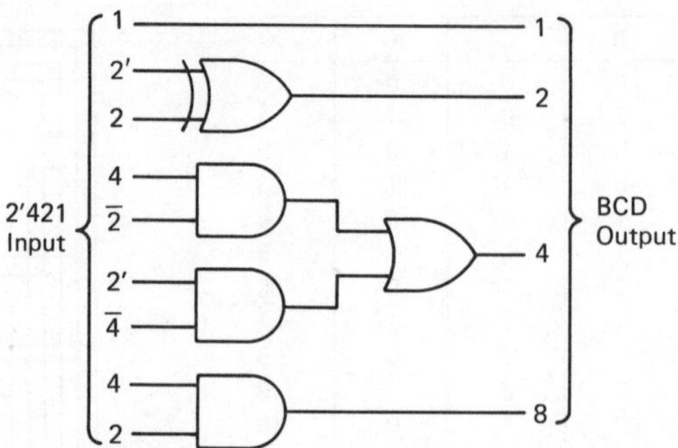

Note 1: The 2′421 code is also called the 2421 code. The 2′ notation signifies that the left-most place in the nibble is denoted.

Note 2: The $\bar{Z}$ input signal to the decoder is provided by passing the 2 signal through an inverter. Similarly, the $\overline{4}$ put signal is obtained by passing the 4 signal through an inverter.

Note 3: The 2421 code is used to obtain the 9′s complement of a binary number. Thus, the 9′s complement is obtained by taking the 1′s complement of the 2421 number.

Figure 6–7 A logic circuit for converting from 2′421 to 8421 BCD code.

encoder in Fig. 6–8 inputs a pulse from a keyboard switch which corresponds to a decimal digit. In turn, the encoder outputs a BCD signal (8421 code) which corresponds to the decimal digit. Four OR gates are employed in this example. They are related to the encoding operation by the logic equations:

$$(1) = 1 + 3 + 5 + 7 + 9$$
$$(2) = 2 + 3 + 6 + 7$$
$$(4) = 4 + 5 + 6 + 7$$
$$(8) = 8 + 9$$

This is just another way of saying that 8421 is the BCD number which corresponds to a combination of the decimal digits 0, 1, 2, 3, 4, 5, 6, 7, 8, 9. The foregoing logic equations state that the BCD LSB output shall be logic-high whenever the decimal digit is 1, or 3, or 5, or 7, or 9, as tabulated in the foregoing truth table. The next-most-significant digit, in turn, shall be logic-high whenever the decimal digit is 2, or 3, or 6, or 7, and so on. Inasmuch as only OR relations are involved in the encoder, the logic diagram comprises only four OR gates.

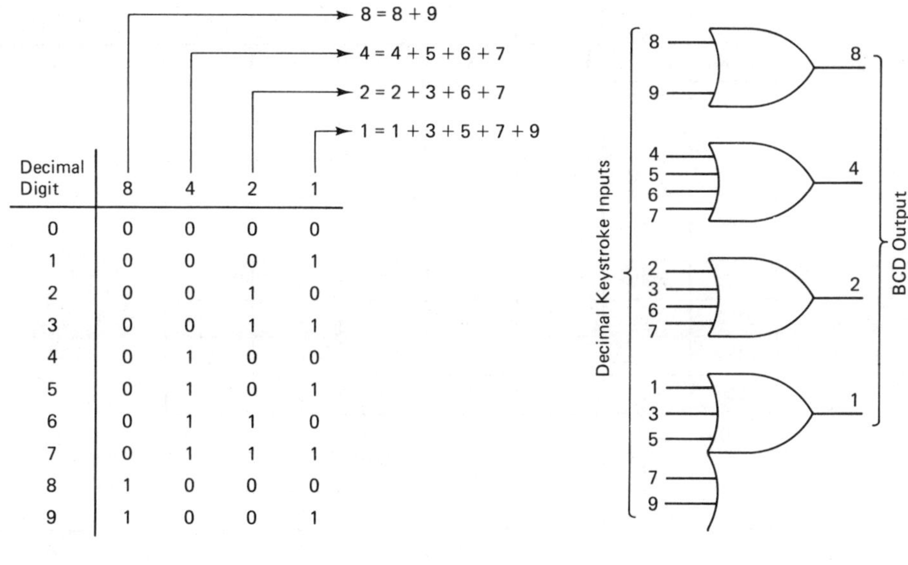

Decimal Digit	8	4	2	1
0	0	0	0	0
1	0	0	0	1
2	0	0	1	0
3	0	0	1	1
4	0	1	0	0
5	0	1	0	1
6	0	1	1	0
7	0	1	1	1
8	1	0	0	0
9	1	0	0	1

(a)

Note: In a typical system, an ASCII decoder would be employed between the keyboard and the bus lines in a microcomputer. Thus, the decimal keystroke signals would be transmitted to the CPU in ASCII code.

Figure 6–8 (a) An encoder that inputs keystrokes from a decimal keyboard and outputs binary coded decimal signals. (b) American Standard Code for Information Exchange (ASCII). (Reproduced by permission of Reston Publishing Co. from *Understanding Microprocessors*.)

BINARY-TO-SEVEN-SEGMENT ENCODER

The digital troubleshooter works with binary-to-seven-segment decoders in a large variety of digital equipment. With reference to the example in Fig. 6–9, ten input lines are provided which correspond to the decimal digits. The encoder has seven output lines (*a* through *g*) which correspond to the segments in a 7-segment display device. The display device reads out a single decimal digit. When an encoder input line is energized, current flows into corresponding output lines associated with the diodes that interconnect the input and output lines.

Observe in Fig. 6–9 that when the decimal 7 input line is energized, the output *a,b,c* lines are in turn energized. Accordingly, the numeral 7 is dis-

	000	001	010	011	100	101	110	111
0000	NULL	① DC_0	ƀ	0	@	P		
0001	SOM	DC_1	!	1	A	Q		
0010	EAO	DC_2	"	2	B	R		
0011	EOM	DC_3	#	3	C	S		
0100	EOT	DC_4 (stop)	$	4	D	T		
0101	WRU	ERR	%	5	E	U		
0110	RU	SYNC	&	6	F	V		
0111	BELL	LEM	'	7	G	W	Unassigned	
1000	FE_0	S_0	(	8	H	X		
1001	HT / SK	S_1	)	9	I	Y		
1010	LF	S_2	*	:	J	Z		
1011	V_{TAB}	S_3	+	;	K	[		
1100	FF	S_4	(comma) ,	<	L	\		ACK
1101	CR	S_5	—	=	M	]		②
1110	SO	S_6	·	>	N	↑		ESC
1111	SI	S_7	/	?	O	←		DEL

Example: | 100 | 0001 | = A

b_7 — — — — — — — b_1

The Abbreviations Used in the Figure Mean:

NULL	Null Idle	DC_0	Device Control ① Reserved for Data Link Escape
SOM	Start of Message	DC_1-DC_3	Device Control
EOA	End of Address	ERR	Error
EOM	End of Message	SYNC	Synchronous Idle
EOT	End of Transmission	LEM	Logical End of Media
WRU	"Who Are You?"	SO_0-SO_7	Separator (Information) Word Separator (Blank, Normally Non-printing)
RU	"Are You . . .?"	ACK	Acknowledge
BELL	Audible Signal	②	Unassigned Control
FE	Format Effector	ESC	Escape
HT	Horizontal Tabulation	DEL	Delete Idle
SK	Skip (Punched Card)		
LF	Line Feed		
V/TAB	Vertical Tabulation		
FF	Form Feed		
CR	Carriage Return		
SO	Shift Out		
SI	Shift In		

(b)

(Reproduced by Permission of Reston Publishing Company From "Understanding Microprocessors")

Figure 6–8 *(cont.)*

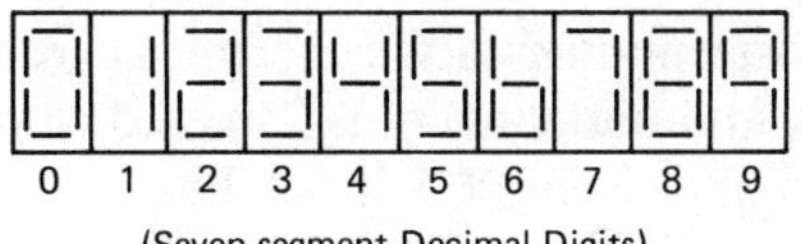

(Seven-segment Decimal Digits)

Decimal Digit-select Lines

0
1
2
3
4
5
6
7
8
9

a b c d e f g

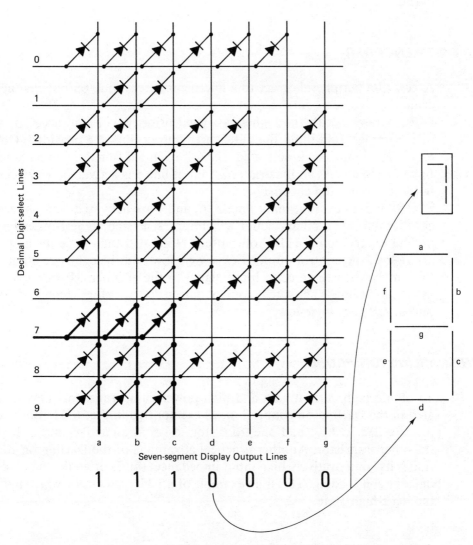

Seven-segment Display Output Lines

1 1 1 0 0 0 0

a
f b
g
e c
d

Note: This diode matrix is an example of a read-only memory (ROM).

Figure 6–9 A decimal/7-segment diode matrix encoder.

played. Note that the encoder diodes permit current flow in only one direction. Consequently, an input current is "steered" into particular output lines. For example, when the decimal 7 input line is energized, forward current flows into the *a,b,c* output lines. Current flow is blocked from the *d,e,f,g* output lines.

PRIORITY ENCODER

Access of a peripheral device to a microcomputer input/output channel, for example, is often controlled by a priority encoder such as shown in Fig. 6–10. A device with the highest assigned priority will gain access first. In this example, eight input lines and three output lines are provided. Observe that if one input line is activated, the arrangement functions as an 8-line-to-binary encoder. As an illustration, if input line $\overline{5}$ is activated, a 101 count is outputted from the encoder.

Suppose next that input lines $\overline{3}$, $\overline{6}$, and $\overline{7}$ are activated. In this case, the encoder outputs a binary count which represents the highest-order line ($\overline{7}$), or, the binary count 111 is outputted. Note also that since all inputs are active-low, the control output $\overline{EO}$ is logic-low and the control output $\overline{GS}$ is logic-high when there is no input signal to the encoder. However, as soon as any input line is activated, control output $\overline{EO}$ goes logic-high and control output $\overline{GS}$ goes logic-low.

THE DARLINGTON PROBE

To obtain high current gain and high power gain, transistors may be operated in the Darlington connection, as exemplified in Fig. 6–11. Both collectors are tied to V_{CC}, and one transistor serves as an active emitter load for the other transistor. Another practical advantage of the Darlington connection is its comparatively high input impedance; this is often desirable to minimize circuit loading. This is an example of a TTL logic probe which indicates the logic-high state.

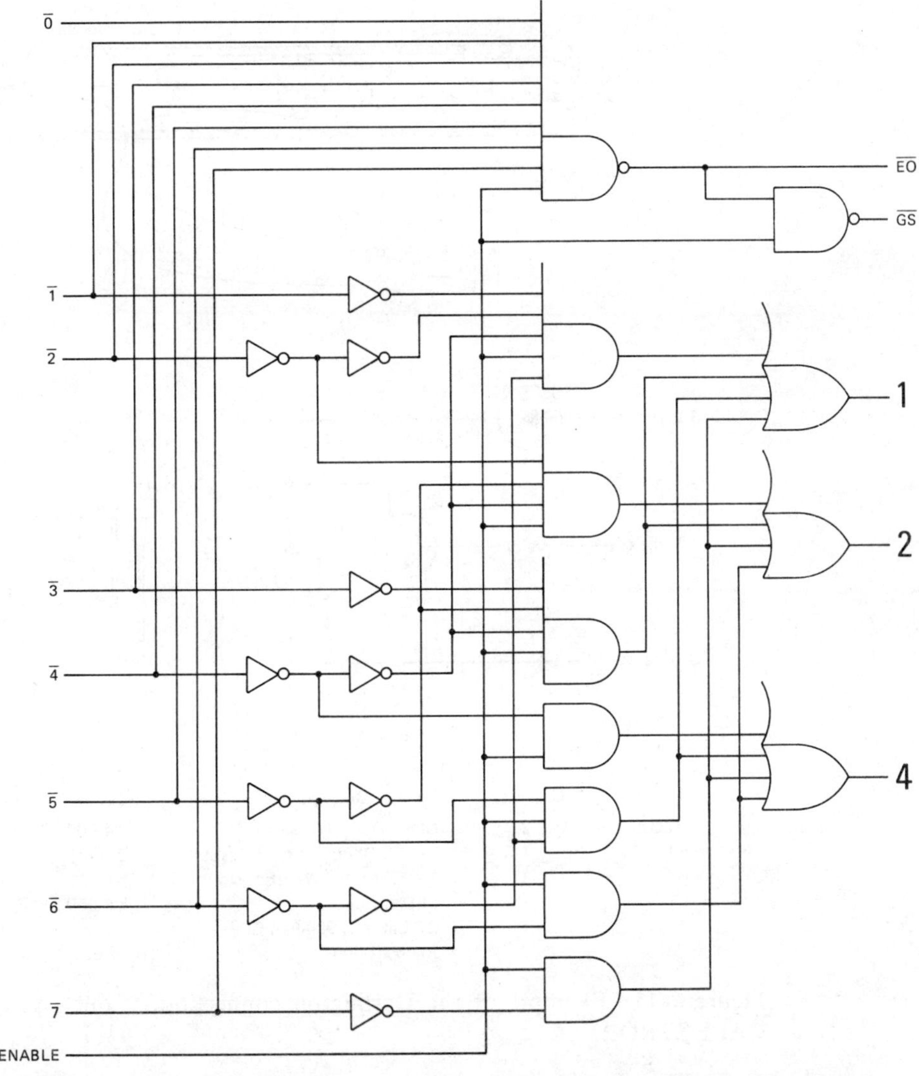

Note: This configuration is a priority encoder arrangement which is an elaborated 8-line-to-binary encoder configuration. The priority encoder includes an Enable control line whereby the input data flow can be either enabled or inhibited. Observe that the eight input lines develop a 421-binary output, and also produce an Enable Output ($\overline{EO}$) signal and a Gate or Strobe signal ($\overline{GS}$) for control of access to the microcomputer input/output channel by peripheral devices.

Figure 6–10 A priority encoder arrangement. (Courtesy, Hewlett-Packard.)

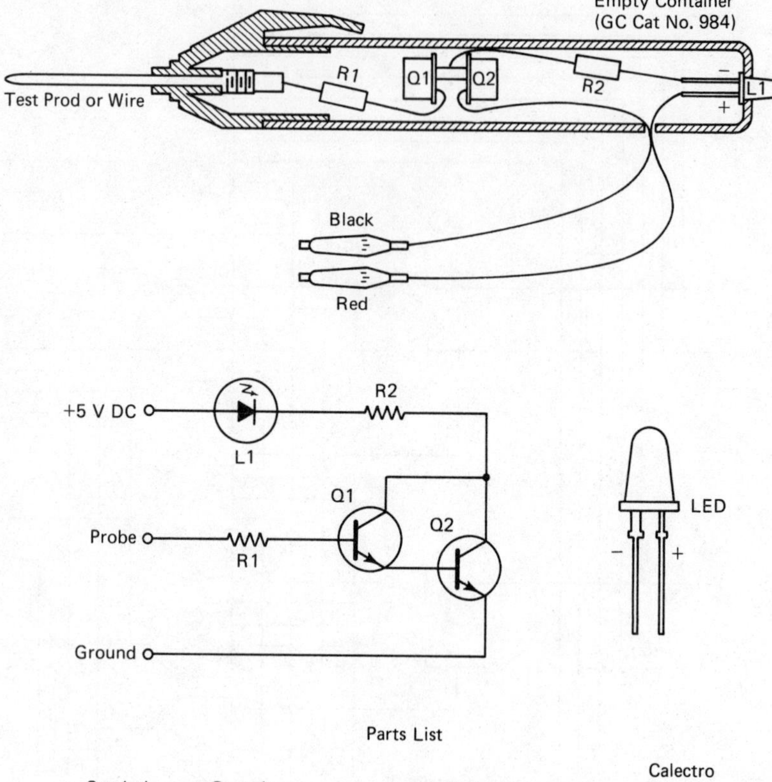

Parts List

Symbol	Quantity	Description	Calectro Cat. No.
Q1, Q2	2	NPN Silicon Transistor	K4-507
R1	1	27,000 Ohm, 1/2 Watt Resistor	B1-401
R2	1	150 Ohm, 1/2 Watt Resistor	B1-374
L1	1	LED Lamp	K4-559
	1	GC Cat. No. 984 Pen Oiler	
	2	Jumper Wires	J4-650

Figure 6–11 Example of the Darlington connection. (Courtesy, CALECTRO.)

I

Time-Saving Computer Programs for the Experimenter, Hobbyist, Technician, Student, and Circuit Designer

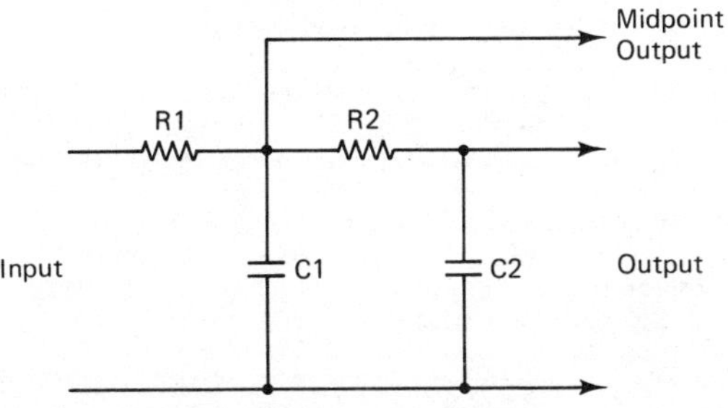

The following programs are written in Microsoft® for the IBM PC computer. However, they will run on various other computers without difficulty. In some cases, a bit of conversion may be required. Note that like many desk-top computers, the IBM PC processes data in terms of radians (not degrees), and in terms of natural logarithms (not base-10 logarithms). Most of the trigonometric functions are provided, although arc-sin and arc-cos must be derived. Observe that if you are using an old type of personal computer, all trigonometric and logarithmic functions must be programmed as subroutines.

```
1 LPRINT "UNSYMMETRICAL TWO-SECTION CASCADED RC INTEGRATING NETWORK"
2 PRINT "UNSYMMETRICAL TWO-SECTION RC INTEGRATING NETWORK"
3 LPRINT "Frequency Response, Phase Characteristic, Slope, Delay, Input Impedanc
e, Output Impedance, Midpoint Frequency Response and Phase Angle"
4 PRINT "Frequency Response, Phase Characteristic, Slope, Delay, Input Impedance
, Output Impedance, Midpoint Frequency Response and Phase Angle"
5 LPRINT "* Survey Program *": PRINT "* Survey Program *":LPRINT:PRINT
6 INPUT "R1 (Ohms)=";RO
7 LPRINT "R1 (Ohms)=";RO
8 INPUT "R2 (Ohms)=";RT
9 LPRINT "R2 (Ohms)=";RT
10 INPUT "C1 (Mfd)=";CO
11 LPRINT "C1 (Mfd)=";CO
12 INPUT "C2 (Mfd)=";CT
13 LPRINT "C2 (Mfd)=";CT
14 INPUT "f (Hz)=";H
15 LPRINT "f (Hz)=";H
16 N=0:G=H:PRINT:LPRINT:A$="########":B$="####.##"
17 F=G+N:LPRINT"f (Hz)=";USING A$;F:PRINT"f (Hz)=";USING A$;F
18 XO=1/(6.2832*F*CO*10^-6):XT=1/(6.2832*F*CT*10^-6)
19 AB=(RO*RO+XO*XO)^.5:BA=-ATN(XO/RO):LK=XO+XT:BC=(RT*RT+LK*LK)^.5
20 CB=-ATN(LK/RT):AC=BC*AB:CA=CB+BA:AD=AC/XO:DA=CA+6.2832/4:AE=AD*COS(DA)
21 EA=AD*SIN(DA):BE=EA+XO:CE=(AE*AE+BE*BE)^.5:EC=-ATN(BE/AE):AN=XT/CE
22 NA=-EC-6.2832/4:N=N+1000
23 LPRINT "Eout/Ein=";USING B$;AN:PRINT "Eout/Ein=";USING B$;AN
24 PRINT "Phase Angle (Degrees)=";USING B$;180+NA*360/6.2832
25 LPRINT "Phase Angle (Degrees)=";USING B$;180+NA*360/6.2832
26 H=F+10:XP=1/(6.2832*H*CO*10^-6):XU=1/(6.2832*H*CT*10^-6)
27 AZ=(RO*RO+XP*XP)^.5:ZA=-ATN(XP/RO):KL=XP+XU:BZ=(RT*RT+KL*KL)^.5
28 CY=-ATN(KL/RT):AQ=BZ*AZ:CQ=CY+ZA:FG=AQ/XP:GF=CQ+6.2832/4
29 BF=FG*COS(GF):FB=FG*SIN(GF):LM=FB+XP:MN=(BF*BF+LM*LM)^.5
30 NM=-ATN(LM/BF):BL=XU/MN:LB=-NM-6.2832/4:PS=(NA-LB)*360/62.832
31 C$="#####.####":LPRINT"Phase Slope (Deg/Hz at f)=";USING C$;-PS
32 PRINT"Phase Slope (Deg/Hz at f)=";USING C$;-PS:PD=NA/(6.2832*F)
33 D$="##.######":LPRINT"Phase Delay (Seconds)=";USING D$;-PD
34 PRINT"Phase Delay (Seconds)=";USING D$;-PD
35 ZA=(RT*RT+XT*XT)^.5:AZ=ATN(XT/RT):ZB=XO*ZA:BZ=AZ-6.2832/4:ZC=XO+XT
36 CZ=-6.2832/4:ZD=(RT*RT+ZC*ZC)^.5:DZ=ATN(ZC/RT):ZE=ZB/ZD:EZ=BZ-DZ
37 ZF=ZE*COS(EZ):ZG=ZE*SIN(EZ):ZH=ZF+RO:ZJ=(ZH*ZH+ZG*ZG)^.5:JZ=ATN(ZG/ZH)
38 LPRINT"Input Impedance (Ohms)=";USING A$;ZJ
39 PRINT"Input Impedance (Ohms)=";USING A$;ZJ
40 LPRINT"Phase Angle (Degrees)=";USING B$;JZ*360/6.2832
41 PRINT"Phase Angle (Degrees)=";USING B$;JZ*360/6.2832
42 AB=RO*XO:BA=-6.2832/4:AC=RO+XO:CA=-6.2832/4:AD=AB/AC:DA=BA-CA:
AE=AD*COS(DA)
43 AF=AD*SIN(DA):AG=RT+AE:AH=(AG*AG+AF*AF)^.5:HA=ATN(AF/AG):AK=AH*XT
44 KA=HA-6.2832/4:AL=(AH*AH+XT*XT)^.5:LA=ATN(XT/AH):AM=AK/AL:MA=-(KA-LA)
45 LPRINT"Output Impedance (Ohms)=";USING A$;AM
46 PRINT"Output Impedance (Ohms)=";USING A$;AM
47 LPRINT "Phase Angle (Degrees)=";USING B$;MA*360/6.2832-180
48 PRINT"Phase Angle (Degrees)=";USING B$;MA*360/6.2832-180
49 TH=ATN(XT/RT):EM=AN/SIN(TH):PH=6.2832/4-TH:PS=NA-PH:PS=PS*360/6.2832
50 LPRINT"Midpoint Ein/Eout=";USING B$;EM
51 PRINT"Midpoint Ein/Eout=";USING B$;EM
52 LPRINT"Phase Angle (Degrees)=";USING B$;180+PS
53 PRINT"Phase Angle (Degrees)=";USING B$;180+PS
54 IF F>20000 THEN 56
55 PRINT:LPRINT:GOTO 17
56 END
```

```
UNSYMMETRICAL TWO-SECTION CASCADED RC INTEGRATING NETWORK
Frequency Response, Phase Characteristic, Slope, Delay, Input Impedance, Output
Impedance, Midpoint Frequency Response and Phase Angle
* Survey Program *

R1 (Ohms)= 100
R2 (Ohms)= 1000
C1 (Mfd)= 1
C2 (Mfd)= .1
f (Hz)= .1

f (Hz)=        0
Eout/Ein=   1.00
Phase Angle (Degrees)=   0.01
Phase Slope (Deg/Hz at f)=     0.0756
Phase Delay (Seconds)= 4.999778
Input Impedance (Ohms)= 1446863
Phase Angle (Degrees)= -90.00
Output Impedance (Ohms)=     1100
Phase Angle (Degrees)=  -0.00
Midpoint Ein/Eout=   1.00
Phase Angle (Degrees)=   0.00

f (Hz)=     1000
Eout/Ein=   0.69
Phase Angle (Degrees)=  65.37
Phase Slope (Deg/Hz at f)=     0.0499
Phase Delay (Seconds)= 0.000318
Input Impedance (Ohms)=      175
Phase Angle (Degrees)= -57.68
Output Impedance (Ohms)=      883
Phase Angle (Degrees)= -33.70
Midpoint Ein/Eout=   0.81
Phase Angle (Degrees)=  33.22

f (Hz)=     2000
Eout/Ein=   0.37
Phase Angle (Degrees)= 102.38
Phase Slope (Deg/Hz at f)=     0.0266
Phase Delay (Seconds)= 0.000108
Input Impedance (Ohms)=      123
Phase Angle (Degrees)= -38.41
Output Impedance (Ohms)=      633
Phase Angle (Degrees)= -52.69
Midpoint Ein/Eout=   0.59
Phase Angle (Degrees)=  50.89


f (Hz)=     3000
Eout/Ein=   0.21
Phase Angle (Degrees)= 122.82
Phase Slope (Deg/Hz at f)=     0.0155
Phase Delay (Seconds)= 0.000053
Input Impedance (Ohms)=      111
Phase Angle (Degrees)= -27.89
Output Impedance (Ohms)=      472
Phase Angle (Degrees)= -62.85
Midpoint Ein/Eout=   0.45
Phase Angle (Degrees)=  60.77
```

```
f (Hz)=     4000
Eout/Ein=    0.13
Phase Angle (Degrees)= 135.21
Phase Slope (Deg/Hz at f)=     0.0098
Phase Delay (Seconds)= 0.000031
Input Impedance (Ohms)=      106
Phase Angle (Degrees)= -21.67
Output Impedance (Ohms)=      371
Phase Angle (Degrees)= -68.85
Midpoint Ein/Eout=    0.36
Phase Angle (Degrees)=  66.91

f (Hz)=     5000
Eout/Ein=    0.09
Phase Angle (Degrees)= 143.36
Phase Slope (Deg/Hz at f)=     0.0067
Phase Delay (Seconds)= 0.000020
Input Impedance (Ohms)=      104
Phase Angle (Degrees)= -17.64
Output Impedance (Ohms)=      304
Phase Angle (Degrees)= -72.73
Midpoint Ein/Eout=    0.30
Phase Angle (Degrees)=  71.01
```

> **Practical Note:** You will not necessarily measure exactly the same values in an experimental circuit as stated by the computer. This depends upon component tolerances, meter accuracy, generator output resistance, and generator wave-form error.

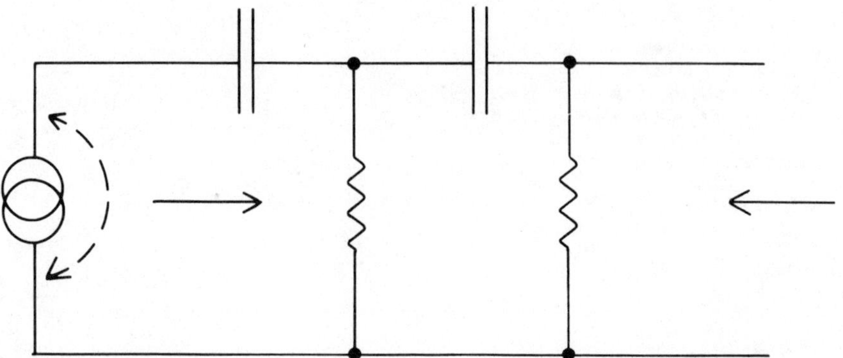

This program is for a constant-voltage source. Thus, the output impedance is computed with respect to a short-circuited input. This program is also for a negligible output load. Accordingly, the input impedance is computed with respect to an open-circuited output. (Subsequent programs are provided for a loaded output and for a source with significant internal resistance.)

```
1 LPRINT "UNSYMMETRICAL TWO-SECTION CASCADED RC LAG (INTEGRATING) NETWORK":
  LPRINT:PRINT "UNSYMMETRICAL TWO-SECTION CASCADED RC LAG (INTEGRATING) NETWORK"
: PRINT
2 LPRINT "Computes Eout/Ein, Phase Angle, Midpoint Eout/Ein & Phase Angle":
  LPRINT:PRINT"Computes Eout/Ein, Phase Angle, Midpoint Eout/Ein, & Phase Angle"
: PRINT:M=0
3 LPRINT "* Computes Eout/Ein and Phase for Off-Tolerance Parameters *":LPRINT:
  PRINT "* Computes Eout/Ein and Phase for Off-Tolerance Parameters *":PRINT
4 INPUT "R1 (Ohms)=";RO
5 LPRINT "R1 (Ohms)=";RO:INPUT "R2 (Ohms)=";RT
6 LPRINT "R2 (Ohms)=";RT:INPUT "C1 (Mfd)=";CO
7 LPRINT "C1 (Mfd)=";CO:INPUT "C2 (Mfd)=";CT
8 LPRINT "C2 (Mfd)=";CT:INPUT "f (Hz)=";F
9 LPRINT"f (Hz)=";F:A$="########":B$="####.##":C$="####.####":INPUT "Given % Tol
erance,=";T
10 LPRINT "Given % Tolerance=";T
11 XO=1/(6.2832*F*CO*10^-6):XT=1/(6.2832*F*CT*10^-6):D$="##.######"
12 AB=(RO^2+XO^2)^.5:BA=-ATN(XO/RO):LK=XO+XT:BC=(RT^2+LK^2)^.5
13 CB=-ATN(LK/RT):AC=BC*AB:CA=CB+BA:AD=AC/XO:DA=CA+6.2832/4:AE=AD*COS(DA)
14 EA=AD*SIN(DA):BE=EA+XO:CE=(AE^2+BE^2)^.5:EC=-ATN(BE/AE):AN=XT/CE
15 NA=-EC-6.2832/4:M=M+1
16 TH=ATN(XT/RT):EM=AN/SIN(TH):PH=6.2832/4-TH:PS=NA-PH:PS=PS*360/6.2832
17 IF M=1 THEN 27
18 IF M=2 THEN 35
19 IF M=3 THEN 44
20 IF M=4 THEN 53
21 IF M=5 THEN 62
22 IF M=6 THEN 71
23 IF M=7 THEN 80
24 IF M=8 THEN 89
25 IF M=9 THEN 98
27 LPRINT:LPRINT "** Normal Characteristics **":LPRINT:PRINT:PRINT "** Normal Ch
aracteristics **":PRINT
28 LPRINT "Eout/Ein=";USING B$;AN:PRINT "Eout/Ein=";USING B$;AN
29 LPRINT"Phase Angle (Degrees)=";USING B$;180+NA*360/6.2832
30 PRINT "Phase Angle (Degrees)=";USING B$;180+NA*360/6.2832
31 LPRINT"Midpoint Eout/Ein=";USING B$;EM:PRINT"Midpoint Eout/Ein=";USING B$;EM
32 LPRINT "Midpoint Phase Angle (Degrees)=";USING B$;180+PS:LPRINT
33 PRINT "Midpoint Phase Angle (Degrees)=";USING B$;180+PS:PRINT
34 RO=RO+.01*T*RO:GOTO 11
35 LPRINT "Eout/Ein, R1 High Tol,=";USING B$;AN
36 PRINT "Eout/Ein, R1 High Tol,=";USING B$;AN
37 LPRINT "Phase Angle (Degrees), R1 High Tol,=";USING B$;180+NA*360/6.2832
38 PRINT "Phase Angle (Degrees), R1 High Tol,=";USING B$;180+NA*360/6.2832
39 LPRINT "Midpoint Eout/Ein, R1 High Tol,=";USING B$;EM
40 PRINT "Midpoint Eout/Ein, R1 High Tol,=";USING B$;EM
41 LPRINT "Phase Angle (Degrees), R1 High Tol,=";USING B$;180+PS:LPRINT
42 PRINT "Phase Angle (Degrees), R1 High Tol,=";USING B$;180+PS:PRINT:RO=RO/(1+.
01*T)
43 RO=RO-.01*T*RO:GOTO 11
44 LPRINT "Eout/Ein, R1 Low Tol,=";USING B$;AN
45 PRINT "Eout/Ein, R1 Low Tol,=";USING B$;AN
46 LPRINT "Phase Angle (Degrees), R1 Low Tol,=";USING B$;180+NA*360/6.2832
47 PRINT "Phase Angle (Degrees), R1 Low Tol,=";USING B$;180+NA*360/6.2832
48 LPRINT "Midpoint Eout/Ein, R1 Low Tol,=";USING B$;EM
49 PRINT "Midpoint Eout/Ein, R1 Low Tol,=";USING B$;EM
50 LPRINT "Phase Angle (Degrees), R1 Low Tol,=";USING B$;180+PS:LPRINT
51 PRINT "Phase Angle (Degrees), R1 Low Tol,=";USING B$;180+PS:PRINT:RO=RO/(1-.0
1*T)
52 RT=RT+.01*T*RT:GOTO 11
53 LPRINT "Eout/Ein, R2 High Tol,=";USING B$;AN
54 PRINT "Eout/Ein, R2 High Tol,=";USING B$;AN
55 LPRINT "Phase Angle (Degrees), R2 High Tol,=";USING B$;180+NA*360/6.2832
56 PRINT "Phase Angle (Degrees), R2 High Tol,=";USING B$;180+NA*360/6.2832
57 LPRINT "Midpoint Eout/Ein, R2 High Tol,=";USING B$;EM
58 PRINT "Midpoint Eout/Ein, R2 High Tol,=";USING B$;EM
59 LPRINT "Phase Angle (Degrees), R2 High Tol,=";USING B$;180+PS:LPRINT
60 PRINT"Phase Angle (Degrees), R2 High Tol,=";USING B$;180+PS:PRINT:RT=RT/(1+.0
1*T)
61 RT=RT-.01*T*RT:GOTO 11
```

```
62 LPRINT "Eout/Ein, R2 Low Tol,=";USING B$;AN
63 PRINT "Eout/Ein, R2 Low Tol,=";USING B$;AN
64 LPRINT "Phase Angle (Degrees), R2 Low Tol,=";USING B$;180+NA*360/6.2832
65 PRINT "Phase Angle (Degrees), R2 Low Tol,=";USING B$;180+NA*360/6.2832
66 LPRINT "Midpoint Eout/Ein, R2 Low Tol,=";USING B$;EM
67 PRINT "Midpoint Eout/Ein, R2 Low Tol,=";USING B$;EM
68 LPRINT "Phase Angle (Degrees), R2 Low Tol,=";USING B$;180+PS:LPRINT
69 PRINT "Phase Angle (Degrees), R2 Low Tol,=";USING B$;180+PS:PRINT:RT=RT/(1-.0
1*T)
70 CO=CO+.01*T*CO:GOTO 11
71 LPRINT "Eout/Ein, C1 High Tol,=";USING B$;AN
72 PRINT "Eout/Ein, C1 High Tol,=";USING B$;AN
73 LPRINT "Phase Angle (Degrees), C1 High Tol,=";USING B$;180+NA*360/6.2832
74 PRINT "Phase Angle (Degrees), C1 High Tol,=";USING B$;180+NA*360/6.2832
75 LPRINT "Midpoint Eout/Ein, C1 High Tol,=";USING B$;EM
76 PRINT "Midpoint Eout/Ein, C1 High Tol,=";USING B$;EM
77 LPRINT "Phase Angle (Degrees), C1 High Tol,=";USING B$;180+PS:LPRINT
78 PRINT "Phase Angle (Degrees), C1 High Tol,=";USING B$;180+PS:PRINT:CO=CO/(1+.
01*T)
79 CO=CO-.01*T*CO:GOTO 11
80 LPRINT "Eout/Ein, C1 Low Tol,=";USING B$;AN
81 PRINT "Eout/Ein, C1 Low Tol,=";USING B$;AN
82 LPRINT "Phase Angle (Degrees), C1 Low Tol,=";USING B$;180+NA*360/6.2832
83 PRINT "Phase Angle (Degrees), C1 Low Tol,=";USING B$;180+NA*360/6.2832
84 LPRINT "Midpoint Eout/Ein, C1 Low Tol,=";USING B$;EM
85 PRINT "Midpoint Eout/Ein, C1 Low Tol,=";USING B$;EM
86 LPRINT "Phase Angle (Degrees), C1 Low Tol,=";USING B$;180+PS:LPRINT
87 PRINT "Phase Angle (Degrees), C1 Low Tol,=";USING B$;180+PS:PRINT:CO=CO/(1-.0
1*T)
88 CT=CT+.01*T*CT:GOTO 11
89 LPRINT "Eout/Ein, C2 High Tol,=";USING B$;AN
90 PRINT "Eout/Ein, C2 High Tol,=";USING B$;AN
91 LPRINT "Phase Angle (Degrees), C2 High Tol,=";USING B$;180+NA*360/6.2832
92 PRINT "Phase Angle (Degrees), C2 High Tol,=";USING B$;180+NA*360/6.2832
93 LPRINT "Midpoint Eout/Ein, C2 High Tol,=";USING B$;EM
94 PRINT "Midpoint Eout/Ein, C2 High Tol,=";USING B$;EM
95 LPRINT "Phase Angle (Degrees), C2 High Tol,=";USING B$;180+PS:LPRINT
96 PRINT "Phase Angle (Degrees), C2 High Tol,=";USING B$;180+PS:PRINT:CT=CT/(1+.
01*T)
97 CT=CT-.01*T*CT:GOTO 11
98 LPRINT "Eout/Ein, C2 Low Tol,=";USING B$;AN
99 PRINT "Eout/Ein, C2 Low Tol,=";USING B$;AN
100 LPRINT "Phase Angle (Degrees), C2 Low Tol,=";USING B$;180+NA*360/6.2832
101 PRINT "Phase Angle (Degrees), C2 Low Tol,=";USING B$;180+NA*360/6.2832
103 LPRINT "Midpoint Eout/Ein, C2 Low Tol,=";USING B$;EM
104 PRINT "Midpoint Eout/Ein, C2 Low Tol,=";USING B$;EM
105 LPRINT "Phase Angle (Degrees), C2 Low Tol,=";USING B$;180+PS:LPRINT
106 PRINT "Phase Angle (Degrees), C2 Low Tol,=";USING B$;180+PS:PRINT:CT=CT/(1-.
01*T)
107 END
```

```
(TYPICAL RUN)


UNSYMMETRICAL TWO-SECTION CASCADED RC LAG (INTEGRATING) NETWORK

Computes Eout/Ein, Phase Angle, Midpoint Eout/Ein & Phase Angle

* Computes Eout/Ein and Phase for Off-Tolerance Parameters *

R1 (Ohms)= 100
R2 (Ohms)= 1000
C1 (Mfd)= 1
C2 (Mfd)= .1
f (Hz)= 1000
Given % Tolerance= 50

** Normal Characteristics **
```

```
Eout/Ein=   0.69
Phase Angle (Degrees)=  65.36
Midpoint Eout/Ein=   0.81
Midpoint Phase Angle (Degrees)=  33.22

Eout/Ein, R1 High Tol,=   0.58
Phase Angle (Degrees), R1 High Tol,=  76.24
Midpoint Eout/Ein, R1 High Tol,=   0.69
Phase Angle (Degrees), R1 High Tol,=  44.10

Eout/Ein, R1 Low Tol,=   0.79
Phase Angle (Degrees), R1 Low Tol,=  50.51
Midpoint Eout/Ein, R1 Low Tol,=   0.94
Phase Angle (Degrees), R1 Low Tol,=  18.37

Eout/Ein, R2 High Tol,=   0.59
Phase Angle (Degrees), R2 High Tol,=  75.98
Midpoint Eout/Ein, R2 High Tol,=   0.82
Phase Angle (Degrees), R2 High Tol,=  32.68

Eout/Ein, R2 Low Tol,=   0.78
Phase Angle (Degrees), R2 Low Tol,=  51.40
Midpoint Eout/Ein, R2 Low Tol,=   0.81
Phase Angle (Degrees), R2 Low Tol,=  33.96

Eout/Ein, C1 High Tol,=   0.59
Phase Angle (Degrees), C1 High Tol,=  75.98
Midpoint Eout/Ein, C1 High Tol,=   0.70
Phase Angle (Degrees), C1 High Tol,=  43.84

Eout/Ein, C1 Low Tol,=   0.78
Phase Angle (Degrees), C1 Low Tol,=  51.40
Midpoint Eout/Ein, C1 Low Tol,=   0.92
Phase Angle (Degrees), C1 Low Tol,=  19.26

Eout/Ein, C2 High Tol,=   0.58
Phase Angle (Degrees), C2 High Tol,=  76.24
Midpoint Eout/Ein, C2 High Tol,=   0.80
Phase Angle (Degrees), C2 High Tol,=  32.93

Eout/Ein, C2 Low Tol,=   0.79
Phase Angle (Degrees), C2 Low Tol,=  50.51
Midpoint Eout/Ein, C2 Low Tol,=   0.83
Phase Angle (Degrees), C2 Low Tol,=  33.07
```

PROGRAM PLANNING

When you start to write a program for computation of the characteristics and response for ac circuits such as exemplified here, the procedural plan is as follows:

1. Apply mesh analysis to obtain an E_{out}/E_{in} equation—e.g., in terms of complex algebra.
2. Assign program variables to break the equation down into sums, differences, products, quotients, powers, roots, and trigonometric functions.
3. Sums and differences are computed with respect to rectangular vectors. Products, quotients, powers, and roots are computed with respect to polar vectors. Angles are computed with respect to arctangents.
4. Many of the computations involve conversion from rectangular vectors to polar vectors, and vice versa.
5. As in the case of pencil-and-paper calculations, 180° ambiguities will be encountered. These must be resolved by observation of a rough sketch of the pertinent vector diagram.
6. Many computers process angles only in terms of radians, and various steps in computation will require conversion from degrees to radians, and vice versa.

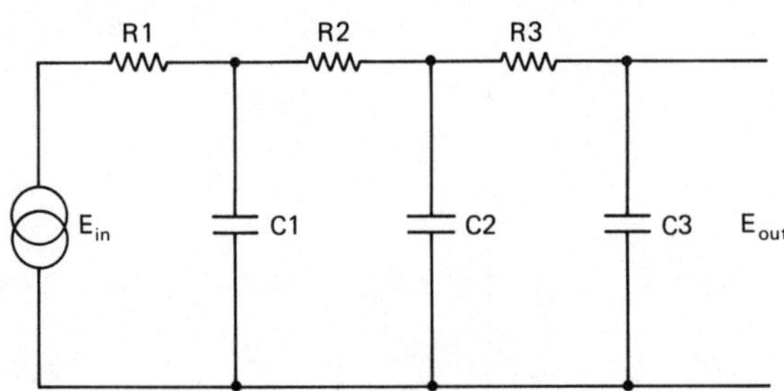

This program is for a constant-voltage source and an unloaded output. Like the preceding programs, equations are based on mesh analysis, and the third RC section is computed with respect to Thevenin's theorem.

```
108 LPRINT "UNSYMMETRICAL 3-SECTION CASCADED RC INTEGRATING NETWORK"
109 PRINT "UNSYMMETRICAL 3-SECTION CASCADED RC INTEGRATING NETWORK"
110 LPRINT "Output Voltage and Phase":PRINT "Output Voltage and Phase"
111 LPRINT "* Survey Program *":PRINT "* Survey Program *":LPRINT:PRINT
112 INPUT "R1 (Ohms)=";RO
113 LPRINT "R1 (Ohms)=";RO
114 INPUT "R2 (Ohms)=";RT
115 LPRINT "R2 (Ohms)=";RT
116 INPUT "R3 (Ohms)=";RQ
117 LPRINT "R3 (Ohms)=";RQ
118 INPUT "C1 (Mfd)=";CO
119 LPRINT "C1 (Mfd)=";CO
120 INPUT "C2 (Mfd)=";CT
121 LPRINT "C2 (Mfd)=";CT
122 INPUT "C3 (Mfd)=";CQ
123 LPRINT "C3 (Mfd)=";CQ
124 INPUT "f (Hz)=";H
125 LPRINT "f (Hz)=";H
126 N=0:G=H:PRINT:LPRINT:A$="#######":B$="####.##"
127 F=G+N:PRINT"f (Hz)=";USING A$;F:LPRINT"f (Hz)=";USING A$;F
128 XO=1/(6.2832*F*CO*10^-6):XT=1/(6.2382*F*CT*10^-6):XQ=1/(6.2832*F*CQ*10^-6)
129 AB=(RO^2+XO^2)^.5:BA=ATN(XO/RO):LK=XO+XT:BC=(RT^2+LK^2)^.5
130 CB=ATN(LK/RT):AC=BC*AB:CA=CB+BA:AD=AC/XO:DA=CA+6.2832/4:AE=AD*COS(DA)
131 EA=AD*SIN(DA):BE=EA+XO:CE=(AE^2+BE^2)^.5:EC=ATN(BE/AE):AN=XT/CE
132 NA=-EC-6.2832/4:NA=NA*360/6.2832:AB=RO*XO:BA=-6.2832/4
133 AF=AD*SIN(DA):AG=AE+RT:AH=(AG^2+AF^2)^.5:HA=-ATN(AF/AG):AJ=AH*XT
134 JA=HA-6.2832/4:AK=AF-XT:KA=FA-6.2832/4:AL=(AG^2+AK^2)^.5:LA=-ATN(AK/AG)
135 AM=AJ/AL:MA=JA-LA:ZN=AM:NZ=MA:NZ=NZ*360/6.2832+180:PQ=-NA*6.2832/360
136 QP=-NZ*6.2832/360:RS=AN*COS(PQ):SR=AN*SIN(PQ):RR=RS+RQ:IT=SR+XQ
137 RU=(RR^2+IT^2)^.5:IL=AN/RU:VO=IL*XQ:ZI=(RQ^2+XQ^2)^.5:IZ=ATN(XQ/RQ)
138 BF=IZ-3.1416/2:NA=NA*6.2832/360:OV=-NA+BF:OV=OV*360/6.2832-180:N=N+1000
139 LPRINT"Eout/Ein=";USING B$;VO:PRINT"Eout/Ein=";USING B$;VO
140 LPRINT"Phase Angle (Degrees)=";USING B$;-OV
141 PRINT"Phase Angle (Degrees)=";USING B$;-OV
142 IF F>20000 THEN 144
143 PRINT:LPRINT:GOTO 127
144 END
```

```
UNSYMMETRICAL 3-SECTION CASCADED RC INTEGRATING NETWORK
Output Voltage and Phase
* Survey Program *

R1 (Ohms)= 100
R2 (Ohms)= 1000
R3 (Ohms)= 10000
C1 (Mfd)= 1
C2 (Mfd)= .1
C3 (Mfd)= .01
f (Hz)= 1

f (Hz)=       1
Eout/Ein=    1.00
Phase Angle (Degrees)=    0.11
```

```
f (Hz)=   1001
Eout/Ein=    0.58
Phase Angle (Degrees)=   97.39

f (Hz)=   2001
Eout/Ein=    0.23
Phase Angle (Degrees)= 153.72

f (Hz)=   3001
Eout/Ein=    0.10
Phase Angle (Degrees)= 184.73

f (Hz)=   4001
Eout/Ein=    0.05
Phase Angle (Degrees)= 203.39

f (Hz)=   5001
Eout/Ein=    0.03
Phase Angle (Degrees)= 215.59

f (Hz)=   6001
Eout/Ein=    0.02
Phase Angle (Degrees)= 224.12

f (Hz)=   7001
Eout/Ein=    0.01
Phase Angle (Degrees)= 230.38

f (Hz)=   8001
Eout/Ein=    0.01
Phase Angle (Degrees)= 235.17
```

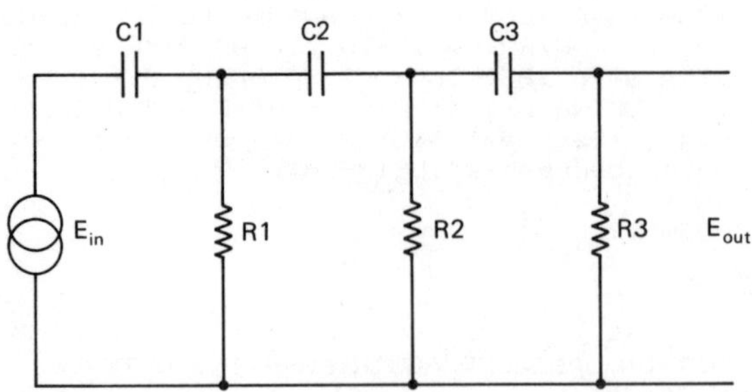

This program is essentially similar to the previous program except that the capacitors and resistors are configured in a differentiating mode. Observe that when reactive circuitry is being computed, programs must be written to take any 180° ambiguities into account. (This is fundamentally no different from solving the problems by pencil and paper—the vector diagram must be followed, and any 180° ambiguities accounted for.)

```
145 PRINT "UNSYMMETRICAL 3-SECTION RC DIFFERENTIATING NETWORK"
146 LPRINT "UNSYMMETRICAL 3-SECTION RC DIFFERENTIATING NETWORK"
147 PRINT "Eout/Ein and Phase Angle":LPRINT "Eout/Ein and Phase Angle"
148 PRINT "* Survey Program *":LPRINT "* Survey Program *":PRINT:LPRINT
149 INPUT "R1 (Ohms)=";RO
150 LPRINT "R1 (Ohms)=";RO
151 INPUT "R2 (Ohms)=";RT
152 LPRINT "R2 (Ohms)=";RT
153 INPUT "R3 (Ohms)=";RQ
154 LPRINT "R3 (Ohms)=";RQ
155 INPUT "C1 (Mfd)=";CO
156 LPRINT "C1 (Mfd)=";CO
157 INPUT "C2 (Mfd)=";CT
158 LPRINT "C2 (Mfd)=";CT
159 INPUT "C3 (Mfd)=";CQ
160 LPRINT "C3 (Mfd)=";CQ
161 INPUT "f (Hz)=";H
162 LPRINT "f (Hz)=";H:N=0:G=H:PRINT:LPRINT
163 F=G+N:A$="########"
164 PRINT "f (Hz)=";USING A$;F:LPRINT "f (Hz)=";USING A$;F
165 XO=1/(6.2832*F*CO*10^-6):XT=1/(6.2832*F*CT*10^-6):XQ=1/(6.2832*F*CQ*10^-6)
166 RS=RO+RT:AB=(RS^2+XT^2)^.5:BA=ATN(XT/RS):AC=(RO^2+XO^2)^.5:CA=ATN(XO/RO)
167 AD=AB*AC:DA=BA+CA:AE=AD/RO:EA=DA:AF=AE*COS(EA):AG=AE*SIN(EA):AH=AF-RO
168 AJ=(AH*AH+AG*AG)^.5:JA=ATN(AG/AH):AK=RT/AJ:KA=-JA:KA=KA*360/6.2832
169 B$="####.##":KA=KA-180:N=N+1000
170 IF ABS(KA)>180 THEN KA=KA+180
171 BC=(RT*RT+XT*XT)^.5:CB=ATN(XT/RT):BD=RO*BC:DB=CB:BE=RO+BD
172 BF=(RO*RO+BC*BC)^.5:FB=ATN(BC/RO):BG=BD/BF:GB=DB-FB:BH=BG*COS(GB)
173 BJ=BG*SIN(GB):BK=BJ+XO:ZL=BG+RQ:LZ=BK+XQ:YL=(ZL*ZL+LZ*LZ)^.5
174 LY=ATN(LZ/ZL):IL=AK/YL:KA=KA*6.2832/360:LI=KA-LY:VO=IL*RQ
175 OV=LI*360/6.2832:PRINT "Eout/Ein=";USING B$;VO
176 LPRINT "Eout/Ein=";USING B$;VO
177 PRINT "Phase Angle (Degrees)=";USING B$;OV
178 LPRINT "Phase Angle (Degrees)=";USING B$;OV
179 IF F>20000 THEN 181
180 PRINT:LPRINT:GOTO 163
181 END
```

```
UNSYMMETRICAL 3-SECTION RC DIFFERENTIATING NETWORK
Eout/Ein and Phase Angle
* Survey Program *

R1 (Ohms)= 100
R2 (Ohms)= 1000
R3 (Ohms)= 10000
C1 (Mfd)= 1
C2 (Mfd)= .1
C3 (Mfd)= .01
f (Hz)= .1

f (Hz)=        0
Eout/Ein=   0.00
Phase Angle (Degrees)=-269.99

f (Hz)=     1000
Eout/Ein=   0.14
Phase Angle (Degrees)=-172.41
```

```
f (Hz)=       2000
Eout/Ein=      0.45
Phase Angle (Degrees)=-115.88

f (Hz)=       3000
Eout/Ein=      0.66
Phase Angle (Degrees)= -84.75

f (Hz)=       4000
Eout/Ein=      0.78
Phase Angle (Degrees)= -66.05

f (Hz)=       5000
Eout/Ein=      0.84
Phase Angle (Degrees)= -53.83

f (Hz)=       6000
Eout/Ein=      0.89
Phase Angle (Degrees)= -45.29

f (Hz)=       7000
Eout/Ein=      0.91
Phase Angle (Degrees)= -39.02
```

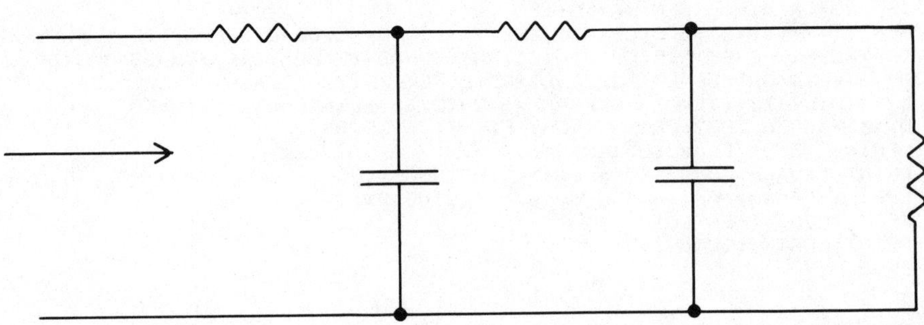

This is a highly practical program, inasmuch as an integrating circuit often works into a resistive load such as the base-emitter circuit of a bipolar transistor. Output loading reduces the output voltage and changes the output phase angle with respect to an unloaded condition of operation.

```
182 LPRINT "UNSYMMETRICAL 2-SECTION INTEGRATING CIRCUIT (With Resistive Load)":P
RINT "UNSYMMETRICAL 2-SECTION INTEGRATING CIRCUIT (With Resistive Load)"
183 LPRINT "* Input Impedance and Phase Angle *":PRINT "* Input Impedance and Ph
ase Angle *":PRINT:LPRINT
184 INPUT "R1 (Ohms)=";RO
185 LPRINT "R1 (Ohms)=";RO
186 INPUT "R2 (Ohms)=";RT
187 LPRINT "R2 (Ohms)=";RT
188 INPUT "C1 (Mfd)=";CO
189 LPRINT "C1 (Mfd)=";CO
190 INPUT "C2 (Mfd)=";CT
191 LPRINT "C2 (Mfd)=";CT
192 INPUT "RL (Ohms)=";RL
193 LPRINT "RL (Ohms)=";RL
194 INPUT "f (Hz)=";F
195 LPRINT "f (Hz)=";F
196 XO=1/(6.2832*F*CO*10^-6):XT=1/(6.2832*F*CT*10^-6)
197 AB=RL*XT:BA=-6.2832/4:AC=(RL*RL+XT*XT)^.5:CA=-ATN(XT/RL):AD=AB/AC
198 DA=BA-CA:AE=AD*COS(DA):AF=AD*SIN(DA):AG=AE+RT:AH=(AG*AG+AF*AF)^.5
199 HA=-ATN(AF/AG):AJ=AH*XO:JA=HA-6.2832/4:AK=AF+XO:KA=-6.2832/4
200 AL=(AG^2+AK^2)^.5:LA=-ATN(AK/AG):AM=AJ/AL:MA=JA-LA:AN=AM*COS(MA)
201 AO=AM*SIN(MA):AP=RO+AN:AQ=(AP^2+AO^2)^.5:QA=-ATN(AO/AP):PF=F*QA
202 LPRINT:PRINT:A$="#######":LPRINT"Zin (Ohms)=";USING A$;AQ
203 PRINT "Zin (Ohms)=";USING A$;AQ
204 LPRINT "Phase Angle (Degrees)=";USING A$;-QA*360/6.2832
205 PRINT "Phase Angle (Degrees)=";USING A$;-QA*360/6.2832
206 END
```

```
JNSYMMETRICAL 2-SECTION INTEGRATING CIRCUIT (With Resistive Load)
 Input Impedance and Phase Angle *

1 (Ohms)= 1000
2 (Ohms)= 2500
1 (Mfd)= .15
2 (Mfd)= .1
L (Ohms)= 15000
 (Hz)= 1200

in (Ohms)=    1592
hase Angle (Degrees)=    -36
```

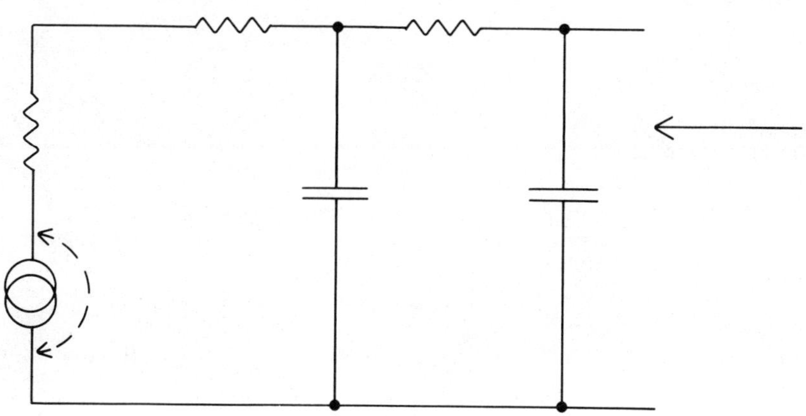

This is another highly practical program, inasmuch as the source voltage for an integrating circuit is often associated with significant internal resistance, such as the output resistance of a bipolar transistor. Internal resistance has the effect of increasing the output impedance and changing its phase angle with respect to the constant-voltage mode of operation.

```
207 LPRINT "UNSYMMETRICAL 2-SECTION INTEGRATING CIRCUIT":PRINT "UNSYMMETRICAL 2-
SECTION INTEGRATING CIRCUIT"
208 LPRINT "* Output Impedance and Phase Angle *":PRINT "* Output Impedance and
Phase Angle *":PRINT:LPRINT
209 INPUT "R1 (Ohms)=";RO
210 LPRINT "R1 (Ohms)=";RO
211 INPUT "R2 (Ohms)=";RT
212 LPRINT "R2 (Ohms)=";RT
213 INPUT "C1 (Mfd)=";CO
214 LPRINT "C1 (Mfd)=";CO
215 INPUT "C2 (Mfd)=";CT
216 LPRINT "C2 (Mfd)=";CT
217 INPUT "Rs (Ohms=";RS
218 LPRINT "Rs (Ohms)=";RS
219 INPUT "f (Hz)=";F
220 LPRINT "f (Hz)=";F
221 XO=1/(6.2832*F*CO*10^-6):XT=1/(6.2832*F*CT*10^-6):AB=(RS+RO)*XO
222 BA=-6.2832/4:AC=((RS+RO)*(RS+RO)+XO*XO)^.5:N=RS+RO:CA=-ATN(XO/N):AD=AB/AC
223 DA=BA-CA:AE=AD*COS(DA):AF=AD*SIN(DA):AG=AE+RT:AH=(AG*AG+AF*AF)^.5
224 HA=-ATN(AF/AG):AJ=AH*XT:JA=HA-6.2832/4:AK=AF-XT:KA=FA-6.2832/4
225 AL=(AG*AG+AK*AK)^.5:LA=-ATN(AK/AG):AM=AJ/AL:MA=JA-LA:AN=AM:NA=MA
226 A$="########":LPRINT:PRINT:LPRINT "Zout (Ohms)=";USING A$;AN
227 PRINT "Zout (Ohms)=";USING A$;AN
228 LPRINT "Phase Angle (Degrees)=";USING A$; NA*360/6.2832
229 PRINT "Phase Angle (Degrees)=";USING A$; NA*360/6.2832
230 END
```

```
UNSYMMETRICAL 2-SECTION INTEGRATING CIRCUIT
* Output Impedance and Phase Angle *

R1 (Ohms)= 1500
R2 (Ohms)= 2000
C1 (Mfd)= .16
C2 (Mfd)= .2
Rs (Ohms)= 950
f (Hz)= 1000

Zout (Ohms)=      693
Phase Angle (Degrees)=     -105
```

11

Time-Saving Program Conversions

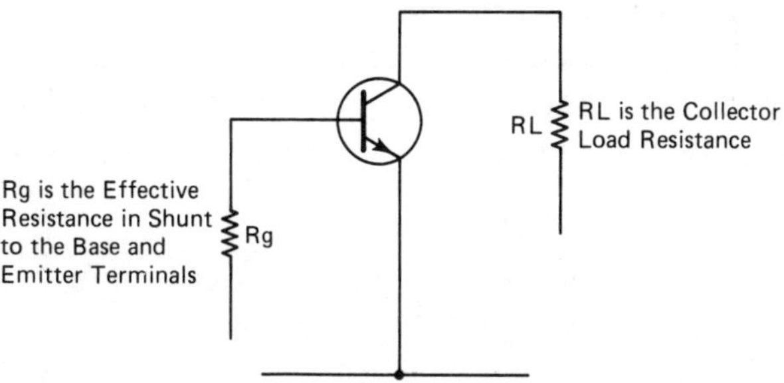
RL
RL is the Collector
Load Resistance
Rg is the Effective
Resistance in Shunt
to the Base and
Emitter Terminals
Rg

Most personal comptuers use the BASIC high-level programming language. However, various computers differ considerably in Input/Output (I/O) functions. You will also find that there is often comparatively limited processing capability in the simpler types of computers. In some cases, limited processing capability entails lack of various mathematical functions; this limitation can be coped with by means of subroutines. In other cases, limited processing capability entails register overflow with loss of INPUTed variable values; limited program memory capacity can be coped with by addition of supplementary RAM. In other words, limited processing capability has various aspects, some of which can be overcome. Other limitations may make it impractical to convert the more involved types of troubleshooting programs.

The first example of program conversion shows how a comparatively simple routine written for the IBM PC computer can be converted to RUN on the Commodore 64 personal computer.

Notes are provided concerning program conversion from the IBM PC to Apple computers. Since many TRS-80 personal computers are still in use, conversion notes are also provided for this group of computers.

NOTES ON PROGRAM CONVERSION

The foregoing example shows how a routine written for the IBM PC computer can be converted to RUN on the Commodore 64 personal computer. Although not absolutely essential, it will save time if you decide at the outset whether the conversion will be made for video display only, or for combined video and printer display. (A program may also be converted for PRINTing hard copy only). Separate conversions for video display and for combined video and printer display are shown next.

You can learn conversion procedure most easily if you examine IBM PC programs and their converted formats; however, this is not always feasible, and in turn, you must learn various conversion "rules" for guidance. Note that an IBM PC program may contain up to 80 characters in a line. However, the Commodore 64 can include a maximum of only 40 characters in a program line. Accordingly, you may occasionally need to divide a program line into two parts, and to convert it as a pair of successive lines for the 64.

TROUBLESHOOTING PROGRAM FOR THE IBM PC

And Compatibles

```
500 LPRINT "AMPLIFIER ANALYSIS BASED ON OFF-TOLERANCE PARAMETERS":
    LPRINT "Generalized Troubleshooting Program No. 1":LPRINT
501 PRINT "AMPLIFIER ANALYSIS BASED ON OFF-TOLERANCE PARAMETERS":
    PRINT "Generalized Troubleshooting Program No. 1":PRINT
502 LPRINT "(Input/Output Values Are Computed from H Parameters, Source Voltage,
              Source Resistance, and Load Resistance)":LPRINT
503 PRINT "(Input/Output Values Are Computed from H Parameters, Source Voltage,
             Source Resistance, and Load Resistance)":PRINT
504 LPRINT "*** CE Configuration ***":PRINT "*** CE Configuration ***":LPRINT:
    PRINT
505 LPRINT "  Computes Eout, Iout, Ein, Iin, Voltage Gain, Current Gain, Power
               Gain, Source Voltage, Input Impedance, Output Impedance":LPRINT
506 PRINT "  Computes Eout, Iout, Ein, Iin, Voltage Gain, Current Gain, Power
              Gain, Source Voltage, Input Impedance, Output Impedance":PRINT
507 LPRINT "Computes Reduced Voltage Gain Resulting from a 30% Off-Tolerance
             on Each Circuit Parameter"
508 PRINT  "Computes Reduced Voltage Gain Resulting from a 30% Off-Tolerance
             on Each Circuit Parameter"
509 LPRINT:PRINT:INPUT "Hfe=";A
510 LPRINT "Hfe=";A:INPUT "Hoe=";B
511 LPRINT "Hoe=";B:INPUT "Hie=";C
```

```
512 LPRINT "Hie=";C:INPUT "Hre=";D
513 LPRINT "Hre=";D:INPUT "RL (Ohms)=";E
514 LPRINT "RL (Ohms)=";E:INPUT "Vg (Volts)=";VG
515 LPRINT "Vg (Volts)=";VG:INPUT "Rg (Ohms)=";F
516 LPRINT "Rg (Ohms)=";F:LPRINT:PRINT:A$="##########"
519 G=A*E/((C*B-A*D)*E+C):H=A/(B*E+1):I=G*H:J=(C+(B*C-A*D)*E)/(B*E+1):B$="######
.##":C$="######.###"
525 K=(C+F)/(B*C-D*A+B*F):LPRINT:PRINT:LPRINT "* Normal Characteristics *":LPRIN
T:PRINT "* Normal Characteristics *":PRINT
526 LPRINT "Voltage Gain=";USING A$;-G:PRINT "Voltage Gain=";USING A$;-G
527 LPRINT "Current Gain=";USING A$;-H:PRINT "Current Gain=";USING A$;-H
528 LPRINT "Power Gain=";USING A$;I:PRINT "Power Gain=";USING A$;I
529 LPRINT "Power Gain (dB)=";USING A$;10*LOG(I)/LOG(10):PRINT "Power Gain (dB)=
";USING A$;10*LOG(I)/LOG(10)
530 LPRINT "Output Impedance (Ohms)=";USING A$;K:PRINT "Output Impedance (Ohms)=
";USING A$;K:IB=VG/(F+J)
531 LPRINT "Input Impedance (Ohms)=";USING A$;J
532 PRINT "Input Impedance (Ohms)=";USING A$;J
533 IC=IB*H:EI=IB*J:VO=IC*E
534 LPRINT "Ib (Microamps)=";USING B$;IB*10^6
535 PRINT "Ib (Microamps)=";USING B$;IB*10^6
536 LPRINT "Ic (Microamps)=";USING B$;-IC*10^6
537 PRINT "Ic (Microamps)=";USING B$;-IC*10^6
538 LPRINT "Eout (Volts)=";USING B$;-VO
539 PRINT "Eout (Volts)=";USING B$;-VO
540 LPRINT "Generator Voltage (Volts)=";USING C$;VG
541 PRINT "Generator Voltage (Voltage)=";USING C$;VG
542 LPRINT "Input Voltage (Volts()=";USING C$;EI
543 PRINT "Input Voltage (Volts)=";USING C$;EI
544 A=A-.3*A:G=A*E/((C*B-A*D)*E+C):H=A/(B*E+1)
545 I=A^2*E/((B*E+1)*((C*B-A*D)*E+C)):LPRINT:PRINT
546 LPRINT "Voltage Gain, Hfe 30% Low,=";USING A$;-G
547 PRINT "Voltage Gain, Hfe 30% Low,=";USING A$;-G
548 LPRINT "Current Gain, Hfe 30% Low,=";USING A$;-H
549 PRINT "Current Gain, Hfe 30% Low,=";USING A$;-H:LPRINT:PRINT
550 A=A*1.428572:B=B+.3*B:G=A*E/((C*B-A*D)*E+C):H=A/(B*E+1)
551 LPRINT "Voltage Gain, Hoe 30% High,=";USING A$;-G
552 PRINT "Voltage Gain, Hoe 30% High,=";USING A$;-G
553 LPRINT "Current Gain, Hoe 30% High,=";USING A$;-H
554 PRINT "Current Gain, Hoe 30% High,=";USING A$;-H:LPRINT:PRINT
555 B=B*.7692308:C=C+.3*C:G=A*E/((C*B-A*D)*E+C):H=A/(B*E+1)
556 LPRINT "Voltage Gain, Hie 30% High,=";USING A$;-G
557 PRINT "Voltage Gain, Hie 30% High,=";USING A$;-G
558 LPRINT "Current Gain, Hie 30% High,=";USING A$;-H
559 PRINT "Current Gain, Hie 30% High,=";USING A$;-H:LPRINT:PRINT
560 C=C*.7692308:D=D-.3*D:G=A*E/((C*B-A*D)*E+C):H=A/(B*E+1)
561 LPRINT "Voltage Gain, Hre 30% Low,=";USING A$;-G
562 PRINT "Voltage Gain, Hre 30% Low,=";USING A$;-G
563 LPRINT "Current Gain, Hre 30% Low,=";USING A$;-H
564 PRINT "Current Gain, Hre 30% Low,=";USING A$;-H:LPRINT:PRINT
565 D=D*1.428572:E=E-.3*E:G=A*E/((C*B-A*D)*E+C):H=A/(B*E+1)
566 LPRINT "Voltage Gain, RL 30% Low,=";USING A$;-G
567 PRINT "Voltage Gain, RL 30% Low,=";USING A$;-G
568 LPRINT "Current Gain, RL 30% Low,=";USING A$;-H
569 PRINT "Current Gain, RL 30% Low,=";USING A$;-H:LPRINT:PRINT
570 E=E*1.428572:F=F+.3*F:G=A*E/((C*B-A*D)*E+C):H=A/(B*E+1)
571 LPRINT "Voltage Gain, Rg 30% High,=";USING A$;-G
572 PRINT "Voltage Gain, Rg 30% High,=";USING A$;-G
573 LPRINT "Current Gain, Rg 30% High,=";USING A$;-H
574 PRINT "Current Gain, Rg 30% High,=";USING A$;-H:LPRINT:PRINT
575 F=F*.7692308:A=A-.3*A:B=B+.3*B:C=C+.3*C:D=D-.3*D
576 G=A*E/((C*B-A*D)*E+C):H=A/(B*E+1)
577 LPRINT "Voltage Gain, all h parameters 30% out of tolerance,=";USING A$;-G
578 PRINT "Voltage Gain, all h parameters 30% out of tolerance,=";USING A$;-G
579 LPRINT "Current Gain, all h parameters 30% out of tolerance,=";USING A$;-H
580 PRINT "Current Gain, all h parameters 30% out of tolerance,=";USING A$;-H
581 LPRINT:PRINT: A=A*1.428572:B=B*.7692308:C=C*.7692308:D=D*1.428572
583 PRINT  "Note: If you do not have the rated h parameters available for a part
icular small-signal transistor, the typical values shown in the sample RUN will
be helpful."
584 END
```

```
(TYPICAL RUN)

AMPLIFIER ANALYSIS BASED ON OFF-TOLERANCE PARAMETERS
Generalized Troubleshooting Program No. 1

(Input/Output Values Are Computed from H Parameters, Source Voltage,
 Source Resistance, and Load Resistance)

*** CE Configuration ***

  Computes Eout, Iout, Ein, Iin, Voltage Gain, Current Gain, Power
  Gain, Source Voltage, Input Impedance, Output Impedance

Computes Reduced Voltage Gain Resulting from a 30% Off-Tolerance
on Each Circuit Parameter

Hfe= 50
Hoe= .00002
Hie= 1500
Hre= .0005
RL (Ohms)= 15000
Vg (Volts)= .004
Rg (Ohms)= 1500


* Normal Characteristics *

Voltage Gain=       -476
Current Gain=        -38
Power Gain=     18315
Power Gain (dB)=         43
Output Impedance (Ohms)=      85714
Input Impedance (Ohms)=       1212
Ib (Microamps)=      1.48
Ic (Microamps)=    -56.74
Eout (Volts)=    -0.85
Generator Voltage (Volts)=      0.004
Input Voltage (Volts()=      0.002

Voltage Gain, Hfe 30% Low,=       -311
Current Gain, Hfe 30% Low,=        -27

Voltage Gain, Hoe 30% High,=       -439
Current Gain, Hoe 30% High,=        -36

Voltage Gain, Hie 30% High,=       -347
Current Gain, Hie 30% High,=        -38

Voltage Gain, Hre 30% Low,=       -444
Current Gain, Hre 30% Low,=        -38

Voltage Gain, RL 30% Low,=       -338
Current Gain, RL 30% Low,=        -41

Voltage Gain, Rg 30% High,=       -476
Current Gain, Rg 30% High,=        -38

Voltage Gain, all h parameters 30% out of tolerance,=       -208
Current Gain, all h parameters 30% out of tolerance,=        -25
```

If you are converting a program to RUN on the 64 printer, the first program line should be coded "OPEN1,4" to switch the printer into action. Each line that contains a variable to be printed must start with the command ""PRINT#1," (wherein the trailing comma is necessary). The program should finally end with the command "CLOSE1,4." Most 64 computers employ "4," although an occasional model may employ "8."

Unless otherwise instructed, any comptuer will print out the decimal values to a comparatively large number of places. To designate printing out to a specified number of decimal places, a suitable print format must be used. As you will see from the foregoing examples, the IBM PC utilizes a character-string (A$) command for this purpose. On the other hand, the Commodore 64 uses the integer (INT) function for this purpose, with associated 100 and .01 multipliers, for example. The INT function alone removes all numerals to the right of the decimal point in the printout.

Observe that the IBM PC uses a caret exponential symbol, whereas the Commodore 64 uses an "up arrow" symbol. Note also that the IBM PC provides both capital and lower-case letters, whereas the 64 provides capital letters only. There are various minor I/O differences to be observed, such as the keys to be used for clearing the screen, for printing a hard copy of the program, and so on.

```
5 REM PRG 1 W/OUT PRINTER
10 PRINT "AMPLIFIER ANALYSIS BASED ON OFF-TOLERANCE PARAMETERS"
15 PRINT "GENERALIZED TROUBLESHOOTING PROGRAM NO.1":PRINT
20 PRINT "(INPUT/OUTPUT VALUES ARE COMPUTED FROM H PARAMETERS,"
25 PRINT "SOURCE CURRENT, SOURCE RESISTANCE, AND LOAD RESISTANCE)":PRINT
30 PRINT "*** CE CONFIGURATION ***"
40 PRINT "COMPUTES E/OUT, I/OUT, E/IN, I/IN, VOLTAGE GAIN, CURRENT GAIN,"
50 PRINT "POWER GAIN, SOURCE VOLTAGE, INPUT IMPEDANCE.":PRINT
60 PRINT "COMPUTES REDUCED VOLTAGE GAIN RESULTING FROM A 30% OFF-TOLERANCE ON"
70 PRINT"EACH CIRCUIT PARAMETER, AND ALL H PARAMETERS 30% OFF-TOLERANCE."
80 PRINT
90 INPUT "HFE=";A
100 INPUT "HOE=";B
110 INPUT "HIE=";C
130 INPUT "HRE=";D
150 INPUT "RL (OHMS)=";E
170 INPUT "IB (MICROAMPS)=";IB
190 INPUT "RG (OHMS)=";F
210 PRINT
220 G=A*E/((C*B-A*D)*E+C):H=A/(B*E+1)
230 I=A↑2*E/((B*E+1)*((C*B-A*D)*E+C))
240 J=(C+(B*C-A*D)*E)/(B*E+1)
250 K=(C+F)/(B*C-D*A+B*F)
255 PRINT "* NORMAL CHARACTERISTICS *":PRINT
258 YA=INT(-G)
260 PRINT "VOLTAGE GAIN=";YA
262 YB=INT(-H)
270 PRINT "CURRENT GAIN=";YB
272 YC=INT(I)
280 PRINT "POWER GAIN=";YC
282 YD=INT(10*LOG(I)/LOG(10))
290 PRINT "POWER GAIN (DB)=";YD
292 YE=INT(J)
310 PRINT "INPUT IMPEDANCE (OHMS)=";YE
330 IC=IB*H:VO=IC*10↑-6*E:EI=IB*10↑-6*J:VG=EI+IB*10↑-6*F
```

```
332 YF=.01*INT(IB*100)
340 PRINT "IB (MICROAMPS)=";YF
342 YG=.01*INT(-IC*100)
360 PRINT "IC (MICROAMPS)=";YG
362 YH=.01*INT(-VO*100)
380 PRINT "E/OUT (VOLTS)=";YH
382 YI=.001*INT(VG*1000)
400 PRINT "GENERATOR VOLTAGE (VOLTS)=";YI
410 YJ=.001*INT(EI*1000)
420 PRINT "INPUT VOLTAGE (VOLTS)=";YJ
440 A=A-.3*A:G=A*E/((C*B-A*D)*E+C):H=A/(B*E+1)
450 I=A↑2*E/((B*E+1)*((C*B-A*D)*E+C)):PRINT
455 YK=INT(-G)
460 PRINT "VOLTAGE GAIN, HFE 30% LOW,=";YK
470 YL=INT(-H)
480 PRINT "CURRENT GAIN, HFE 30% LOW,=";YL
500 A=A*1.428572:B=B+.3*B:G=A*E/((C*B-A*D)*E+C):H=A/(B*E+1):PRINT
505 YM=INT(-G)
510 PRINT "VOLTAGE GAIN, HOE 30% HIGH,=";YM
515 YN=INT(-H)
530 PRINT "CURRENT GAIN, HOE 30% HIGH,=";YN
550 B=B*.7692308:C=C+.3*C:G=A*E/((C*B-A*D)*E+C):H=A/(B*E+1):PRINT
555 YO=INT(-G)
560 PRINT "VOLTAGE GAIN, HIE 30% HIGH,=";YO
570 YP=INT(-H)
580 PRINT "CURRENT GAIN, HIE 30% HIGH,=";YP
600 C=C*.7692308:D=D-.3*D:G=A*E/((C*B-A*D)*E+C):H=A/(B*E+1):PRINT
605 YQ=INT(-G)
610 PRINT "VOLTAGE GAIN, HRE 30% LOW,=";YQ
620 YR=INT(-H)
630 PRINT "CURRENT GAIN, HRE 30% LOW,=";YR
650 D=D*1.428572:E=E-.3*E:G=A*E/((C*B-A*D)*E+C):H=A/(B*E+1):PRINT
655 YS=INT(-G)
660 PRINT "VOLTAGE GAIN, RL 30% LOW,=";YS
670 YT=INT(-H)
680 PRINT "CURRENT GAIN, RL 30% LOW,=";YT
700 E=E*1.428572:F=F+.3*F:G=A*E/((C*B-A*D)*E+C):H=A/(B*E+1):PRINT
705 TU=INT(-G)
710 PRINT "VOLTAGE GAIN, RG 30% HIGH,=";TU
720 TV=INT(-H)
730 PRINT "CURRENT GAIN, RG 30% HIGH,=";TV
750 F=F*.7692308:A=A-.3*A:B=B+.3*B:C=C+.3*C:D=D-.3*D
760 G=A*E/((C*B-A*D)*E+C):H=A/(B*E+1):PRINT
765 YU=INT(-G)
770 PRINT "VOLTAGE GAIN, ALL H PARAMETERS 30% OUT OF TOLERANCE,=";YU
780 YV=INT(-H)
790 PRINT "CURRENT GAIN, ALL H PARAMETERS 30% OUT OF TOLERANCE,=";YV
810 A=A*1.428572:B=B*.7692308:C=C*.7692308:D=D*1.428572:PRINT
820 PRINT "NOTE: IF YOU DO NOT HAVE THE RATED H PARAMETERS AVAILABLE"
830 PRINT "FOR A PARTICULAR SMALL-SIGNAL TRANSISTOR, THE TYPICAL"
840 PRINT "VALUES SHOWN IN THE SAMPLE RUN WILL BE HELPFUL
850 END

READY.
```

CONVERTED TROUBLESHOOTING PROGRAM

For the Commodore 64

(With Printer)

```
5 REM PRG 1 WITH PRINTER
10 PRINT "AMPLIFIER ANALYSIS BASED ON OFF-TOLERANCE PARAMETERS"
12 OPEN1,4:PRINT#1,"AMPLIFIER ANALYSIS BASED ON OFF-TOLERANCE PARAMETERS"
13 CLOSE1,4
15 PRINT "GENERALIZED TROUBLESHOOTING PROGRAM NO.1":PRINT
17 OPEN1,4:PRINT#1,"GENERALIZED TROUBLESHOOTING PROGRAM NO.1":PRINT#1,
18 CLOSE1,4
20 PRINT "(INPUT/OUTPUT VALUES ARE COMPUTED FROM H PARAMETERS,"
22 OPEN1,4:PRINT#1,"(INPUT/OUTPUT VALUES ARE COMPUTED FROM H PARAMETERS,"
23 CLOSE1,4
25 PRINT "SOURCE CURRENT, SOURCE RESISTANCE, AND LOAD RESISTANCE)":PRINT
26 OPEN1,4:PRINT#1,"SOURCE CURRENT, SOURCE RESISTANCE, AND LOAD RESISTANCE)"
27 PRINT#1,:CLOSE1,4
30 PRINT "*** CE CONFIGURATION ***"
32 PRINT
35 OPEN1,4:PRINT#1,"*** CE CONFIGURATION ***":PRINT#1,:CLOSE1,4
40 PRINT "COMPUTES E/OUT, I/OUT, E/IN, I/IN, VOLTAGE GAIN, CURRENT GAIN,"
45 OPEN1,4:PRINT#1,"COMPUTES E/OUT, I/OUT, E/IN, I/IN, VOLTAGE GAIN,"
46 PRINT#1,"CURRENT GAIN,":CLOSE1,4
50 PRINT "POWER GAIN, SOURCE VOLTAGE, INPUT IMPEDANCE.":PRINT
55 OPEN1,4:PRINT#1,"POWER GAIN, SOURCE VOLTAGE, INPUT IMPEDANCE.":PRINT#1,
56 CLOSE1,4
60 PRINT "COMPUTES REDUCED VOLTAGE GAIN RESULTING FROM A 30% OFF-TOLERANCE ON"
65 OPEN1,4
67 PRINT#1,"COMPUTES REDUCED VOLTAGE GAIN RESULTING FROM A 30% OFF-TOLERANCE"
68 CLOSE1,4
70 PRINT"EACH CIRCUIT PARAMETER, AND ALL H PARAMETERS 30% OFF-TOLERANCE."
72 OPEN1,4
75 PRINT#1,"ON EACH CIRCUIT PARAMETER, AND ALL H PARAMETERS 30% OFF-TOLERANCE."
77 PRINT#1,:CLOSE1,4
80 PRINT
90 INPUT "HFE=";A
95 OPEN1,4:PRINT#1,"HFE=";A:CLOSE1,4
100 INPUT "HOE=";B
105 OPEN1,4:PRINT#1,"HOE=";B:CLOSE1,4
110 INPUT "HIE=";C
120 OPEN1,4:PRINT#1,"HIE=";C:CLOSE1,4
130 INPUT "HRE=";D
140 OPEN1,4:PRINT#1,"HRE=";D:CLOSE1,4
150 INPUT "RL (OHMS)=";E
160 OPEN1,4:PRINT#1,"RL (OHMS)=";E:CLOSE1,4
170 INPUT "IB (MICROAMPS)=";IB
180 OPEN1,4:PRINT#1,"IB (MICROAMPS)=";IB:CLOSE1,4
190 INPUT "RG (OHMS)=";F
200 OPEN1,4:PRINT#1,"RG (OHMS)=";F:PRINT#1,:CLOSE1,4
210 PRINT
220 G=A*E/((C*B-A*D)*E+C):H=A/(B*E+1)
230 I=A↑2*E/((B*E+1)*((C*B-A*D)*E+C))
240 J=(C+(B*C-A*D)*E)/(B*E+1)
250 K=(C+F)/(B*C-D*A+B*F)
255 PRINT "* NORMAL CHARACTERISTICS *":PRINT
256 OPEN1,4:PRINT#1,"* NORMAL CHARACTERISTICS *":PRINT#1,:CLOSE1,4
258 YA=INT(-G)
260 PRINT "VOLTAGE GAIN=";YA
261 OPEN1,4:PRINT#1,"VOLTAGE GAIN=";YA:CLOSE1,4
262 YB=INT(-H)
270 PRINT "CURRENT GAIN=";YB
271 OPEN1,4:PRINT#1,"CURRENT GAIN=";YB:CLOSE1,4
272 YC=INT(I)
280 PRINT "POWER GAIN=";YC
281 OPEN1,4:PRINT#1,"POWER GAIN=";YC:CLOSE1,4
```

```
282 YD=INT(10*LOG(I)/LOG(10))
290 PRINT "POWER GAIN (DB)=";YD
291 OPEN1,4:PRINT#1,"POWER GAIN (DB)=";YD:CLOSE1,4
292 YE=INT(J)
310 PRINT "INPUT IMPEDANCE (OHMS)=";YE
315 OPEN1,4:PRINT#1,"INPUT IMPEDANCE (OHMS)=";YE:CLOSE1,4
330 IC=IB*H:VO=IC*10↑-6*E:EI=IB*10↑-6*J:VG=EI+IB*10↑-6*F
332 YF=.01*INT(IB*100)
340 PRINT "IB (MICROAMPS)=";YF
341 OPEN1,4:PRINT#1,"IB (MICROAMPS)=";YF:CLOSE1,4
342 YG=.01*INT(-IC*100)
360 PRINT "IC (MICROAMPS)=";YG
361 OPEN1,4:PRINT#1,"IC (MICROAMPS)=";YG:CLOSE1,4
362 YH=.01*INT(-VO*100)
380 PRINT "E/OUT (VOLTS)=";YH
381 OPEN1,4:PRINT#1,"E/OUT (VOLTS)=";YH:CLOSE1,4
382 YI=.001*INT(VG*1000)
400 PRINT "GENERATOR VOLTAGE (VOLTS)=";YI
405 OPEN1,4:PRINT#1,"GENERATOR VOLTAGE (VOLTS)=";YI:CLOSE1,4
410 YJ=.001*INT(EI*1000)
420 PRINT "INPUT VOLTAGE (VOLTS)=";YJ
430 OPEN1,4:PRINT#1,"INPUT VOLTAGE (VOLTS)=";YJ:PRINT#1,:CLOSE1,4
440 A=A-.3*A:G=A*E/((C*B-A*D)*E+C):H=A/(B*E+1)
450 I=A↑2*E/((B*E+1)*((C*B-A*D)*E+C)):PRINT
455 YK=INT(-G)
460 PRINT "VOLTAGE GAIN, HFE 30% LOW,=";YK
465 OPEN1,4:PRINT#1,"VOLTAGE GAIN, HFE 30% LOW,=";YK:CLOSE1,4
470 YL=INT(-H)
480 PRINT "CURRENT GAIN, HFE 30% LOW,=";YL
490 OPEN1,4:PRINT#1,"CURRENT GAIN, HFE 30% LOW,=";YL:PRINT#1,:CLOSE1,4
500 A=A*1.428572:B=B+.3*B:G=A*E/((C*B-A*D)*E+C):H=A/(B*E+1):PRINT
505 YM=INT(-G)
510 PRINT "VOLTAGE GAIN, HOE 30% HIGH,=";YM
512 OPEN1,4:PRINT#1,"VOLTAGE GAIN, HOE 30% HIGH,=";YM:CLOSE1,4
515 YN=INT(-H)
530 PRINT "CURRENT GAIN, HOE 30% HIGH,=";YN
540 OPEN1,4:PRINT#1,"CURRENT GAIN, HOE 30% HIGH,=";YN:PRINT#1,:CLOSE1,4
550 B=B*.7692308:C=C+.3*C:G=A*E/((C*B-A*D)*E+C):H=A/(B*E+1):PRINT
555 YO=INT(-G)
560 PRINT "VOLTAGE GAIN, HIE 30% HIGH,=";YO
565 OPEN1,4:PRINT#1,"VOLTAGE GAIN, HIE 30% HIGH,=";YO:CLOSE1,4
570 YP=INT(-H)
580 PRINT "CURRENT GAIN, HIE 30% HIGH,=";YP
590 OPEN1,4:PRINT#1,"CURRENT GAIN, HIE 30% HIGH,=";YP:PRINT#1,:CLOSE1,4
600 C=C*.7692308:D=D-.3*D:G=A*E/((C*B-A*D)*E+C):H=A/(B*E+1):PRINT
605 YQ=INT(-G)
610 PRINT "VOLTAGE GAIN, HRE 30% LOW,=";YQ
615 OPEN1,4:PRINT#1,"VOLTAGE GAIN, HRE 30% LOW,=";YQ:CLOSE1,4
620 YR=INT(-H)
630 PRINT "CURRENT GAIN, HRE 30% LOW,=";YR
640 OPEN1,4:PRINT#1,"CURRENT GAIN, HRE 30% LOW,=";YR:PRINT#1,:CLOSE1,4
650 D=D*1.428572:E=E-.3*E:G=A*E/((C*B-A*D)*E+C):H=A/(B*E+1):PRINT
655 YS=INT(-G)
660 PRINT "VOLTAGE GAIN, RL 30% LOW,=";YS
665 OPEN1,4:PRINT#1,"VOLTAGE GAIN, RL 30% LOW,=";YS:CLOSE1,4
670 YT=INT(-H)
680 PRINT "CURRENT GAIN, RL 30% LOW,=";YT
690 OPEN1,4:PRINT#1,"CURRENT GAIN, RL 30% LOW,=";YT:PRINT#1,:CLOSE1,4
700 E=E*1.428572:F=F+.3*F:G=A*E/((C*B-A*D)*E+C):H=A/(B*E+1):PRINT
705 TU=INT(-G)
710 PRINT "VOLTAGE GAIN, RG 30% HIGH,=";TU
715 OPEN1,4:PRINT#1,"VOLTAGE GAIN, RG 30% HIGH,=";TU:CLOSE1,4
720 TV=INT(-H)
730 PRINT "CURRENT GAIN, RG 30% HIGH,=";TV
740 OPEN1,4:PRINT#1,"CURRENT GAIN, RG 30% HIGH,=";TV:PRINT#1,:CLOSE1,4
750 F=F*.7692308:A=A-.3*A:B=B+.3*B:C=C+.3*C:D=D-.3*D
760 G=A*E/((C*B-A*D)*E+C):H=A/(B*E+1):PRINT
765 YU=INT(-G)
770 PRINT "VOLTAGE GAIN, ALL H PARAMETERS 30% OUT OF TOLERANCE,=";YU
772 OPEN1,4
775 PRINT#1,"VOLTAGE GAIN, ALL H PARAMETERS 30% OUT OF TOLERANCE,=";YU
```

```
777 CLOSE1,4
780 YV=INT(-H)
790 PRINT "CURRENT GAIN, ALL H PARAMETERS 30% OUT OF TOLERANCE,=";YV
792 OPEN1,4
795 PRINT#1,"CURRENT GAIN, ALL H PARAMETERS 30% OUT OF TOLERANCE,=";YV
800 PRINT#1,:CLOSE1,4
810 A=A*1.428572:B=B*.7692308:C=C*.7692308:D=D*1.428572:PRINT
820 PRINT "NOTE: IF YOU DO NOT HAVE THE RATED H PARAMETERS AVAILABLE"
822 OPEN1,4
825 PRINT#1,"NOTE: IF YOU DO NOT HAVE THE RATED H PARAMETERS AVAILABLE"
827 CLOSE1,4
830 PRINT "FOR A PARTICULAR SMALL-SIGNAL TRANSISTOR, THE TYPICAL"
832 OPEN1,4
835 PRINT#1,"FOR A PARTICULAR SMALL-SIGNAL TRANSISTOR, THE TYPICAL"
837 CLOSE1,4
840 PRINT "VALUES SHOWN IN THE SAMPLE RUN WILL BE HELPFUL
842 OPEN1,4
845 PRINT#1,"VALUES SHOWN IN THE SAMPLE RUN WILL BE HELPFUL.":CLOSE1,4
850 END

READY.
```

```
AMPLIFIER ANALYSIS BASED ON OFF-TOLERANCE PARAMETERS
GENERALIZED TROUBLESHOOTING PROGRAM NO.1

(INPUT/OUTPUT VALUES ARE COMPUTED FROM H PARAMETERS,
SOURCE CURRENT, SOURCE RESISTANCE, AND LOAD RESISTANCE)

*** CE CONFIGURATION ***

COMPUTES E/OUT, I/OUT, E/IN, I/IN, VOLTAGE GAIN,
CURRENT GAIN,
POWER GAIN, SOURCE VOLTAGE, INPUT IMPEDANCE.

COMPUTES REDUCED VOLTAGE GAIN RESULTING FROM A 30% OFF-TOLERANCE
ON EACH CIRCUIT PARAMETER, AND ALL H PARAMETERS 30% OFF-TOLERANCE.

HFE= 50
HOE= 2E-05
HIE= 1500
HRE= 5E-04
RL (OHMS)= 15000
IB (MICROAMPS)= 5
RG (OHMS)= 1500

* NORMAL CHARACTERISTICS *

VOLTAGE GAIN=-477
CURRENT GAIN=-39
POWER GAIN= 18315
POWER GAIN (DB)= 42
INPUT IMPEDANCE (OHMS)= 1211
IB (MICROAMPS)= 5
IC (MICROAMPS)=-192.31
E/OUT (VOLTS)=-2.89
GENERATOR VOLTAGE (VOLTS)= .013
INPUT VOLTAGE (VOLTS)= 6E-03

VOLTAGE GAIN, HFE 30% LOW,=-312
CURRENT GAIN, HFE 30% LOW,=-27

VOLTAGE GAIN, HOE 30% HIGH,=-439
CURRENT GAIN, HOE 30% HIGH,=-36

VOLTAGE GAIN, HIE 30% HIGH,=-348
CURRENT GAIN, HIE 30% HIGH,=-39

VOLTAGE GAIN, HRE 30% LOW,=-445
CURRENT GAIN, HRE 30% LOW,=-39

VOLTAGE GAIN, RL 30% LOW,=-339
CURRENT GAIN, RL 30% LOW,=-42
```

```
VOLTAGE GAIN, RG 30% HIGH,=-477
CURRENT GAIN, RG 30% HIGH,=-39

VOLTAGE GAIN, ALL H PARAMETERS 30% OUT OF TOLERANCE,=-208
CURRENT GAIN, ALL H PARAMETERS 30% OUT OF TOLERANCE,=-26

NOTE: IF YOU DO NOT HAVE THE RATED H PARAMETERS AVAILABLE
FOR A PARTICULAR SMALL-SIGNAL TRANSISTOR, THE TYPICAL
VALUES SHOWN IN THE SAMPLE RUN WILL BE HELPFUL.
```

Common Conversion "Bugs"

Error messages are frequently vague, and the programmer must be alert to the probabilities and possibilities for "illegal" entries. Some typical "bugs" are:

1. Numeral 0 typed in instead of a capital O, or vice versa. Note in passing that the numeral 0 will be displayed as Ø on the IBM video screen, although it may be (correctly) displayed as a capital letter in hard copy; the programmer must accordingly be aware of his computer's idiosyncrasies.
2. A two-letter variable may have been inadvertently reversed, such as PQ for QP.
3. A semicolon may have been entered instead of a colon (or vice versa).
4. An entire program line may have been omitted.
5. A tricky difficulty involves "illegal" word processing in the course of correcting a program line that leaves a "bug" hidden down in the program memory. If this trouble is suspected, type the entire line over again.
6. Reserved variables cannot be used arbitrarily. Thus, RO is an acceptable variable, whereas OR is reserved and may be employed only in logical statements (no numerical value can be assigned to OR).
7. In the case of the IBM PC, program lines can be renumbered in a chosen sequence by means of the RENUM function (this function is not provided on the 64). Whenever the RENUM command is given, follow it up to make certain that all GOTO or GOSUB references are still correct.
8. Remember that the computer provides ony natural logarithms; base-10 logs must be called by typing 10*LOG(P)/LOG(10).
9. Remember also that the computer provides only radian angular values. Degrees must be called by typing 360/6.2832 (and it is easy to type 6.2832/360 instead).

10. Too, a program may appear to have a "bug" when the fault is actually an INPUT error. For example, if the programmer inadvertently INPUTs 15.5 instead of 1.55, it will appear that there is a "bug" in the program. In other words, if an incorrect answer is printed out, first double-check your INPUTs.

One Line at a Time

Sometimes a converted program will RUN with elusive incorrect answers. In such a case, check out the equations for a possible "slip." Since anyone is prone to "pet" errors, it is helpful to have someone else re-check equations in difficult circumstances. Note that you can follow each program line, one at a time, with the print message PRINT "Line 56 X=";X, and in turn this information will be displayed on the RUN. Thereby, the programmer can follow the processing action one line at a time, and zero-in on the fault.

One of the common elusive errors is failure to correctly increment and decrement variables. As an illustration, the INPUTed value for R1 might be 1000 ohms. If you subsequently increment its value with the coding "RO=RO+.01*T*RO, this increment must be followed at an appropriate point with the coding RO=RO/(1+.01*T). It is easy to forget to restore the variable's value, or to type in a minus sign instead of a plus sign (or vice versa).

IBM PC Jr. In general, programs written for the IBM PC will RUN without conversion on the IBM PC Jr.

APPLE IIe AND II+. A simple Commodore 64 program may RUN without conversion on the APPLE computer. Just as different keys are utilized to call the same function in the IBM PC versus the Commodore 64, so are different keys employed to call the same function in the APPLE IIe or II+ versus the 64. As an illustration, the command CLS clears the screen in the IBM PC, whereas the command "HOME" must be typed into the APPLE IIe. After "HOME" is typed as a direct command, the RETURN key must then be pressed.

When using arrays, the APPLE computer is limited to a maximum of 88 dimensions, compared with 255 dimensions for the IBM PC or the 64. However, a long IBM array can be sectioned into two or more shorter arrays in converting to the APPLE. To make a hard copy of a program with the APPLE computer, you can command 100: PR#1:LIST:PRO#0 where 1 is (ordinarily) the slot number of the printer. In the case of the IBM PC, you type LLIST and press the ENTER key. To hard-copy a program on the 64, the command OPEN4,4:CMD4:LIST:PRINT#4:CLOSE4 may be used. (In most cases, the slot number will be 4).

Just as you must open and close files to PRINT hard copy on the 64, so is it necessary to redefine output slots on the APPLE computer. For example, an LPRINT command in an IBM PC program could be converted to an APPLE format by means of the following lines:

```
10 PRINT CHR$(4);''PR#1"
20 PRINT ''THIS IS HARD COPY"
30 PRINT ''CHR$(4);PR#0"
```

A BASIC program cannot be RENUMbered from within itself on the APPLE IIe or II+. ATN returns the angle (in radians) corresponding to the ATN value, just as for the IBM PC and the 64. Exclusive OR (XOR) is employed as a command on the IBM PC, but must be simulated on the APPLE (and 64) computers. AND NOT OR NOT AND formulation is utilized. As previously noted, the easiest way to become familiar with conversion procedures is to compare equivalent programs that have been written for the IBM PC and the APPLE computers.

TRS-80 COMPUTERS. Many TRS-80 personal computers are still in use. Among them, the PC-2 Pocket Computer is the most capable, and provides some functions that are lacking in the IBM PC. For example, the PC2 provides angular computations in radians, degree, or grads; it also provides base-10 logarithms, and has arc-sin and arc-cos functions. On the other hand, the Model 1 has virtually no mathematical functions available, with the result that square root, log, trig, and exponential functions must be programmed as subroutines.

The PC-2 printer (actually an XY plotter) provides various sizes of type, with a choice of four colors. As might be anticipated, the plotter is considerably slower than the matrix print heads in the IBM PC and Commodore 64 computers. However, the owner of a PC-2 has a machine that focuses more on mathematical operations than on graphics. The chief drawback noted by most users is its inability to display more than one line at a time on the video "screen."

III

Time-Saving Program for Computation of RC High-Pass Filter Bootstrap Circuit Output Voltage and Phase versus F

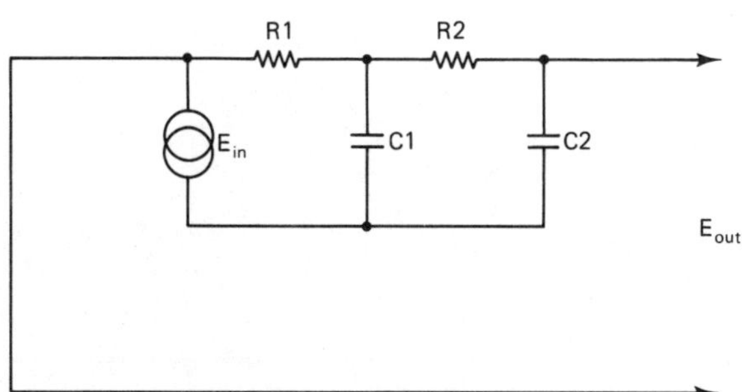

This network is an ''experimenter's curiosity.'' It develops an output voltage that is greater than its input voltage at its resonant frequency, and at all frequencies higher than its resonant frequency. The circuit has high-pass filter action. The experimenter can cascade and/or ''piggy-back'' additional sections to obtain increased voltage bootstrapping.

```
1 LPRINT "UNSYMMETRICAL 2-SECTION RC NETWORK WITH VOLTAGE GAIN"
2 PRINT "UNSYMMETRICAL 2-SECTION RC NETWORK WITH VOLTAGE GAIN"
3 LPRINT "(Integrating Sections)":PRINT"(Integrating Sections)"
4 LPRINT "* High-Pass Filter Action *":PRINT "* High-Pass Filter Action *"
5 LPRINT "Eout/Ein and Phase vs. F":PRINT "Eout/Ein and Phase vs. F"
6 LPRINT "(Survey Program)":PRINT "(Survey Program)":LPRINT:PRINT
7 INPUT "R1 (Ohms)=";RO
8 LPRINT "R1 (Ohms)=";RO
9 INPUT "R2 (Ohms)=";RT
10 LPRINT "R2 (Ohms)=";RT
11 INPUT "C1 (Mfd)=";CO
12 LPRINT "C1 (Mfd)=";CO
13 INPUT "C2 (Mfd)=";CT
14 LPRINT "C2 (Mfd)=";CT
15 INPUT "Starting f (Hz)=";H
16 LPRINT "Starting f (Hz)=";H
17 F=0:N=0:G=H:LPRINT:PRINT
18 F=G+N:A$="#######":LPRINT"f (Hz)=";USING A$;F:PRINT"f (Hz)=";USING A$;F
19 XO=1/(6.2832*F*CO*10^-6):XT=1/(6.2832*F*CT*10^-6)
20 AB=(RO*RO+XO*XO)^.5:BA=-ATN(XO/RO):B$="####.###"
21 LK=XO+XT:BC=(RT*RT+LK*LK)^.5:CB=-ATN(LK/RT):AC=BC*AB:CA=CB+BA
22 AD=AC/XO:DA=CA+6.2832/4:AE=AD*COS(DA):EA=AD*SIN(DA)
23 BE=EA+XO:CE=(AE*AE+BE*BE)^.5:EC=-ATN(BE/AE):AN=XT/CE:NA=-EC-6.2832/4
24 SH=AN*COS(NA):HS=AN*SIN(NA):BS=1+SH
25 PU=(BS*BS+HS*HS)^.5:UP=-ATN(HS/BS):N=N+100:C$="####.#"
26 LPRINT "Eout/Ein=";USING B$;PU:PRINT "Eout/Ein=";USING B$;PU
27 LPRINT "Phase Angle (Degrees)=";USING C$;-UP*360/6.2832
28 PRINT "Phase Angle (Degrees)=";USING C$;-UP*360/6.2832:LPRINT:PRINT
29 IF N>20000 THEN 31
30 GOTO 18
31 END
```

IV

Getting It All Together

Although this topic is not really new, it is a real time-saver and deserves mention. A good workbench with shelves and drawers for storage of test equipment, tools, components, devices, circuit boards and miscellany is not a luxury—it is a necessity for anyone who is going into electronics. A good example of "getting it all together" is shown below.

Electrical outlets are required. A test panel is also a major time-saver. The panel should be provided with terminals and jacks from variable ac and dc power supplies, battery chargers, isolation transformer, test speakers, audio output transformer, audio amplifiers, and antenna connections for AM, FM, CB, VHF TV, UHF TV, and auto radio.

If you are "starting from scratch," you will have immediate need for the following hand tools: Various pliers such as long needle-nosed, long-nose, slim-nose, linesman, and diagonal cutting pliers, various screwdrivers such as flat-blade, Phillip's-head and Allen-head screwdrivers, an assortment of nut drivers, a dual-blade wire stripper, soldering gun, small soldering iron, soldering aid, and an antiwicking tool.

If you plan to get into record changer servicing, you will need a chassis cradle, hexs and spline key wrenches, a set of tweezers with various sizes, a dental-type inspection mirror, a set of feeler gauges, a set of small box end wrenches, and a stylus inspection microscope. It is also advisable to have a stroboscope disc handy, for checking turntable speed.

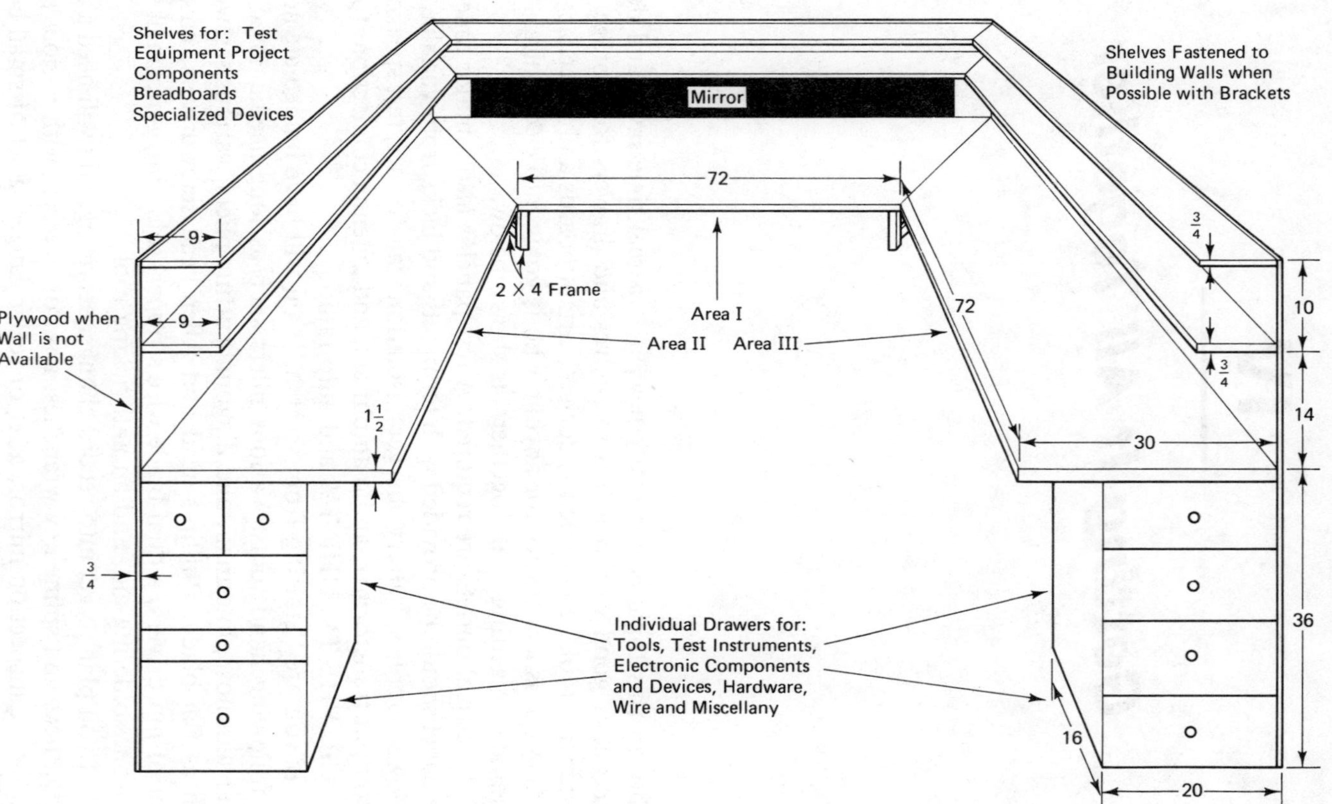

(Reproduced by Permission of Parker Publishing Co., Inc., From "Electronic Workshop Manual and Guide," by Carl G. Grolle)

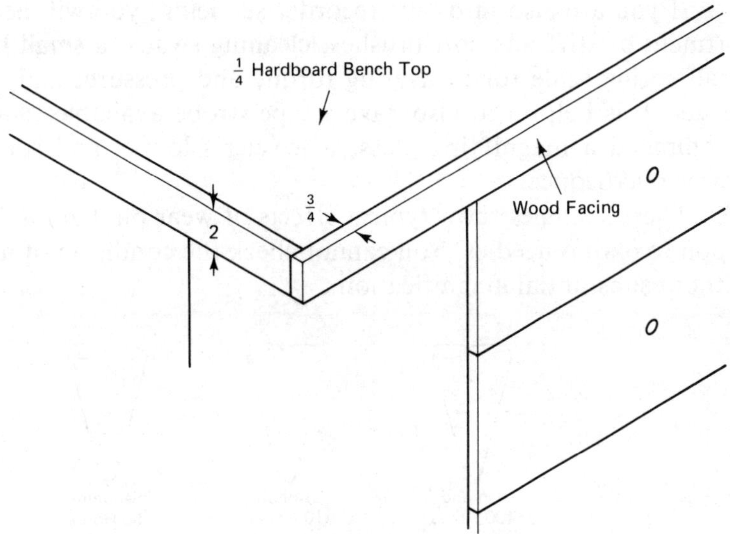

Wood trim for the workbench top front edge

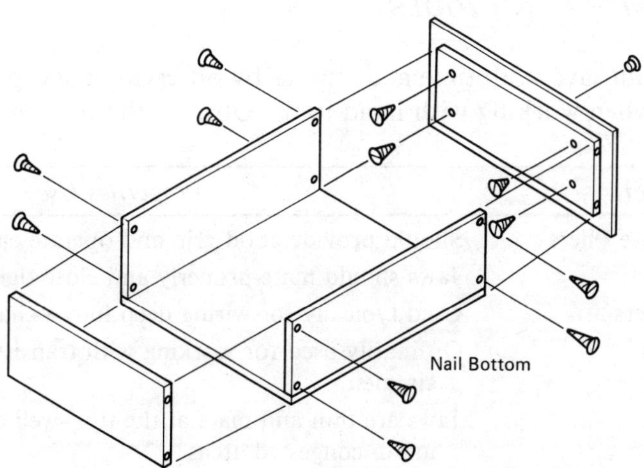

Plan for tool and parts drawer

If you are also into tape-recorder servicing, you will need a varied assortment of stiff and soft brushes, cleaning swabs, a small hand blower, a small spring scale for measuring torque and pressure, and a head demagnetizer. It is helpful to also have a tape strobe available. Some technicians recommend a magnifying glass, a jeweler's loupe, and sometimes a low-power microsopce.

These sketches show typical effects of wear on diamond, osmium, and sapphire phono needles. You cannot check the condition of a needle (stylus) without substantial magnification.

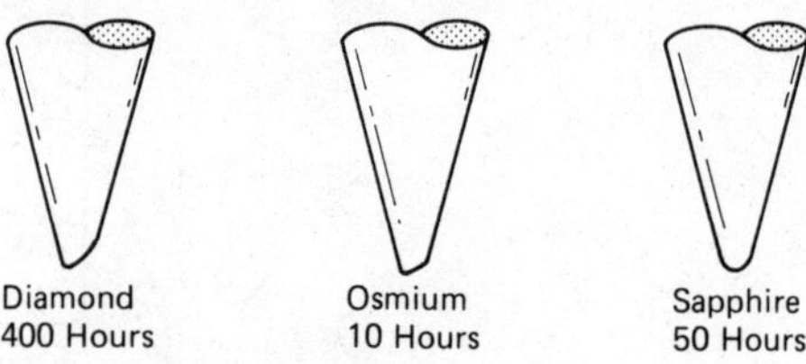

(Reproduced by Special Permission of Reston Publishing Co., Inc., and Derek Cameron from "Audio Technology Systems")

WORKING WITH HAND TOOLS

You can save both time and temper by observing good practices and shortcuts when working with hand tools. Observe the following basic points:

Tool	*Properties Use*
General-purpose Pliers	Should provide good grip and operate easily
	Jaws should mate properly and close tightly
Eight-Inch Pliers	Used typically for wiring deep into a chassis
Six-Inch Pliers	Ordinarily used for working with transistors or integrated circuits.
	Jaws are thin and mate at the tip—well adapted for working in congested areas
Six-Inch Round-Nose Pliers	Generally used for bending wire and forming loops
Six-inch Heavy-Duty Combination Long-Nose Side-Cutting Pliers	Typically used for twisting wire around terminal strips and cutting wire
Side-Cutting Diagonal Pliers	Used to cut component leads and light-gauge wire

Rugged Side-Cutting Diagonal Pliers	Cuts up to #8 gauge bolts; can cut heavy-gauge wire or rods
End Nippers (5-Inch)	Used on congested PC boards where standard "dykes" will not go
Combination Pliers-Clamp	Jaws will lock together; use as a heat sink, miniature vise, solder holder, nut starter, glue clamp, or tool jig
Wire Stripper	Use for stripping insulation without danger of damaging the wire
Razor Blade	Handy "tool" for opening up a multiconductor cable
Quick-Heating Soldering Gun	Basic type of electronic soldering tool
Soldering Iron	It is good practice to have several sized available, including a miniature iron Cordless solding irons are handy for jobs where a power outlet is not available
Desoldering Tool	Facilitates desoldering procedures, and leaves one hand free
Extractor for Integrated Circuits	Saves time and damage to PC board; use with a slotted-bar tip on a soldering iron
Drills	Sizes 1 though 60; a drill index indicates: Drill diameter Common tap sizes Tap drill numbers Body drill numbers Other useful bits are 1/4″, 5/16″, 3/8″, 7/16″, and 1/2″
Countersink, Deburring Tool	Used to "clean up" holes drilled in sheet metal.
Files	Keep an assortment of sizes at hand; use a double-cut second-cut tooth type for sheet metal; a half-round type is needed in addition to a round file and a square file; for delicate work use a set of 4″ jeweler's or Swiss-pattern files.

WORKBENCH TEST PANEL

A workbench test panel is a significant time-saver. As shown below, a basic test panel for general electronics requirements consists of a 4-ohm speaker and an 8-ohm speaker, UHF/VHF/FM/CB/Auto antenna facilities, AC and DC power supplies, an isolation transformer, an audio output transformer, a utility audio amplifier, and a stereo amplifier. The stereo amplifier has tape, phono, and tuner inputs. If you do much stereo testing, you will probably wish to add a pair of stereo speakers to the complement.

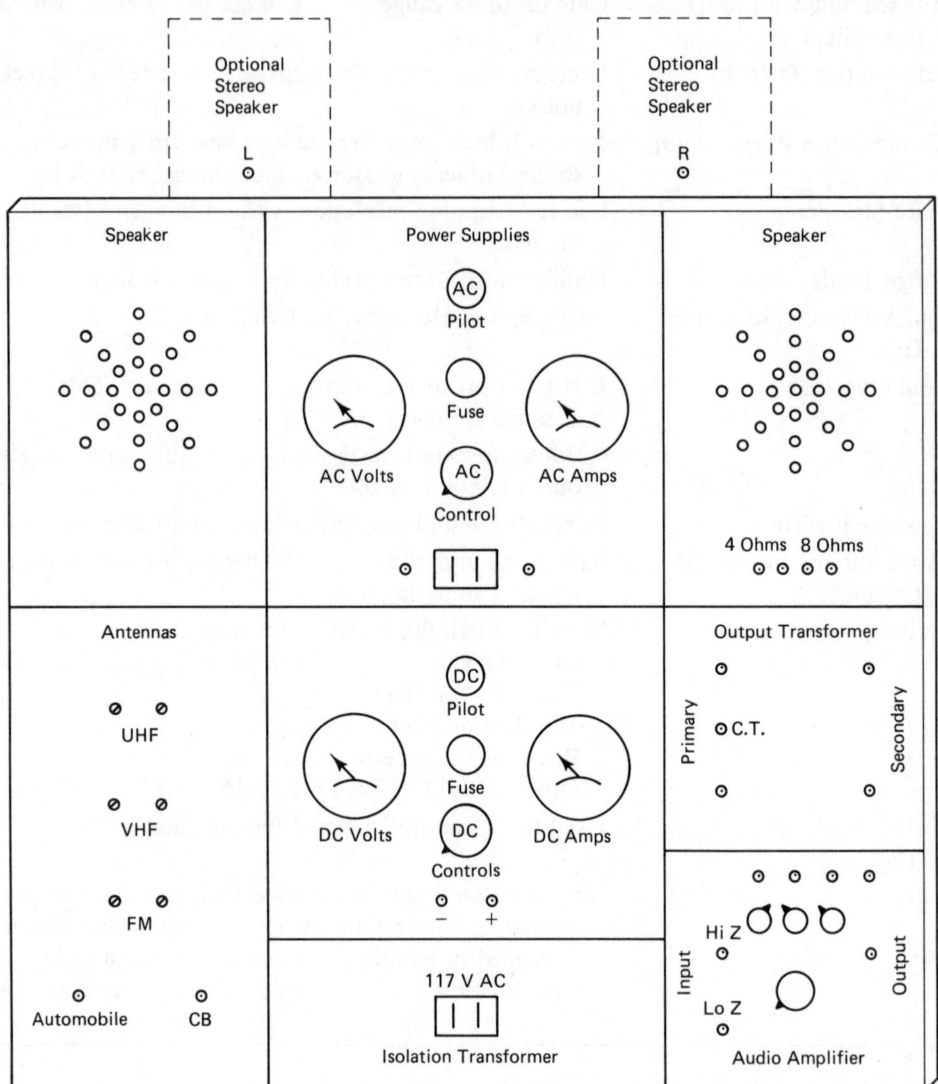

Optional Stereo Speaker
L
Optional Stereo Speaker
R
Speaker
Power Supplies
AC
Pilot
Fuse
AC Volts
AC
AC Amps
Control
Speaker
4 Ohms
8 Ohms
Antennas
UHF
VHF
FM
Automobile
CB
DC
Pilot
Fuse
DC Volts
DC
DC Amps
Controls
−
+
117 V AC
Isolation Transformer
Output Transformer
Primary
C.T.
Secondary
Hi Z
Lo Z
Input
Output
Audio Amplifier

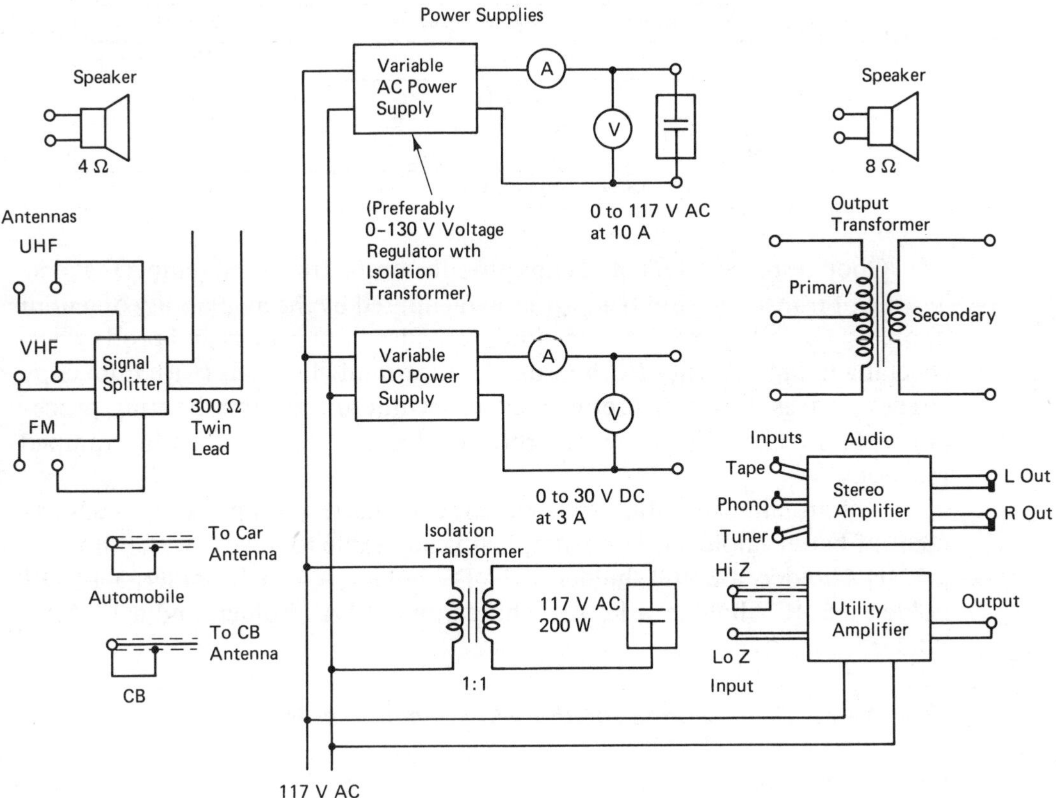

Note that the number of controls provided in the audio amplifier section will depend on the particular amplifiers that you build (or purchase) for the workbench test panel.

Basic Test Panel Schematic

This is a basic test panel schematic. You are likely to wish to include some specialized items. For example, if you are a radio amateur, you will probably add several antenna facilities. If you are into computers, you may wish to add a telephone interface. If you are into instrument maintenance or repair, you will probably add a variety of calibrating voltage sources and precision resistors.

You will also require a set of basic test leads. A list of essential types is as follows:

Quantity	End Connector	Type of Conductor	End Connector
2	AC Plug	Zip Cord (Line Cord)	Two Insulated Alligator Clips
2	AC Plug	"Cheater Cord"	Female Receptacle
1	Motorola Type Plug	Coaxial Antenna Lead-in	Motorola Type Plug
2	Red Banana Plug	Test Lead Wire	Red Banana Plug

Good test leads, kits, and clips are substantial time (and temper) savers. Testing of transistors and ICs is greatly facilitated by the availability of a mini hook test clip adapter for connection to standard test probes and prods when checking in high-density PC boards. A set of mini clip leads (with mini clips at each end) is also very useful in experimental and troubleshooting procedures. Micro test clips may be preferred for text connections in cramped locations.

You are almost certain to need some 9V battery snaps, and an assortment of battery holders. For example it is advisable to have a 1 "C" holder, a 1 "D" holder, a 2 "C" holder, a 2 "D" holder, a 4 "C" holder, a 4 "D" holder, a 6 "C" holder, a 1 "AA" holder, a 2 "AA" holder, and a 4 "AA" holder.

Time-Saver Applications for the Workbench Test Panel

Utility Audio Amplifier:

1. Checking microphones and PA equipment.
2. Audio signal tracer.
3. Patch amplifier when troubleshooting the audio section in a radio or TV receiver, tape recorder, record player, or intercom.
4. Output boost amplifier for a low-level audio oscillator.
5. Simple high-frequency signal tracer (when used with a series input diode).

Stereo Amplifier:

1. Patch L amplifier or patch R amplifier when troubleshooting a hi-fi system.
   If the workbench stereo amplifier has individual pre-amp, driver, and output sections, individual sections can also be used as patch amplifiers.

2. L and R stereo speakers can be used in substitution tests when troubleshooting a hi-fi system.
3. The workbench stereo amplifier can be used as a reference in comparison tests of malfunctioning hi-fi systems to determine normal dB separation, noise levels, and so on.

DC Power Supply

1. Substitute V_{CC} supply when troubleshooting any unit of electronic equipment.
2. Power source for breadboards and other experimental units.
3. Bias voltage source for use in tests and measurements.
4. Substitute for battery charger.
5. Determination of dc power demand by a unit of electronic equipment (volts times amps equals watts).

AC Power Supply

1. Adjustable line-voltage supply when troubleshooting any unit of line-operated electronic equipment.
2. Determination of volt-ampere power demand by a unit of electronic equipment (volts times amps equals volt-amps).
3. Facilitation of intermittent troubleshooting by checking equipment operation at low line voltage and at high line voltage.
4. Source of 60-Hz ac current when localizing ground-loop hum problems in audio equipment.
5. Source of pulsating dc (when combined with the adjustable dc power supply) when checking power-supply hum problems in any unit of electronic equipment.

Output Transformer

1. Besides its conventional use in matching an audio amplifier to a speaker, the output transformer can serve as a handy a-f step-up or step-down transformer in experimental arrangements.
2. The output level from an audio oscillator can be stepped up by this means (distortion may be increased, however).

3. The output transformer can also function as a practical isolation device to minimize capacitance unbalance to ground in "floating" audio-frequency test arrangements.

Speakers

1. A speaker will serve as a practical temporary microphone.

Components Storage Cabinet

One or more components storage cabinet is needed in addition to a workbench in order to "get it all together." You can easily build a components storage cabinet as shown below, or you can purchase a similar commercial cabinet. Both shelves and drawers are provided; each drawer can be partitioned as desired.

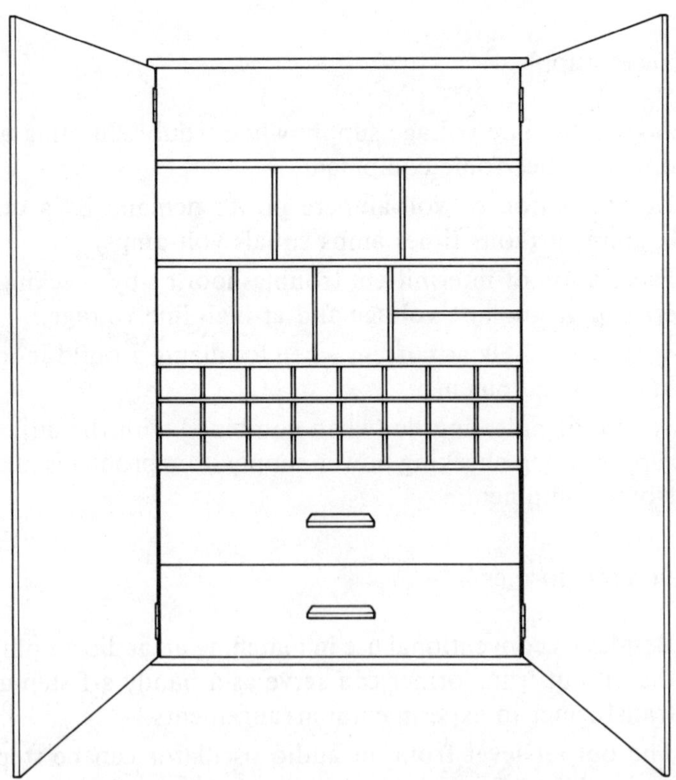

This Type of Components Storage Cabinet is a Major Time-saver

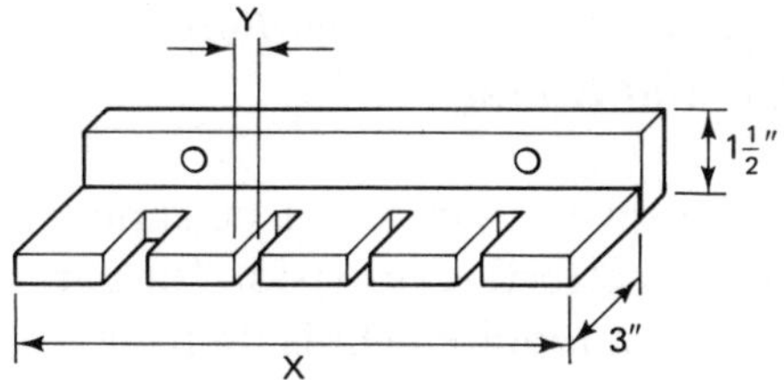

Wooden probe holder for the workbench or storage cabinet

An entire drawer can be used for a particular type of components such as digital logic boards and associated devices. Or, a large drawer may be partitioned into as many as 10 compartments for segregation of related components such as paper capacitors, mica capacitors, electrolytic capacitors, ceramic capacitors, adjustable capacitors, and so on. Similarly, fixed resistors with various values and power ratings or specialized resistors such as metaloxide and wirewound resistors may be stored in another partitioned drawer.

Observe that you can install horizontal poles above the shelves to support spools of wire. In turn, any spool is immediately accessible, and you need only unroll as much wire as required for a particular job. You may provide spools of test-lead wire, solid wire, hook-up wire, coax cable, twin lead, line cord, and heater appliance cord, for example.

A partitioned drawer may be assigned to machine screws, nuts, and washers. Most screws and nuts used in electronics projects are gauge numbers 4, 6, 8, and 10. (The gauge number denotes the thread diameter). Gauge numbers, diameters, and threads per inch may be tabulated as follows:

Gauge #	*Diameter (inches)*	*Threads per inch*	
		Coarse	*Fine*
4	0.112	40	48
6	0.138	32	40
8	0.164	32	36
10	0.190	24	32

Machine screws may also be segregated according to length. You will usually be concerned with 3/8″, 1/2″ and 3/4″ lengths. A few longer machine screws in larger diameters may also be included for special requirements. Be sure to include #6-32 hex head nuts, #6 flat washers, and #6 internal lock washers.

Workbench Tips on Parts and Hardware Organization and Storage

More time can be saved than might be supposed by the short-timers if the workbench area has effective provision for parts and hardware organization and storage. The first requirement is "a place for everything, and everything in its place." The second and equally important requirement is human-engineered organization of the "places."

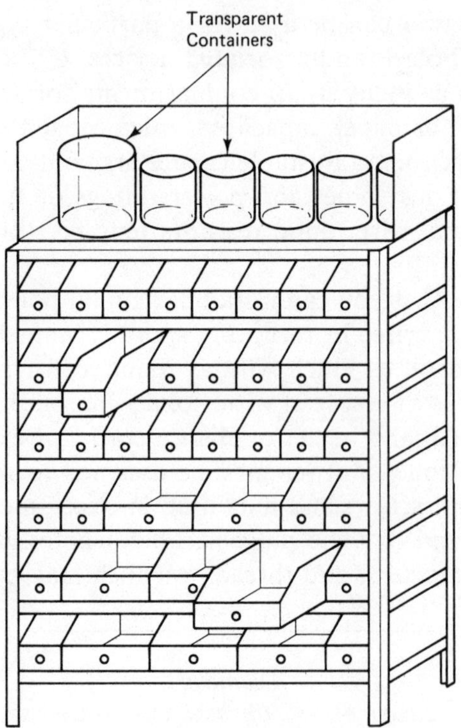

As an illustration, you will find that a shelf of transparent containers on top of a storage rack provides faster location of most-often-used hardware items. Your most-often-used items will vary from project to project, and from one phase of experimentation to another. Possible candidates for storage in transparent containers include:

Solder	Terminal Strips	Clips
Flux	Spacers, Posts, and Rods	Fasteners
Fuses	Switch and Potentiometer	Clamps

Machine Screws	Hardware	Grommets
Nuts	Connectors	Handles
Washers	Lugs	Knobs
Sheet Metal Screws	Insulators	PC Boards

You will also find it helpful to organize the larger drawers in a storage rack by means of half a dozen or more transparent containers. Possible candidates for storage in these containers include:

PC Drill Bits	Twine
Miniature Soldering Iron Tips	White Grease
#0, 1, 2 Screws	Silicone Grease
C Rings	Petroleum Jelly
Rivets	Acetone
Recessed Hex-Head Screws	Alcohol
Switch Contacts	Cabinet Polish
Miniature Springs	Liquid Cleaner
Ferrite Cores	Coolant Liquid
Spray Can Valves	Color Paints

WORKBENCH TIME-SAVER CHEMICALS

Unexpected time and effort can be saved by having basic chemicals available at the electronics workbench, and knowing how to use them. For example, freeze sprays can sometimes save a headache, and the right lubricant can make all the difference in repairing a tape recorder. Basic chemicals are as follows:

Chemical	*Properties Usage*
Cleaning Agents	Evaporative cleaners, and lubricative cleaners, often supplied in aerosol cans; for switches, relays, potentiometers and similar devices
Degreasers	For removal of dirt and grease buildup on moving parts; follow up with a cleaner and lubricant
Aerosol Hand Cleaner	To clean hands on tape recorders; may be used with an extension spray brush
Aerosol Vacuum Cleaner-Blower System	For removing dust in congested locations
Freeze Spray	For troubleshooting intermittent circuits; can pinch-hit for a heat sink in difficult areas

Epoxy Cement	To securely bond parts to metal, glass, plastic, or coated surfaces
Silicone Grease	Provides good heat transfer from a power transistor to a metallic sink, for example
Silicone Rubber	Use in same applications as epoxy; silicone rubber provides a good insulating film in addition to bonding
Cyanoacrylate Adhesive	For quick-set bonding without heat or pressure to rubber, plastic, porcelain, metal, wood, or glass
Aerosol Spray Paint	To finish project chassis-cabinets
Zinc Chromate Primer Paint	To prepare aluminum surfaces for final painting
Finish Preparations	For obtaining a dull flat, glossy, two- or three-tone, crackle, spatter, hammertone, metallic-flake, or transparent finish

Time-Saver Tips on Practical Soldering Techniques

An unexpectedly large amount of time and vexation (and money) can be saved by observing good practices in soldering procedures. As a case in point, electronics kit suppliers have stated that 90 percent of the kits returned for factory advice and repair do not work or exhibit trouble symptoms because of poor soldering procedures.

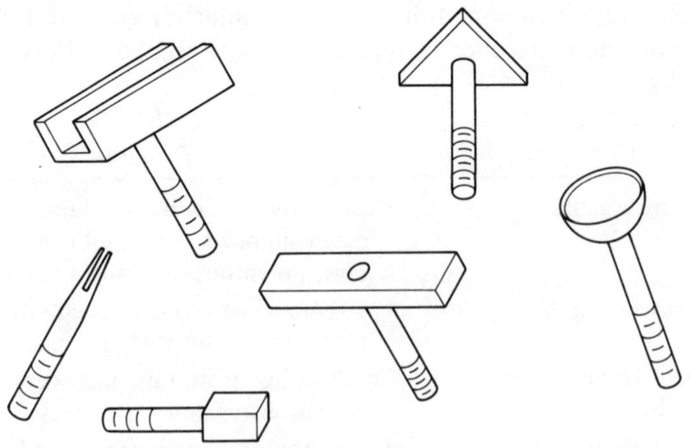

Desoldering Tips

(Reproduced by Permission of Parker Publishing Company, Inc., From "Electronic Workshop Manual and Guide," by Carl G. Grolle)

Desoldering is often more difficult than soldering techniques. Congested PC boards with integrated circuits and miniaturized components and easily-damaged conductors are real gremlins.

Basic requirements include proper soldering and desoldering tools. Most of us start out by under-rating the importance of the right tools and building up practical experience in the art of soldering.

A fast-heating solder gun is convenient and effective for making a few solder joints. However, if you have a number of solder joints to make, you will save time and annoyance by using a soldering iron. Cold-solder joints are a classic source of intermittent operation and tough-dog trouble situations. A good solder joint is smooth and shiny; all of the joint connections are visible, and appear to be a single unit. Excessive heat can cause damage to the insulation and structures associated with the connection.

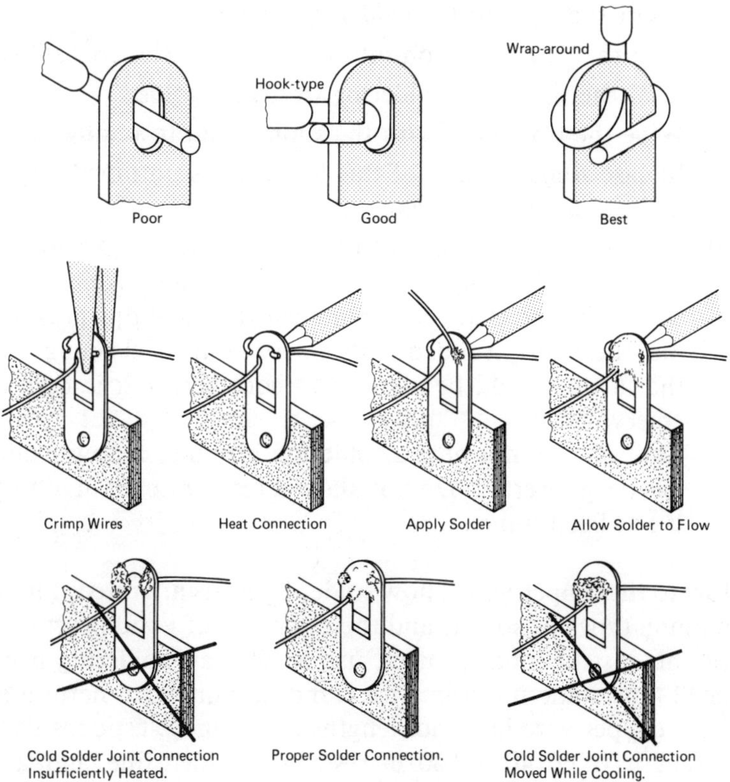

Step-by-step Soldering Process
and Typical Completed Joints

(Reproduced by Permission of Prentice-Hall, Inc., From "Electronics Shop Practices, Equipment, and Materials")

Good Practices Familiarization:
How to clean a soldering iron, to dress the tip, and to tin the tip, is essential basic knowledge. For this instructional project, use a soldering iron with a copper tip and with a supplementary iron tip; obtain a stand for the soldering iron, a natural sponge pad, some soft solder (wire-type solder with resin core), and have a 117-volt 60-Hz power source available. Proceed as follows:

1. Remove the tip from the soldering iron.
2. Scrape all loose oxide from the tip. This ensures that the tip will not "freeze" into the iron and it also permits improved heat flow into the tip.
3. Scrape any loose oxide from the tip.
4. Insert the tip into the soldering iron.
5. Plug the soldering iron into the power outlet and allow it to heat up.
6. Coat the tip of the soldering iron with solder.
7. Wipe the tip lightly with the sponge pad to remove any excess solder.
8. In case some portion of this tip remains untinned, repeat the cleaning and tinning procedure.
9. If necessary, file the tip to facilitate tinning. (An iron-clad tip should not be filed.) If a copper tip is filed, the angles of the flat surfaces should not be changed. Permanently tinned tips must not be filed, but may be lightly rubbed with emery cloth if cleaning is required. Note that a tip should be allowed to cool before it is filed or rubbed with emery cloth.
   Caution: Do not plug a soldering iron into a power outlet unless a tip has been inserted. Do not allow a tip to remain in a heated state unless it has been tinned.

How to tin copper wires, how soldering irons are rated, familiarization with common types of solder, and the essentials of good practices are also basic. For this instructional project, use a 100-watt soldering iron with stand, a pencil-type soldering iron with stand, a dampened natural sponge, No. 12 gauge copper wire in 1-foot lengths (solid and plastic-insulated types), sample lengths of 60/40 and 40/60 resin-core solder, fine sandpaper, a wire stripper, and have a 117-volt 60-Hz power source available. Proceed as follows:

1. Using the 60/40 solder, tin the tips of both soldering irons.
2. Strip both ends of the insulated copper wire.

3. Observe the exposed surface; if it is oxidized, remove the oxide with sandpaper.
4. Wipe the tip of the 100-watt soldering iron with the damp sponge and, if advisable, add a small amount of solder.
5. With the soldering iron placed on its stand, hold the stripped and cleaned end of the copper wire against the tinned tip for transfer of heat.
6. Touch the 60/40 solder to the heated wire, but do not permit the solder to touch the tip of the iron.
7. Revolve the wire rapidly to obtain a uniformly tinned surface. Avoid burning the insulation on the wire.
8. Observe the result to determine whether the film of solder is continuous, smooth, and shiny. If necessary, reheat the end of the copper wire.
9. Then repeat the foregoing procedure, using the other end of the copper wire, using the pencil soldering iron and 40/60 solder.

How to solder joints, and familiarization with good mechanical joints and good soldering practices are also basic. For this instructional project, obtain a 50 watt soldering iron, a dampened natural sponge, No. 18 gauge 2-foot lengths of bare solid copper wire, a sample length of 60/40 resin-core solder, fine sandpaper, a bending tool (or long-nose pliers with jaws covered by plastic tape), and have available a 117-volt 60Hz power source. Proceed as follows:

1. Tin the tip of the solding iron.
2. Clean and tip two ends on the copper wires.
3. Use a bending tool or taped pliers and form the ends of wires into a hook splice. Then make a good mechanical joint, avoiding damage to the wires.
4. Apply heat to one side of the splice with the soldering iron and touch the solder to the other side of the splice. A small amount of solder may be added to the tip.
5. Hold the wires steady while the joint is cooling; otherwise, a cold-solder joint may result.
6. Do not attempt to use artificial cooling to speed up the job.
7. Observe the completed joint, looking for insufficient or excess solder; surfaces should be completely covered without visible projections or holes. The solder surface should be shiny and smooth.
8. If the joint is defective in any way, reheat the splice.

How to unsolder a joint between two wires, familiarization with the mechanical properties of a soldered joint, and acquiring practical experience in working with hot metals are also basic. For this instructional project, obtain a 50-watt soldering iron and stand, dampened natural sponge, a sample length of 60/40 resin-core solder, long-nose pliers, soldering aid, and have a 117-volt 60-Hz power outlet available. Proceed as follows:

1. Tin the tip of the soldering iron.
2. Wipe the tip with the sponge and place the soldered splice on the tip.
3. Drain away as much of the solder as is practical and again wipe the tip.
4. Again place the splice on the tip and loosen the joint with the soldering aid and long-nose pliers.
5. After the ends of the wires have been separated, straighten them as well as feasible with the pliers.
6. Complete the job by tinning the ends of the wires in preparation for a new splice

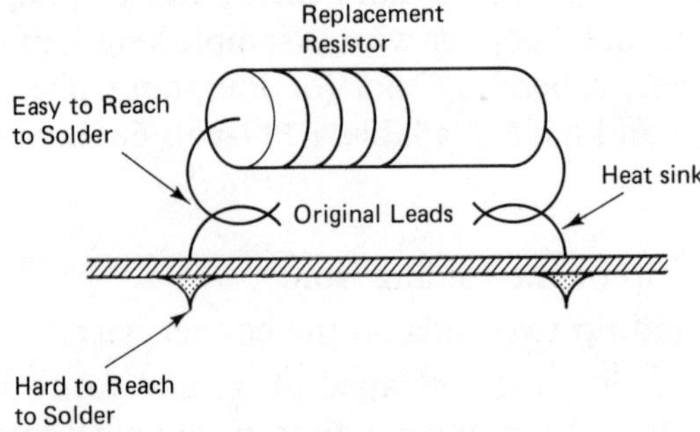

Replacing a Resistor on a Printed Circuit Board
Using the Original Resistor's Leads

Here are the essential time-saving factors in all soldering procedures:

1. The joint to be soldered should be clean and mechanically tight.
2. Tin the cleaned soldering tip; for example, dip it in a small puddle of 60/40 tin-lead resin-core solder.
3. Apply the soldering tip, if possible, at the bottom of the joint so that gravity will help the solder flow.
4. Maintain as much contact as possible between the soldering tip and the connection surface for optimum heat transfer.

5. Apply the solder to the top of the joint. It should melt almost instantly, flowing down and over the joint toward the soldering tip. Use only sufficient solder to thoroughly wet the connection.
6. Remove the tip and the solder, and keep the joint immobile until the solder has hardened.

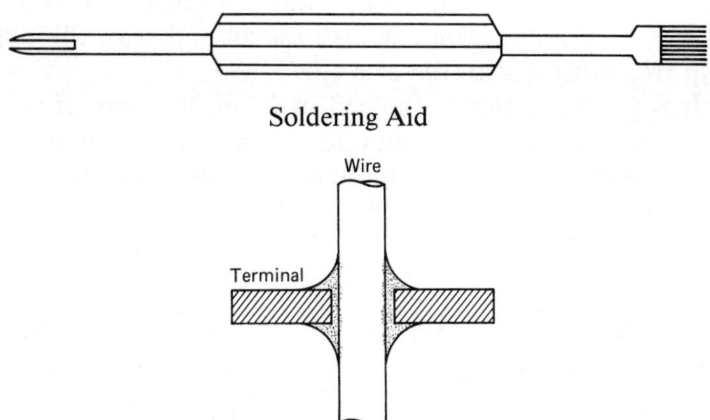

Soldering Aid

A Solder Joint Made in Accordance with Good Practices

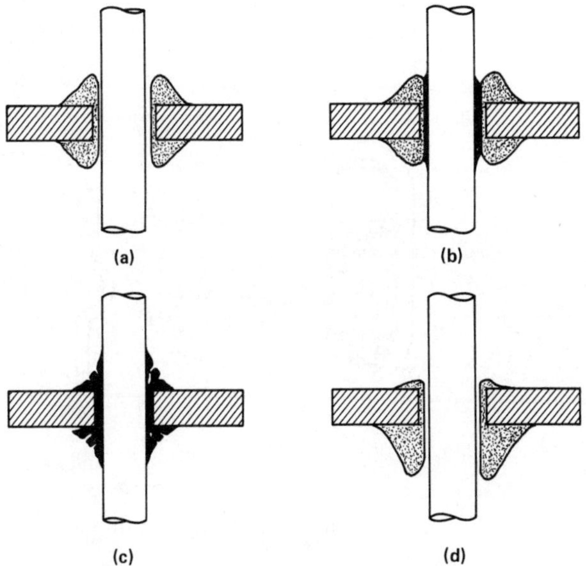

Defective solder joints: (a) cold-solder formation; (b) resin joint; (c) fractured solder; (d) excessively heated joint.

Low-wattage irons are essential for printed circuits and integrated circuits. Use a 20 to 30 watt iron. Observe that many of these low-wattage irons have interchangeable tips for various kinds of close work. A 3/64 inch tip is common. You can obtain straight tips, offset tips, pencil, pyramid, screwdriver, needle, and spade tips.

If your iron fails to develop normal heat, there is probably thermal contact resistance between the tip and the element. Loosen and tighten the tip a couple of times. You can also use an anti-seize compound to prevent the tip from fusing with the element.

It is good practice at the end of 20 or 30 hours of use to remove the tip and reapply anti-seize compound to the threads. Note the tips often oxidize where they are secured to the gun (the gun body then heats up, instead of the tip).

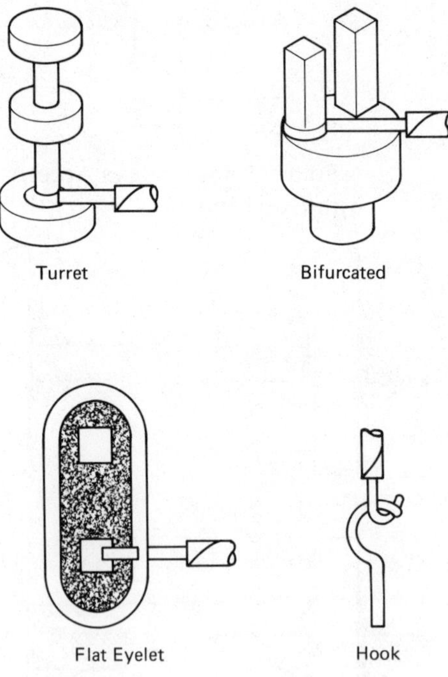

Common Types of Terminals

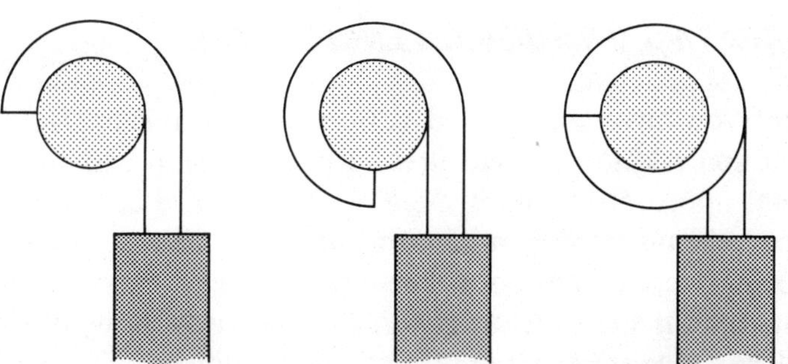

Type of Wraps: 180° or $\frac{1}{2}$, $\frac{3}{4}$, and $1\frac{1}{2}$ (Maximum)

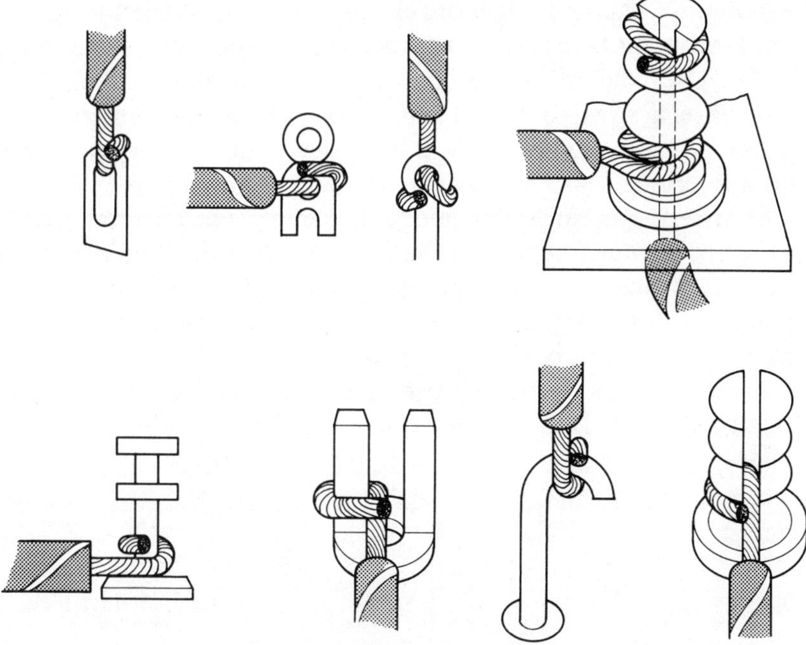

Typical Wraps of Leads to Lugs, Terminals, and the Like, Prior to Soldering. The Style of Wrap may Vary with Wire Diameter.

(Reproduced by Permission of Prentice-Hall, Inc., From "Electronics Shop Practices, Equipment, and Materials")

TIME-SAVING TIPS ON BUILDING ENCLOSURES

After you have constructed various testers and projects, you are likely to want some kind of enclosures to protect them and to give them a professional appearance. Sheet-metal enclosures are desirable, and you can fabricate them with a minimum of tools. Sheet metal is comparatively inexpensive, and has good shielding action. Controls and parts can be easily mounted on sheet metal. If we have the know-how, it is easy to work with, and has maximum strength with minimum bulk. Sheet metal can also be given a good appearance with black vinyl finish.

It is advisable to use 22-gauge cold-rolled steel or 16-20 gauge semihard aluminum. Aluminum is more expensive, but it is easier to work with. Start by sketching a layout on cardboard that is about the same thickness as the metal. Use your metal-working tools to form the cardboard enclosure. If you have a small box and pan brake and a bench-mounted sheet-metal squaring shear, your work will be easier—but ordinary hand tools can also do a fine job.

After your cardboard model is completed, a scratch awl can be used to scribe layout lines on the metal. (These lines should fall inside of the finished enclosure, so that they will not mar the outer appearance.) It is good practice to include flaps on all edges of the chassis for structural strength as well as for improved appearance.

Strain relief during the bending operations can be provided by drilling small holes with 1/8″ diameter at all corners. It is good pratice to lay out

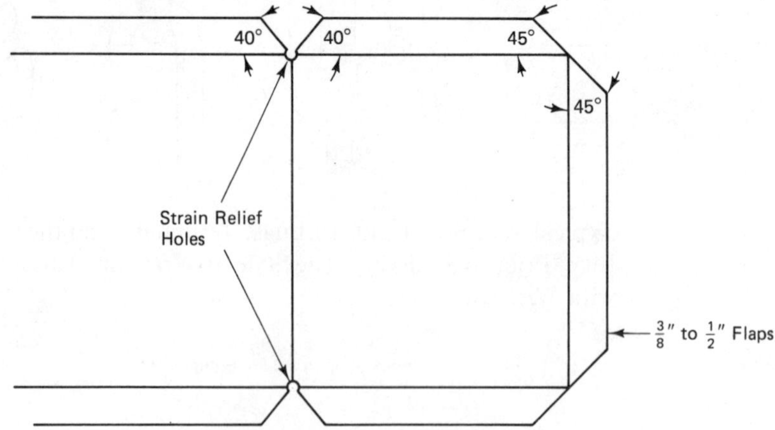

Typical Chassis Pattern (Partial View)

(Reproduced by Permission of Parker Publishing Company, Inc., From "Electronic Workshop Manual and Guide," by Carl G. Grolle)

the corner angles less than 45°, because when the metal is bent 90° the flaps will not touch each other and cause a buckle. (Don't start bending the metal before all cuts, openings, and holes have been completed).

The bending operation should be started by first bending all of the flaps. After the flaps are bent into position, it is advisable to bend the other angles less than 90° at first, and complete the operation later with full 90° bends. Note that hardwood strips and C clamps can be utilized as banding tools, if you do not have a box and pan brake available. The hardwood strips may be 1″ square, and cut to the exact length of the band. The strips are then clamped securely above and below the bend as illustrated below. Use another strip of wood and bend the metal to the required angle; use the same proceudre for all of the remaining bends.

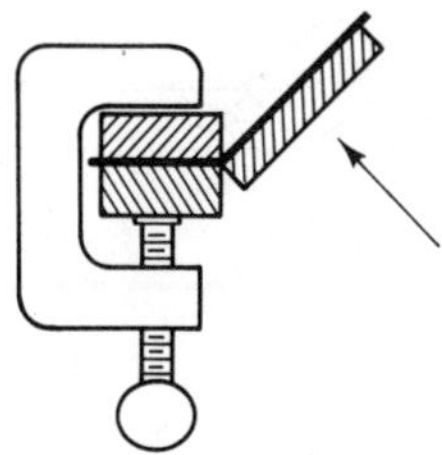

Bending Sheet Metal Using Wood Strips

Next, you may wish to construct a cowl-type minibox. It's not difficult to fabricate, and prevents overhead glare; it also protects the front-panel controls from accidental mechanical damage. A layout for a cowl-type minibox is shown below.

Start by transferring the dimensions of the minibox base to the sheet metal. It is good practice to not make the cover until after the base is completed. You can cut the metal with squaring shears, sheet-metal shears, a nibbler, or even a saber saw if you use a very fine-toothed blade. Cutting will be facilitated by clamping the metal to 1/4″ hardboard before you start cutting with a saber saw. After cutting, any rough spots may be removed with a fine-toothed file. Then sand the edges with fine-grit wet-dry abrasive paper. At this point, drill holes and cut openings; it's easier at this time than after bending the sheet metal.

Flaps A, B, and C may be bent simultaneously at a 90° angle, followed by flaps D, E, and F. Flaps G and H may then be bent, followed by bending at I with a 90° angle. Finally, bend at line J.

Before laying out the cowl cover, check the outside dimensions of the minibox base. Then, if you find that the bends are precise and square, the

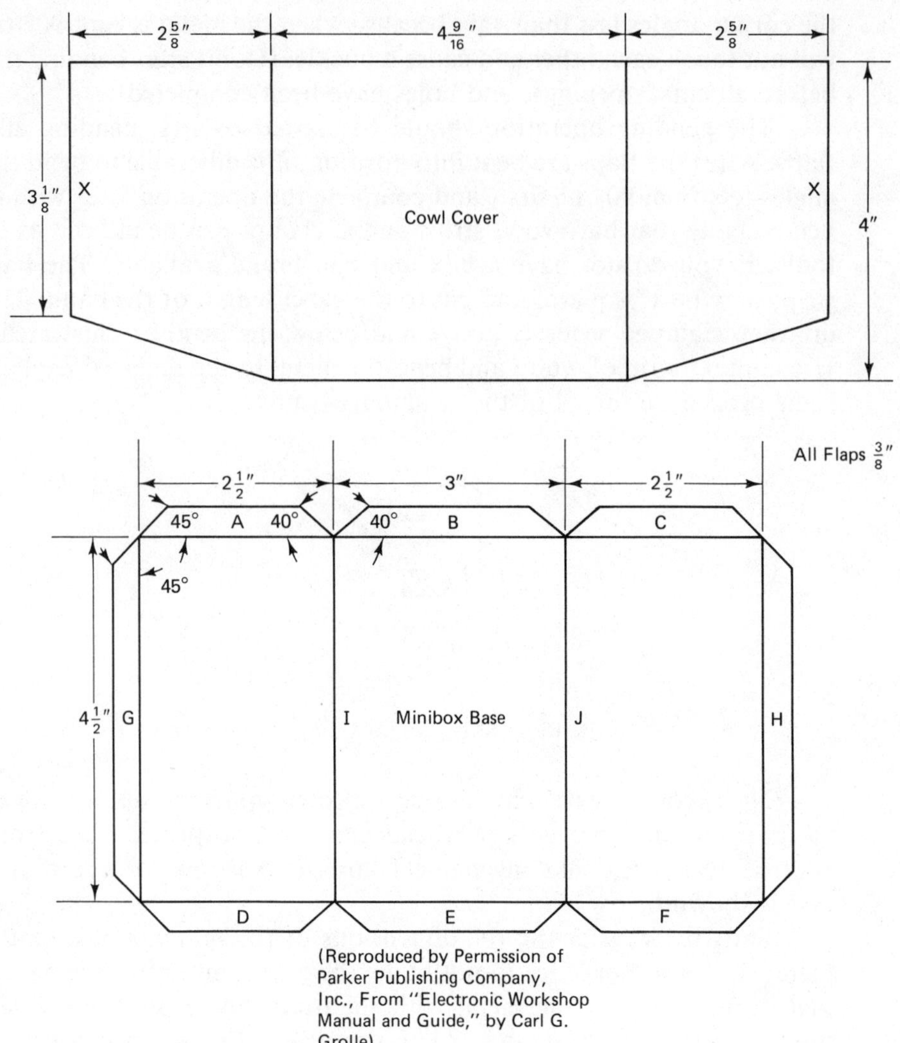

(Reproduced by Permission of Parker Publishing Company, Inc., From "Electronic Workshop Manual and Guide," by Carl G. Grolle)

planned measurements for the cover will be ok. On the other hand, if the base did not turn out quite as planned, adjust the cover dimensions as required to provide a custom structural fit.

Now the cowl cover may be cut, and the edges smoothed up if necessary. The two sides are then bent to 90°; snap the cover over the base for optimum fit. Then clamp or tape the cover in place while lightly center-punching holes at the points marked X. Follow up by drilling a 1/16" hole through the cover and base flap at each of the X points. Then, after the pilot holes have been drilled, separate the two sections of the minibox. Only the cover holes are then redrilled to 1/8″. The cowl cover can then be secured to the base with No. 4 × 3/8″ sheet-metal screws through these holes.

TIME-SAVING TIPS ON CHASSIS BUILDING

Instead of using an experimenter socket, you may wish to construct your favorite project or tester on a chassis for installation in an enclosure. A layout for an all-purpose chassis is shown below.

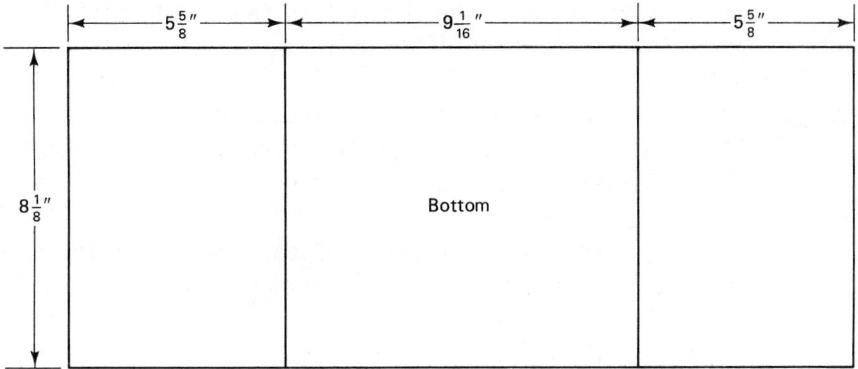

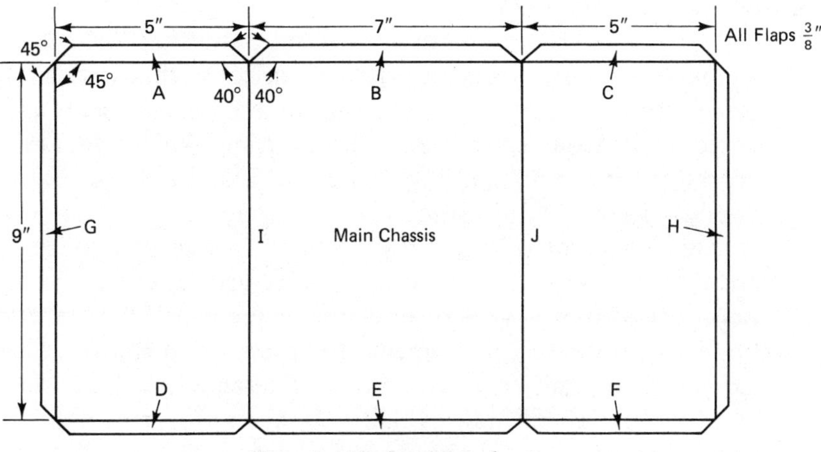

(Reproduced by Permission of Parker Publishing Company, Inc., From "Electronic Workshop Manual and Guide," by Carl G. Grolle)

The main section dimensions may be laid out on sheet metal, such as 18 gauge semihard aluminum. Next, cut out the pattern with squaring shears, sheet-metal shears, nibblers, or a saber saw with a very fine-tooth blade. Use a smooth flat file and abrasive paper to smooth up any irregularities that might remain. Then cut all of the holes and openings before bending the sheet metal.

Now, bend flaps A, B, and C together at a right angle. Bend flaps D, E, and F similarly. Then finish the flaps with a 90° bend at G and finally H. Bend a right angle at I, and also at J. "Size up" the chassis, and if any surfaces are not quite square and perpendicular, bend the surfaces slightly as required.

Before proceeding to lay out the bottom section, measure the outside dimensions of the main chassis. If there is a noticeable departure from the planned dimensions, you can change the cover measurements slightly to compensate for the deviation. Then, after you are certain of the exact size, proceed to scribe the bottom pattern on the sheet metal, and cut the pattern out. Bend the sides perpendicular to the middle section, and smooth any rough edges.

You are now ready for "the moment of truth." Slide the bottom section over the main chassis. Unless you have miscalculated, the fit will be satisfactory. If you wish, the main chassis may be spray painted in a light color. A rubber foot may be secured to each corner at the bottom.

CLEANING UP THE ACTION . . .

Beginners are sometimes surprised to discover that a cleaning of an electronic unit that has been in use for some time may serve as a shortcut to correction of trouble symptoms. This is a well-known fact in regard to tape recorders, for example, although we will find that it applies to various other types of electronic units also.

As a general rule, preventive maintenance procedures, including removal of dirt, grime, corrosion, excessive grease, and so on, will extend the useful life of many types of electronic equipment. Before starting to recalibrate any electronic instrument, for example, a thorough cleaning job is very desirable, and may save you considerable time in the long run.

The foreign substances that gain access to printed-circuit boards and mechanisms such as tape heads and transports include carbon, sulphur, metallic oxides, lead, and other substances. In areas near the ocean, salt can be anticipated. These foreign substances absorb moisture when the humidity is high, and erratic leakage resistance paths are the inevitable result.

Customary cleaning procedures employ brush and solvent techniques, pressure-spray methods, compressed-air blasts, and ultrasonic baths. A small vacuum nozzle is recommended for preliminary collection of comparatively loose debris. This procedure may be followed by rubbing localized areas with a cotton swab that has been dipped in alcohol.

Although it is not applicable to all electronic equipment, ultrasonic baths, using a solvent agitated by ultrasonic waves can be very effective in removing hard-crusted contaminants. This method is generally used for small parts.

As illustrated below, the pressure-spray method can be employed using an aerosol spray can with an extension spray brush. (A cobra brush is shown in the illustration). This arrangement has a three-foot nylon tube attached to the valve on the aerosol can.

A professional cleaning job can be accomplished with minimum workbench equipment, provided that we know how to use it to best advantage. A vacuum clearner, water, liquid detergent (dishwashing type), a pressure sprayer, and a heat lamp or hair dryer will do the trick. Frist, we vacuum

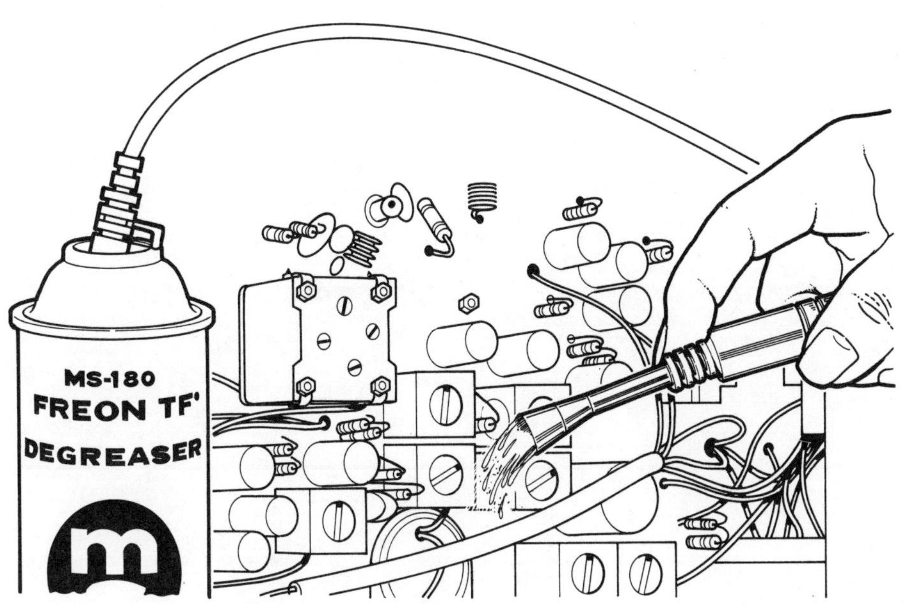

Courtesy of Miller-Stephenson Chemical Co.
(Reproduced by Permission of Parker Publishing Company, Inc., From "Electronic Workshop Manual and Guide," by Carl G. Grolle)

all of the loose dust, dirt, and debris from the electronic unit. Then we mix a cleaning solution containing one part of dishwashing detergent to nine parts of water. The cleaning procedure is as follows:

We set the electronic unit in a laundry tube, making certain that the detergent solution cannot come into contact with items that could be damaged, such as speaker cones, wooden cabinetry, meters, cardboard-cased capacitors, paper labels, and so on. An electric air compressor sprayer may be used for spraying. Or, a garden-type pressurized sprayer is suitable. In a pinch, even a pump-action oil can will serve the purpose.

We proceed to spray the detergent water cleaning solution throughout the unit until it starts to look like new. Then, we follow up with a rinse spray using clean water until all of the detergent bubbles have disappeared. Next, air spray will remove excess water and moisture, followed by a heat lamp or a blow-gun hair dryer. Avoid letting the unit stand moist for any extended length of time, and be careful not to damage any heat-sensitive devices with the heat source.

Index

A

ac amplifier, 94–95
ac current-voltage relations, 97–98
ac voltage monitor/recorder, 89–91
ac voltmeter destortion checks, 83–85
ac voltmeter operation, 100
ac voltmeter response to pure dc, 82
ac voltmeter with full-wave instrument rectifier, 81
Amplifier construction, 18–21
Amplifier input impedance, 117–18
Amplifier output impedance, 119–20
Amplifier power output:
  amplifier construction, 18
  measuring, 18
AM rejection by FM detector, 172–73
AM signal generator, 71
AND-OR gate operation, 219
AND-OR INVERT gate operation, 219–21
Asynchronous binary up counter, 262–65
Asynchronous 4-bit down counter, 266
Audible indications, 21–22
Audio current "sniffer," 95–98
  ac current-voltage relations, 97–98
Audio impedance checker, 108
Audio oscillators, 71
Audio-oscillator waveform, 98–100
Audio signal-tracer/level indicator, 143–44
Audio signal-tracer/residue analyzer, 135f
Automatic internal resistance ohmmeter, 66–68

B

Bad-level voltage, 203
Bad logic levels, 183–85
Barrier potentials, 278
Basic ac waveforms, 74
Basic D latch, 215

Battery-clip terminals, 108–11
for radios, 121–26
BCD-to-decimal decoder, 301–2
Bias boxes:
used in electronic test procedures, 13
Binary coded decimal counter, 293–95
Binary number decoder, 302–6
Binary-to-seven-segment encoder, 309–12
Bipolar transistor checkout with ohmmeter, 61–63

C

Capacitor substitution box, 89
Carrier components, 73
CB radio troubleshooting, 102–4
Checkout gated latch, 215–19
Circuit-board wiring, 243–45
Circuit loading, 7–13
Citizen's-band radio (CB), 102–4
Clearing out the "garbage," 265–66
Clock line identification, 235–36
Clock pulses, 234
Clock skew, 297
Clock subber, 234
CMOS current demands, 247–51
CMOS gates, 245–47
inverter circuits, 247
Color-bar generator, 72
Combination deode probe, 91–95
Commercial capacitor testers, 89
Common-collector complementary-symmetry amplifier, 46
Common-emitter complementary-symmetry amplifier, 45
Common faults, 227
Comparative TTL demands, 247–51
Complementary-symmetry compound-connected configuration, 46
Components of pulsating dc waveforms, 77
Constant-current ohmmeter, 224
Continuity wipe test, 242
"Crash off in the weeds," 236
Critical vs. noncritical dc voltages, 37–39
Crossover frequency, 126
Crossover operation, 126–30
Cumulative barrier potentials, 278
Current "sniffer," 95–98
Current spike quick checker, 199–200
Current troubleshooting procedures, 261–62

D

Darlington probe:
advantages, 312–14
Data-domain display, 204
Data flow, 293
Data storage, 185
Data tracing, 256–59
dc current monitor:
audible indication, 30–31
dc current "sniffer," 40
dc voltage monitors, 21–25
reference level indicator, 28–30
dc voltages:
addition, 3–4
critical vs. noncritical, 37–39
subtraction, 3–4
Decibels, 10–13
Digital circuits, 183
Digital counters, 260–62
Digital-logic probe test facilities, 69
Digital pacaphase indicator, 213
Digital stethescope, 299–300
Digital troubleshooters, 195
Digital troubleshooting:
intermittent monitoring, 227–30
procedures, 235–36
Digital waveforms, 230–33
Diode probe with zero barrier potential, 91
Diode switch experiment, 276
Direction of data flow, 293

Distorting amplifiers, 15
Distortion checks, 83–85
Distortion ratings, 17
Double-Darlington emitter-follower, 40–42, 100
Down counters, 286–88
DVM capabilities, 111–14

E

Edge-triggering flip-flops, 239
8/12 ground rule, 201–2
E/I resistance measurements, 55
Electrolytic capacitors, 109
Electromotive force, 69
Emitter-follower for dc voltage monitors, 25–27
Encoders, 306–9
Even-harmonic amplifier distortion, 13
Experimental IC asynchronous up counter, 266–69

F

False alarm nodes, 227
Fan-in and fan-out, 195–97
Fan-out test, 276–78
Finger test, 64
Flip-flops, 209
Flip-flop troubleshooting, 234–42
Follow-up digital troubleshooting, 235–36
FM/TV sweep-and-marker generator, 72
4-bit asynchronous binary up counter, 262–65
4-bit down counter, 266
14-pin IC packages, 225
Frequency counters, 288–93
Frequency-distortion localization, 166
Frequency response, 17
  two-tone signal injector, 164–66
Full-wave instrument rectifier, 81

G

Gate combinations, 183
Gated counter arrangement, 290
Gated latch, 215–19
Gate recognition, 181–83
Gates in 16-pin IC packages, 225
Germanium diode, 91

H

Half-wave ac voltmeter, 77–81
Harmonic-distortion meter, 72
High-frequency ac probes, 91–95
High-impedance circuitry:
  increasing input resistance, 25
High logic levels, 183–85

I

IC asynchronous up counter, 266–69
IC comparison quick checker, 297–98
IFM signal injector, 166
Impedance at battery-clip terminals, 108–11
Impedance check:
  battery clip terminals, 108–11
Impedance measurements, 105, 106
  amplifier input, 117–18
  amplifier output, 119–20
  DVM capabilities, 111–14
  internal, 114–17
  low-level circuits, 117
  microphone output, 120–21
  radio battery-clip terminals, 121–26
  speakers, 120
Impedance measurement shortcuts, 105–8
Impedance of mini-speaker, 126
Impedance range, 4
In-circuit transistor quick-check, 31–33
Inductance measurement:
  DVM, 130–33
  precision, 132

Integrated circuits, 158–62
quick checks, 161–62
soldering tips, 160–61
Intermittent monitoring, 227–30
Internal impedance, 114–17
Internal resistance ohmmeter, 66–68
Inverter implementation, 251–55

J

Junction characteristics, 221–24

K

Keyboard-to-BCD encoder, 306–9

L

Latches, 209
in 14-pin IC packages, 225
in 16-pin IC packages, 225
Level-triggering flip-flops, 239
Linear taper, 58
Localized strayfield quick checker, 299–300
Logic comparator troubleshooting, 269–76
Logic-high or -low, 183
"Loose" tolerance, 37
Low-impedance electronic equipment, 30–31
Low-level circuit impedance, 117
Low-logic levels, 183–85
Low-power ohmmeters, 61

M

Mapout procedures, 279
Maximum power output, 17
Measurement of amplifier:
input impedance, 117–18
output impedance, 119–20
Measurement of internal impedance, 114–17
Measuring inductance, 130–33
Measuring stage gain, 85–87
Microphone output impedance, 120–21
Mini-speaker, 126
Modified emitter-follower with zero insertion loss, 100–2
Monitoring voltages, 91
Multimeter, 8
Multitesters, 8

N

NAND implementation, 242
NC pins, 202–3
Negative fan-out test, 276–78
Negative-peak high-frequency ac probe, 91–95
Negative-peak voltage, 83
"Normal" bad level, 203

O

Ohmmeter operation, 49–53
Ohmmeter quick check of D latch, 225
Ohmmeter test polarity, 49
1-out-of 16 decoder, 306
Open collector gates, 247–51
Operation of devices, 296–97
Oscilloscopes basics, 204–8
Out-of-place signal voltages, 171–72

P

Parallel-resonant circuits, 106–7
Parallel-T RC filters, 99
Peak clipping, 83
Peak-response ac voltmeters, 81
Peak-to-peak high-frequency ac probe, 91–95
Piezo buzzers, 215
Pinout standards, 198–99

Positive-peak high-frequency ac probe, 91–95
Positive-peak voltage, 83
Potentiometers' resistance values, 58
Power amplifier quick checks, 45–48
Preliminary troubleshooting procedure, 211–12
Principles of data tracing, 256–59
Priority encoder, 312
Probe preamp, 92–94
Programmable down counter, 286–88
Pulsating dc waveforms, 77

Q

Quick check comparisons, 42–43
Quick checker:
  current spike, 199–200
  IC comparison, 297–98
  two-tone signal injector, 164–66
Quick check recycling, 43–45
Quick checks:
  AM rejection by FM detector, 172–73
  D latch, 225
  power amplifiers, 45–48

R

Radio battery-clip terminals, 121–26
Radio frequencies, 145–49
RC differentiating circuit, 74
RC integrating circuits, 74
Recording dc voltage monitor, 22–25
Reference level indicators, 28–30
  dc voltage monitor, 28–29
  supplementary tape recorder, 29
Replacement IC flip-flop packages, 239
Resistance-based troubleshooting procedures, 261–62
Resistance measurement with conventional VOM, 64–66
Resistance measurements, 55, 105
Resistance of stacked diodes, 55–60
Resistors, 56–58
Resonance probe, 166–68
RFI quick checker, 299
RF signals made audible, 149–57
RMS values of basic pulsating dc waveforms, 77–81
Rule of 2, 204
Running out the "garbage," 265

S

Semiconductors, 136–43
Semiconductor tester, 72
Sensitive amplifier distortion test, 13–18
Series operation of devices, 296–97
Series-resonant circuits, 106–7
7/11 ground rule, 198–99
Short-circuit current demands, 247
Shortcuts in impedance measurements, 105–8
Signal injection, 170–71
Signal substituters, 163
Signal tracing:
  at higher frequencies, 145, 149
  at radio frequencies, 145–49
Signal voltages, 171–72
Silicon diodes, 50
  reverse resistance, 52
Single-shot pulsers, 242
16-pin IC packages, 225
Soldering tips, 160–61
Source-impedance step-up advantage, 101
Speaker impedance, 120
Speedy stage identifier, 166–67
Square wave input, 77
Stacked diodes, 55–60
  dc level shifts, 59
Stage gain, 85–87
Stage-gain measurement, 170–71
Strobed latch, 215
Synchronous counters, 279–86

T

Tape-recorder recycling quick test, 43–45
10/12/14 ground rule, 225–26
Test equipment requirements, 68–72
Third-harmonic filters, 99
"Tight" tolerance, 37
Timbre, 74
  analysis, 74
  analyzer, 135
Toggling, 269
Tolerances in comparison quick checks, 42–43
Total harmonic distortion, 17
"Tough-dog" troubleshooting procedures, 280
Tracing circuit-board wiring, 243–45
Transistor click test, 169f
Transparent latch, 215–19
Trouble localization, 103
Troubleshooting:
  basic D latch, 215
  CMOS gates, 245–47
  digital counters, 260–62
  flip-flops, 234–42
  latches, 209
  with the logic comparator, 269–76
  procedures, 185
  shortcuts, 102–4
  synchronous counters, 279–86
True RMS ac voltmeter, 81–82
Turn-off and turn-on quick tests, 31–37
Turnover check, 83
TV analyzer, 173–78
Two-color probe, 213–15
2′421-to-8421 code converter, 306
2-to-4 line decoder, 300-301
Two-tone signal injector, 164–66

U

Unexpected grounds, 203
Unmodulated RF signals made audible, 149–57

V

Voltage-bootstrap action, 97
Voltage measurements, 4–7, 68
Voltage troubleshooting procedures, 261–62
Voltmeter sensitivity, 13

W

"Wipe" test, 242
Wired-AND, 247–51
Wired-OR, 247–51
Workbench test equipment requirements, 68–72
Working with capacitors, 87–89
Working with integrated circuits, 158–62

Z

Zero barrier potential, 91